Nutraceutical Approaches to Cardiovascular and Metabolic Diseases: Evidence and Opportunities

Nutraceutical Approaches to Cardiovascular and Metabolic Diseases: Evidence and Opportunities

Editors

Paolo Magni
Andrea Baragetti
Andrea Poli

MDPI • Basel • Beijing • Wuhan • Barcelona • Belgrade • Manchester • Tokyo • Cluj • Tianjin

Editors
Paolo Magni
Università degli Studi di Milano
Italy

Andrea Baragetti
University of Milan
Italy

Andrea Poli
Nutrition Foundation of Italy
Italy

Editorial Office
MDPI
St. Alban-Anlage 66
4052 Basel, Switzerland

This is a reprint of articles from the Special Issue published online in the open access journal *Nutrients* (ISSN 2072-6643) (available at: https://www.mdpi.com/journal/nutrients/special_issues/ Nutraceutical_Cardiovascular_Metabolic_Diseases).

For citation purposes, cite each article independently as indicated on the article page online and as indicated below:

LastName, A.A.; LastName, B.B.; LastName, C.C. Article Title. *Journal Name* **Year**, *Volume Number*, Page Range.

ISBN 978-3-0365-7130-0 (Hbk)
ISBN 978-3-0365-7131-7 (PDF)

Contents

About the Editors . . . . **vii**

Paolo Magni, Andrea Baragetti and Andrea Poli
Special Issue: Nutraceutical Approaches to Cardiovascular and Metabolic Diseases: Evidence
and Opportunities
Reprinted from: *Nutrients* **2022**, *14*, 5399, doi:10.3390/nu14245399 . . . . **1**

**Hammad Ullah, Eduardo Sommella, Cristina Santarcangelo, Danilo D'Avino,
Antonietta Rossi, Marco Dacrema, Alessandro Di Minno, et al.**
Hydroethanolic Extract of *Prunus domestica* L.: Metabolite Profiling and In Vitro Modulation of
Molecular Mechanisms Associated to Cardiometabolic Diseases
Reprinted from: *Nutrients* **2022**, *14*, 340, doi:10.3390/nu14020340 . . . . **3**

**Achille Parfait Atchan Nwakiban, Anna Passarelli, Lorenzo Da Dalt, Chiara Olivieri,
Tugba Nur Demirci, Stefano Piazza, Enrico Sangiovanni, et al.**
Cameroonian Spice Extracts Modulate Molecular Mechanisms Relevant to Cardiometabolic
Diseases in SW 872 Human Liposarcoma Cells
Reprinted from: *Nutrients* **2021**, *13*, 4271, doi:10.3390/nu13124271 . . . . **23**

**Inés Domínguez-López, Isabella Parilli-Moser, Camila Arancibia-Riveros,
Anna Tresserra-Rimbau, Miguel Angel Martínez-González, Carolina Ortega-Azorín,
Jordi Salas-Salvadó, et al.**
Urinary Tartaric Acid, a Biomarker of Wine Intake, Correlates with Lower Total and LDL
Cholesterol
Reprinted from: *Nutrients* **2021**, *13*, 2883, doi:10.3390/nu13082883 . . . . **39**

Emilio Ros, Annapoorna Singh and James H. O'Keefe
Nuts: Natural Pleiotropic Nutraceuticals
Reprinted from: *Nutrients* **2021**, *13*, 3269, doi:10.3390/nu13093269 . . . . **51**

Drake W. Lem, Dennis L. Gierhart and Pinakin Gunvant Davey
A Systematic Review of Carotenoids in the Management of Diabetic Retinopathy
Reprinted from: *Nutrients* **2021**, *13*, 2441, doi:10.3390/nu13072441 . . . . **79**

**Elena Tragni, Luisella Vigna, Massimiliano Ruscica, Chiara Macchi, Manuela Casula,
Alfonso Santelia, Alberico L. Catapano and Paolo Magni**
Reduction of Cardio-Metabolic Risk and Body Weight through a Multiphasic Very-Low Calorie
Ketogenic Diet Program in Women with Overweight/Obesity: A Study in a Real-World Setting
Reprinted from: *Nutrients* **2021**, *13*, 1804, doi:10.3390/nu13061804 . . . . **103**

**Andrea Baragetti, Marco Severgnini, Elena Olmastroni, Carola Conca Dioguardi, Elisa
Mattavelli, Andrea Angius, Luca Rotta, et al.**
Gut Microbiota Functional Dysbiosis Relates to Individual Diet in Subclinical Carotid
Atherosclerosis
Reprinted from: *Nutrients* **2021**, *13*, 304, doi:10.3390/nu13020304 . . . . **117**

Elisa Mattavelli, Alberico Luigi Catapano and Andrea Baragetti
Molecular Immune-Inflammatory Connections between Dietary Fats and Atherosclerotic
Cardiovascular Disease: Which Translation into Clinics?
Reprinted from: *Nutrients* **2021**, *13*, 3768, doi:10.3390/nu13113768 . . . . **135**

Peter E. Penson and Maciej Banach
Nutraceuticals for the Control of Dyslipidaemias in Clinical Practice
Reprinted from: *Nutrients* **2021**, *13*, 2957, doi:10.3390/nu13092957 . . . . **157**

Stefania Cicolari, Chiara Pavanello, Elena Olmastroni, Marina Del Puppo, Marco Bertolotti, Giuliana Mombelli, Alberico L. Catapano, et al.
Interactions of Oxysterols with Atherosclerosis Biomarkers in Subjects with Moderate Hypercholesterolemia and Effects of a Nutraceutical Combination (*Bifidobacterium longum* BB536, Red Yeast Rice Extract) (Randomized, Double-Blind, Placebo-Controlled Study)
Reprinted from: *Nutrients* **2021**, *13*, 427, doi:10.3390/nu13020427 . . . . **167**

Andrea Poli, Franca Marangoni, Alberto Corsini, Enzo Manzato, Walter Marrocco, Daniela Martini, Gerardo Medea, et al.
Phytosterols, Cholesterol Control, and Cardiovascular Disease
Reprinted from: *Nutrients* **2021**, *13*, 2810, doi:10.3390/nu13082810 . . . . **181**

Manuela Casula, Alberico Luigi Catapano and Paolo Magni
Nutraceuticals for Dyslipidaemia and Glucometabolic Diseases: What the Guidelines Tell Us (and Do Not Tell, Yet)
Reprinted from: *Nutrients* **2022**, *14*, 606, doi:10.3390/nu14030606 . . . . **195**

About the Editors

Paolo Magni

Paolo Magni, MD-PhD, is professor at the Università degli Studi di Milano (Milan, Italy) and collaborates with IRCCS MultiMedica hospital (Milan, Italy). He is interested in experimental and clinical research of cardio-metabolic diseases, including the identification and validation of novel biomarkers using omics approaches and the experimental and clinical assessment of the safety and efficacy of innovative nutraceutical products. He is currently the coordinator of two European projects on the assessment of atherosclerotic cardiovascular disease risk using multiomic technologies and artificial intelligence/machine learning strategies.

Andrea Baragetti

Andrea Baragetti, PhD, is professor at the Università degli Studi di Milano (Milan, Italy) and collaborates with IRCCS MultiMedica hospital (Milan, Italy). He is interested in experimental and clinical research of lipid disorders in the context of cardio-metabolic diseases, including the validation of biomarkers using omics approaches and their relationship with nutritional features and nutraceutical approaches. He is currently the holder of research projects aimed at assessing specific aspects of the pathophysiology of cardiometabolic diseases.

Andrea Poli

Andrea Poli, MD, is President of the Nutrition Foundation of Italy (NFI). In this capacity, he promotes scientific studies in the fields of nutrition and nutraceutical products, with specific interests regarding lipoprotein disorders and atherosclerosis, population interventions for atherosclerosis prevention, and the study of the distribution of risk factors in the Italian population.

Editorial

Special Issue: Nutraceutical Approaches to Cardiovascular and Metabolic Diseases: Evidence and Opportunities

Paolo Magni [1,2,*], **Andrea Baragetti** [1,2] and **Andrea Poli** [3]

1   Department of Pharmacological and Biomolecular Sciences, Università degli Studi di Milano, 20133 Milan, Italy
2   IRCCS Multimedica Hospital, 20099 Milan, Italy
3   Nutrition Foundation of Italy, 20124 Milan, Italy
*   Correspondence: paolo.magni@unimi.it

Citation: Magni, P.; Baragetti, A.; Poli, A. Special Issue: Nutraceutical Approaches to Cardiovascular and Metabolic Diseases: Evidence and Opportunities. *Nutrients* **2022**, *14*, 5399. https://doi.org/10.3390/nu14245399

Received: 3 November 2022
Accepted: 22 November 2022
Published: 19 December 2022

Publisher's Note: MDPI stays neutral with regard to jurisdictional claims in published maps and institutional affiliations.

The effective prevention and treatment of cardiovascular and metabolic diseases is a major task for health systems since these pathological conditions are still major causes of mortality, morbidity, and disability worldwide. To pursue this aim, strategies should be adopted to take advantage of lifestyle changes, including, when appropriate, pharmacological approaches and considering the nutraceutical option, which, according to an increasing research and practical interest, appears to be an additional and effective asset in this biomedical field. This Special Issue of *Nutrients* has thus been devoted to the discussion of the current experimental and clinical evidence regarding the efficacy and the safety of nutraceutical products for managing cardiometabolic diseases, also taking into consideration critical issues such as the quality of nutraceutical products, the related regulatory aspects, and the quality of evidence required to inform guidelines.

Among papers reporting experimental studies, Ullah et al. approached the potential nutraceutical properties of a hydroethanolic extract of *Prunus domestica* L. The authors were able to identify the extract's active components and explore its anti-inflammatory activity and ability to inhibit digestive enzymes involved in some features of the metabolic syndrome [1]. Moreover, Atchan Nwakiban et al. assessed the potential anti-obesity nutraceutical activities of 11 Cameroonian medicinal spice extracts in a human-cell-based model of human adipocytes [2].

Shifting to clinical studies, Domínguez-López et al. highlighted that the level of urinary tartaric acid, a biomarker of wine intake, correlated with lower total and low-density lipoprotein (LDL) cholesterol in postmenopausal women, suggesting that wine consumption may have a positive effect on the cardiovascular risk profile, thus exhibiting some nutraceutical properties [3]. Another paper discusses the beneficial effects of the dietary use of nuts in terms of cardiometabolic diseases, leading Ros et al. to suggest that nuts can be considered natural pleiotropic nutraceuticals [4]. The role of carotenoids, another family of compounds with nutraceutical properties, is then reviewed by Lem et al. in the context of managing diabetic retinopathy [5]. Moreover, a real-world study by Tragni et al. showed that a very-low-calorie ketogenic diet results in a significant reduction in cardio-metabolic risk, in addition to weight loss, in women with overweight/obesity [6], while a study by Baragetti et al. identified a significant relationship between gut microbiota functional dysbiosis and the individual diet in subjects in primary cardiovascular disease (CVD) prevention [7]. Another aspect of the relationship between diet and CVD was addressed by Mattavelli et al., who discussed the complex physiopathological relationship between specific dietary fats, inflammation, and cardiovascular disease and the contrasting data from the literature regarding observational studies and interventional trials [8]. The most common and robust nutraceutical approaches used in clinical practice for cardiovascular and metabolic disease prevention and treatment are presented in a paper by Banach and Penson [9]. Some of these approaches are then discussed in greater detail in other papers in this issue. Cicolari et al. show data on the interactions of oxysterols with atherosclerosis biomarkers in subjects with moderate hypercholesterolemia in relationship with a nutraceutical combination including a probiotic and red yeast rice extract [10]. Phytosterols are a family of nutraceutical compounds with well-established functional properties: their use and effectiveness in cholesterol control and cardiovascular disease management are discussed in detail by Poli et al. [11]. This Special Issue then closes with a critical evaluation of the inclusion (or not) criteria of nutraceutical options in the most relevant guidelines for obesity, diabetes mellitus, and dyslipidemias, highlighting the strengths and limitations of the available evidence (Casula et al.) [12].

This Special Issue of *Nutrients* aims to implement a qualified and open evidence-based discussion on the use of nutraceutical products for cardiometabolic health, thus providing an up-to-date set of information useful for readers involved in experimental/clinical research as well as clinical practice in this biomedical area.

Author Contributions: Conceptualization, P.M. and A.B.; methodology, P.M. and A.P.; resources, P.M. and A.B.; writing—original draft preparation, P.M.; writing—review and editing, P.M., A.B. and A.P.; supervision, P.M. and A.B. All authors have read and agreed to the published version of the manuscript.

Funding: The work of P. Magni is supported by: Universita' degli Studi di Milano (Transition grant PSR2015-1720PMAGN_01; PSR 2021), Ministry of Health-IRCCS MultiMedica GR-2016-02361198, European Union (AtheroNET COST Action CA21153; HORIZON-MSCA-2021-SE-01-01—MSCA Staff Exchanges 2021 CardioSCOPE 101086397). The work of A. Baragetti is supported by: Ministry of Health—Ricerca Corrente—IRCCS MultiMedica; "Cibo, Microbiota, Salute" by "Vini di Batasiolo S.p.A" AL_RIC19ABARA_01; a research award (2021) from "the Peanut Institute"; Ministry of Health—Ricerca Corrente—IRCCS MultiMedica; PRIN 2017H5F943; ERANET ER-2017-2364981; "PNRR M4C2-Investimento 1.4-CN00000041", European Union—NextGeneration EU.

Conflicts of Interest: The authors declare no conflict of interest.

References

1. Ullah, H.; Sommella, E.; Santarcangelo, C.; D'Avino, D.; Rossi, A.; Dacrema, M.; Minno, A.D.; Di Matteo, G.; Mannina, L.; Campiglia, P.; et al. Hydroethanolic Extract of *Prunus domestica* L.: Metabolite Profiling and In Vitro Modulation of Molecular Mechanisms Associated to Cardiometabolic Diseases. *Nutrients* **2022**, *14*, 340. [CrossRef] [PubMed]

2. Atchan Nwakiban, A.P.; Passarelli, A.; Da Dalt, L.; Olivieri, C.; Demirci, T.N.; Piazza, S.; Sangiovanni, E.; Carpentier-Maguire, E.; Martinelli, G.; Shivashankara, S.T.; et al. Cameroonian Spice Extracts Modulate Molecular Mechanisms Relevant to Cardiometabolic Diseases in SW 872 Human Liposarcoma Cells. *Nutrients* **2021**, *13*, 4271. [CrossRef] [PubMed]

3. Domínguez-López, I.; Parilli-Moser, I.; Arancibia-Riveros, C.; Tresserra-Rimbau, A.; Martínez-González, M.A.; Ortega-Azorín, C.; Salas-Salvadó, J.; Castañer, O.; Lapetra, J.; Arós, F.; et al. Urinary Tartaric Acid, a Biomarker of Wine Intake, Correlates with Lower Total and LDL Cholesterol. *Nutrients* **2021**, *13*, 2883. [CrossRef] [PubMed]

4. Ros, E.; Singh, A.; O'Keefe, J.H. Nuts: Natural Pleiotropic Nutraceuticals. *Nutrients* **2021**, *13*, 3269. [CrossRef] [PubMed]

5. Lem, D.W.; Gierhart, D.L.; Davey, P.G. A Systematic Review of Carotenoids in the Management of Diabetic Retinopathy. *Nutrients* **2021**, *13*, 2441. [CrossRef] [PubMed]

6. Tragni, E.; Vigna, L.; Ruscica, M.; Macchi, C.; Casula, M.; Santelia, A.; Catapano, A.L.; Magni, P. Reduction of Cardio-Metabolic Risk and Body Weight through a Multiphasic Very-Low Calorie Ketogenic Diet Program in Women with Overweight/Obesity: A Study in a Real-World Setting. *Nutrients* **2021**, *13*, 1804. [CrossRef] [PubMed]

7. Baragetti, A.; Severgnini, M.; Olmastroni, E.; Dioguardi, C.C.; Mattavelli, E.; Angius, A.; Rotta, L.; Cibella, J.; Caredda, G.; Consolandi, C.; et al. Gut Microbiota Functional Dysbiosis Relates to Individual Diet in Subclinical Carotid Atherosclerosis. *Nutrients* **2021**, *13*, 304. [CrossRef] [PubMed]

8. Mattavelli, E.; Catapano, A.L.; Baragetti, A. Molecular Immune-Inflammatory Connections between Dietary Fats and Atherosclerotic Cardiovascular Disease: Which Translation into Clinics? *Nutrients* **2021**, *13*, 3768. [CrossRef] [PubMed]

9. Penson, P.E.; Banach, M. Nutraceuticals for the Control of Dyslipidaemias in Clinical Practice. *Nutrients* **2021**, *13*, 2957. [CrossRef]

10. Cicolari, S.; Pavanello, C.; Olmastroni, E.; Puppo, M.D.; Bertolotti, M.; Mombelli, G.; Catapano, A.L.; Calabresi, L.; Magni, P. Interactions of Oxysterols with Atherosclerosis Biomarkers in Subjects with Moderate Hypercholesterolemia and Effects of a Nutraceutical Combination. *Nutrients* **2021**, *13*, 427. [CrossRef]

11. Poli, A.; Marangoni, F.; Corsini, A.; Manzato, E.; Marrocco, W.; Martini, D.; Medea, G.; Visioli, F. Phytosterols, Cholesterol Control, and Cardiovascular Disease. *Nutrients* **2021**, *13*, 2810. [CrossRef] [PubMed]

12. Casula, M.; Catapano, A.L.; Magni, P. Nutraceuticals for Dyslipidaemia and Glucometabolic Diseases: What the Guidelines Tell Us (and Do Not Tell, Yet). *Nutrients* **2022**, *14*, 606. [CrossRef] [PubMed]

Article

Hydroethanolic Extract of *Prunus domestica* L.: Metabolite Profiling and In Vitro Modulation of Molecular Mechanisms Associated to Cardiometabolic Diseases

Hammad Ullah [1,†], Eduardo Sommella [2,†], Cristina Santarcangelo [1,†], Danilo D'Avino [1], Antonietta Rossi [1], Marco Dacrema [1], Alessandro Di Minno [1,3], Giacomo Di Matteo [4], Luisa Mannina [4], Pietro Campiglia [2,5], Paolo Magni [6,7,*] and Maria Daglia [1,8,*]

[1] Department of Pharmacy, University of Napoli Federico II, Via D. Montesano 49, 80131 Naples, NA, Italy; hammad.ullah@unina.it (H.U.); cristina.santarcangelo@unina.it (C.S.); dani.davino@studenti.unina.it (D.D.); antrossi@unina.it (A.R.); marco.dacrema@unina.it (M.D.); alessandro.diminno@unina.it (A.D.M.)

[2] Department of Pharmacy, University of Salerno, 84084 Fisciano, SA, Italy; esommella@unisa.it (E.S.); pcampiglia@unisa.it (P.C.)

[3] CEINGE-Biotecnologie Avanzate, Via Gaetano Salvatore 486, 80145 Naples, NA, Italy

[4] Department of Chemistry and Technology of Drugs, Sapienza University of Rome, Piazzale Aldo Moro 5, 00185 Rome, RM, Italy; giacomo.dimatteo@uniroma1.it (G.D.M.); luisa.mannina@uniroma1.it (L.M.)

[5] European Biomedical Research Institute of Salerno, Via De Renzi 50, 84125 Salerno, SA, Italy

[6] Department of Pharmacological and Biomolecular Sciences, Università degli Studi di Milano, 20133 Milan, MI, Italy

[7] IRCCS MultiMedica, Sesto San Giovanni, 20099 Milan, MI, Italy

[8] International Research Center for Food Nutrition and Safety, Jiangsu University, Zhenjiang 212013, China

* Correspondence: paolo.magni@unimi.it (P.M.); maria.daglia@unina.it (M.D.)

† These three authors share the first authorship.

Citation: Ullah, H.; Sommella, E.; Santarcangelo, C.; D'Avino, D.; Rossi, A.; Dacrema, M.; Minno, A.D.; Di Matteo, G.; Mannina, L.; Campiglia, P.; et al. Hydroethanolic Extract of *Prunus domestica* L.: Metabolite Profiling and In Vitro Modulation of Molecular Mechanisms Associated to Cardiometabolic Diseases. *Nutrients* **2022**, *14*, 340. https://doi.org/10.3390/nu14020340

Academic Editors: Flávio Reis and Winston Craig

Received: 2 December 2021
Accepted: 10 January 2022
Published: 14 January 2022

Publisher's Note: MDPI stays neutral with regard to jurisdictional claims in published maps and institutional affiliations.

Abstract: High consumption of fruit and vegetables has an inverse association with cardiometabolic risk factors. This study aimed to chemically characterize the hydroethanolic extract of *P. domestica* subsp. syriaca fruit pulp and evaluate its inhibitory activity against metabolic enzymes and production of proinflammatory mediators. Ultra-high-performance liquid chromatography high-resolution mass spectrometry(UHPLC-HRMS) analysis showed the presence of hydroxycinnamic acids, flavanols, and glycoside flavonols, while nuclear magnetic resonance(NMR) analysis showed, among saccharides, an abundant presence of glucose. *P. domestica* fruit extract inhibited α-amylase, α-glucosidase, pancreatic lipase, and HMG CoA reductase enzyme activities, with IC50 values of 7.01 mg/mL, 6.4 mg/mL, 6.0 mg/mL, and 2.5 mg/mL, respectively. *P. domestica* fruit extract inhibited lipopolysaccharide-induced production of nitrite, interleukin-1 β and PGE$_2$ in activated J774 macrophages. The findings of the present study indicate that *P. domestica* fruit extracts positively modulate in vitro a series of molecular mechanisms involved in the pathophysiology of cardiometabolic diseases. Further research is necessary to better characterize these properties and their potential application for human health.

Keywords: *Prunus domestica* L.; chemical characterization; digestive enzyme inhibition; HMG-CoA reductase inhibition; anti-inflammatory activity

1. Introduction

The metabolic syndrome (MS), (also called as "Reaven's syndrome", "Syndrome X", "Insulin resistance syndrome", and "the deadly quartet") [1–3], is a cluster of pathological conditions including visceral obesity, hyperglycemia, or type 2 diabetes mellitus (T2DM), hypertension, and dyslipidemia, and is associated with higher cardiovascular disease (CVD) risk. A prolonged and persistent condition of MS may indeed silently progress towards serious complications such as T2DM, when not yet present, coronary heart disease (CHD), heart failure, stroke, cerebrovascular accidents (CVA) and stroke, as well as liver complications (nonalcoholic fatty liver disease (NAFLD) and nonalcoholic steatohepatitis (NASH)) [4,5]. Adipose tissue dysfunction is commonly associated with MS, and dysregulated secretion of pro- or anti-inflammatory adipokines may contribute towards MS-induced complications [6,7]. Indeed, elevation of circulating tumor necrosis factor-alpha (TNF-α), interleukin-6 (IL-6), and C-reactive protein (CRP) are predominantly observed in patients with dysregulated metabolism [8,9].

Increasing age, sedentary lifestyle, female gender, lower socioeconomic class, positive family history, excessive alcohol intake, unhealthy dietary patterns, and intake of numerous medications are some of the risk factors associated with the prevalence of MS [10,11]. In this context, prevention is essential. Adopting a healthy lifestyle, aimed at weight normalization, increasing physical activity and adoption of a healthy diet, including increased intake of fruits, vegetables, and whole grains and a low consumption of salt, trans fatty acids, and cholesterol-rich foods [12]. However, as maintaining a healthy lifestyle on a daily basis is challenging, specific pharmacological interventions (i.e., statins, renin–angiotensin–aldosterone system (RAAS) inhibitors, and insulin-sensitizing agents, etc.) are often prescribed to target some selected pathogenetic mechanisms [12–14]. The main limitations of the pharmacological approach are the increase in the occurrence of adverse and side effects without beneficial effects on low-grade inflammation [15]. In this context, there is a growing attention to food supplements able to decrease the modifiable risk factors of MS [16]. One of the most widely used food supplement ingredients is red yeast rice (RYR), made by fermentation of rice with *Monascus purpureus*, producing monacolin K, which, in lactone form, is identical to lovastatin. RYR is used alone or in combination with plant extracts, coenzyme Q10, chromium, and vitamins. In 2018, the Panel on Food Additives and Nutrient Sources added to Food of the European Food Safety Authority (EFSA) has expressed doubts about the actual safety of RYR and concluded that EFSA is "unable to identify a dietary intake of monacolins from RYR that does not give rise to concerns about harmful effects to health" [17]. Besides RYR, other natural compounds could be effective strategies in the treatment of MS and prevention of MS progression towards serious complications through the inhibitory activity of the enzymes associated with carbohydrate/lipid digestion and cardiometabolic diseases (i.e., α-amylase, α-glucosidase, HMG-CoA reductase, and pancreatic lipase) [18–20].

Research analysis has shown an inverse association of increased consumption of fruits and vegetables with the incidence of MS risk factors [21,22], though there is a strong need to identify the specific effects of fruits and vegetables on the risk of MS development [23]. *Prunus domestica* L. (European plum) is known for its health benefits, which may be the result of its antioxidant potential and anti-inflammatory effects [24,25]. Plums are fruits with low glycemic index, thus their consumption in adequate amount and on a regular basis could be a potential preventive strategy against MS [25,26]. In addition, plums are commonly consumed when on a diet, they are commercially available at low cost, and as happens for many fruits, some of the plums produced are discarded as they do not reach the size requirements to be placed on the market.

As finding new food supplement ingredients which are safe and effective in the prevention of MS is an unmet need, especially after the concern of EFSA regarding RYR, the present study aimed to (1) characterize the metabolite profile of a hydroethanolic extract of *Prunus domestica*, using a multimethodological approach that was previously used for the study of different food matrices [27,28], and two analytical techniques, untargeted nuclear magnetic resonance (NMR) spectroscopy and targeted ultra-high-performance liquid chromatography high-resolution mass spectrometry (UHPLC-HRMS), and (2) study the in vitro modulatory effects of this extract on the activity of a panel of enzymes differentially involved in the pathophysiology of MS and the modulation of proinflammatory mediator release.

2. Materials and Methods

2.1. Preparation of Fruit Extracts

Two different varieties of *Prunus domestica* L. fruits (*P. domestica* L. subsp. domestica, also known as common plum, and *P. domestica* L. subsp. syriaca, also known as Mirabelle plum) (Figure 1) were collected from a local cultivator in the Campania Region (Italy) in October 2020. Eight fruits were sampled for each variety (Common Plum and Mirabelle Plum). All fruits were first washed with water to eliminate every dirt residue and were separated into skin and pulp. Both parts were cut into small pieces with a ceramic knife. To control oxidation during preparation, the samples were cut in an ice bath. The samples were freeze-dried and then ground into fine powder using a mortar and pestle. Aliquots of 1 and 2 g of the powdered skin and pulp were added to 20 mL and 40 mL of 50%, 70% and 99% ethanolic solution acidified with 0.1% HCL solution, respectively. The sample pH values were adjusted to 2.0, and the samples were subjected to magnetic stirring for 3 h at room temperature, followed by centrifugation at 6000 rpm for 10 min. The precipitate was separated from the supernatant. The same procedure was repeated three times and the supernatants of each sample were collected and filtered through Whatman cellulose filter paper. The filtrate was concentrated in a rotary evaporator at a temperature lower than 30 °C and submitted to freeze drying. The dry extracts were kept at −20 °C for subsequent determination of total polyphenol content, antioxidant activity, untargeted NMR spectroscopy, and targeted UHPLC-HRMS.

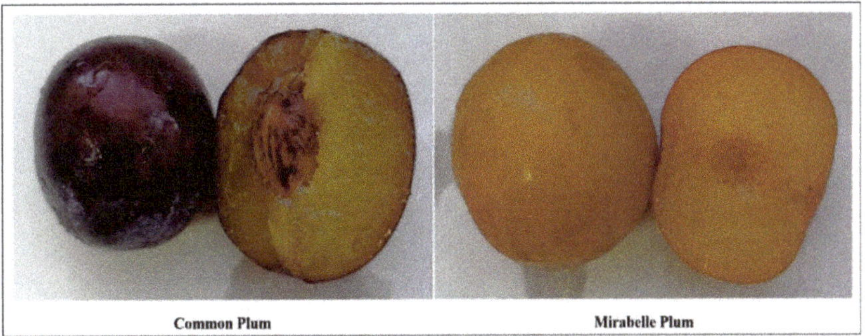

Figure 1. Plum varieties, collected from a local cultivator in the Campania Region (Italy). *P. domestica* L. subsp. domestica (Common plum) and *P. domestica* L. subsp. syriaca (Mirabelle plum).

Based on the assessment of total polyphenol content and antioxidant activity, the fruit pulp extract of *P. domestica* subsp. syriaca obtained with 50% hydroethanolic solution was selected, and in view of the high content of glucose and sucrose determined via NMR, it was subjected to the chemical precipitation of its sugars by treatment with absolute ethanol, followed by an ultra-freezing temperature. The organic solvent was removed under reduced pressure by a rotary evaporator and the dry extract obtained from the fruit pulp extract of *P. domestica* subsp. syriaca without sugars was kept at −20 °C for subsequent biological assays.

2.2. Total Phenolic Contents

Total phenolic content (TPC) was determined using a colorimetric assay (Folin-Ciocalteu method), following the same protocol as set by Singleton et al. with some modifications [29]. An aliquot (10 µL) of the samples (50 mg/mL) or gallic acid standard solutions (200–1000 µg/mL) was taken and added to 50 µL of Folin-Ciocalteu reagent. The solutions were cyclomixed for 4 min and added to 200 µL Na_2CO_3 (15%). The final volume was made to 1 mL with distilled water and allowed to incubate for 2 h at room temperature and under dark conditions. The solutions were read spectrophotometrically at 750 nm. Gallic acid was used as standard compound, with serial dilutions being prepared with known concentrations ranging from 200 to 1000 µg/mL. The results were expressed as mg equivalent to gallic acid/g of extract on dry weight basis.

2.3. Antioxidant Assay

ABTS (2,2′-azino-bis3-ethylbenzthiazoline-6- sulfonic) assay was performed to evaluate the antioxidant potential of the fruit extracts following protocols set by Kok et al. with slight modifications [30]. The assay was conducted by placing 1 mL ABTS solution in a microtube, to which 10 µL of sample (50 mg/mL) or Trolox (0, 15, 20, 25, 30 and 35 µM) was added. The mixture was allowed to incubate for 2.5 min, and the absorbance was read at 734 nm. Thus, the antioxidant compounds present in the fruit extracts quench the color and produce a decoloration of the solution which is proportional to their antioxidant activity. The results were expressed as Trolox equivalent concentration (µM/g of extract on dry weight basis).

2.4. Metabolic Profiling of P. domestica Fruit Pulp Extract

The metabolic profiling of *P. domestica* fruit pulp extract was evaluated using UHPLC-HRMS and NMR analysis.

2.4.1. RP-UHPLC-HRMS Analysis

P. domestica fruit pulp extract was solubilized in methanol/water (50:50 *v/v*). The sample was then filtered through a cellulose acetate/cellulose nitrate mixed esters membrane (0.45 µm; Millipore Corporation, Billerica, MA, USA), and analyzed by RP-UHPLC-HRMS. UHPLC-HRMS analysis was performed on a Shimadzu Nexera UHPLC system, consisting of a CBM-20A controller, two LC-30AD dual-plunger parallel-flow pumps, a DGU-20 AR5 degasser, an SPD-M20A photo diode array detector (PDA), a CTO-20A column oven, and a SIL-30AC autosampler. The system was coupled online to a hybrid Ion trap Time of Flight Mass spectrometer (LCMS-IT-TOF, Shimadzu, Duisburg, F.R. Germany) equipped with an electrospray source (ESI). For RP-UHPLC analysis, a Kinetex Biphenyl 100 mm × 2.1 mm, 2.6 µm (L × I.D, particle size, Phenomenex®, Bologna, Italy) column was employed at a flow rate of 0.4 mL/min. The mobile phases consisted of (A) 0.1% CH_3COOH in H_2O and (B) ACN plus 0.1% CH_3COOH. Analysis was performed in gradient as follows: 0–20.0 min, 2–20% B; 20.01–22.0 min, 20.01–99% B; 99% B hold for 1 min; returning to initial conditions in 0.1 min. The column oven was set to 40 °C and 5 µL sample was injected. PDA detection parameters were sampling rate 12 Hz, time constant 0.160 s and chromatograms were extracted at 280 and 330 nm. LC data elaboration was performed by the LCMS solution® software (Version 3.50.346, Shimadzu, Duisburg, F.R. Germany). MS detection was performed in negative mode ionization as follows: curve desolvation line (CDL), 250 °C; Block Heater, 250 °C; Nebulizing and Drying gas, 1.5 and 10 L/min; ESI⁻ Capillary Voltage, −3.5 kV; MS range, m/z 150–1500; ion accumulation time, 30 ms; ion trap repeat, 3. MS/MS was performed in a data-dependent acquisition (DDA), precursor ions selection was based on the base peak chromatogram (BPC) intensity of 700,000. Collision-induced dissociation (CID), 50%, ion trap repeat. For analysis, the instrument was tuned using sodium trifluoroacetate (NaTFA). Metabolite annotation was based on accurate mass measurement, MS/MS fragmentation pattern and comparison within silico spectra with MS database searching [31,32]. "Formula Predictor" software (Shimadzu, Duisburg, F.R. Germany) was used for the prediction of the molecular formula using the following settings: maximum deviation from mass accuracy: 5 ppm, fragment ion information, and nitrogen rule.

2.4.2. NMR Analysis

P. domestica fruit pulp extract (500 mg) was solubilized in 10 mL of 400 mM phosphate buffer/D2O containing 3-(trimethylsilyl)-propionic-2,2,3,3-d4 acid sodium salt (TSPA), used as internal standard for quantitative measurements, and EDTA, used as a complexing agent for metal ions. Then, an aliquot of 0.7 mL was transferred in a 5 mm NMR tube. The NMR spectra were recorded at 25 °C on a JNM-ECZ 600R (JEOL Ltd., Tokyo, Japan) spectrometer operating at the proton frequency of 600.17 MHz equipped with an autosampler and the SuperCOOL cryogenic probe (JEOL Ltd., Tokyo, Japan). The ^{1}H spectrum was acquired using a presaturation pulse sequence to suppress water signal, a 90° pulse of 12.8 μs and 65 K data points. All the NMR spectra were processed using the JEOL Delta v5.3.1. software (JEOL Ltd., Tokyo, Japan). The ^{1}H spectrum after Fourier transformation was manually phased, automatically base-corrected and referred to the β-glucose CH-1 signal set at 4.66 ppm. 2D NMR experiments, namely ^{1}H-^{1}H COSY, ^{1}H-^{1}H TOCSY, ^{1}H-^{13}C HSQC and ^{1}H-13C HMBC, were performed using the following experimental conditions: ^{1}H-^{1}H COSY and ^{1}H-^{1}H TOCSY experiments were carried out with water presaturation during relaxation delay and 9 kHz of spectral width in both dimensions. ^{1}H-^{1}H COSY was acquired using 4 k × 256 points in F1 and F2, respectively, a relaxation delay of 2.5 s and 44 scans, whereas in the case of ^{1}H-^{1}H TOCSY experiment, 8 k × 256 points in F1 and F2 dimensions, respectively, a mixing time of 80 ms, a relaxation delay of 2 s and 52 scans were used. ^{1}H-^{13}C HSQC experiment was carried out using a 90° ^{1}H pulse of 12.8 μs and 90° ^{13}C pulse of 14.0 μs, a spectral width of 9 kHz and 33 kHz for the ^{1}H and ^{13}C dimensions, respectively, 8 k × 256 points, a relaxation delay of 2 s, 80 scans and a coupling constant 1JC–H of 150 Hz. ^{13}C spectra were referenced to the CH-1 resonance of β-glucose at 97.00 ppm. ^{1}H-13C HMBC experiment was carried out with 12.8 μs for ^{1}H and 14.0 μs for ^{13}C 90° pulse, a spectral width of 9 kHz and 38 kHz for the ^{1}H and ^{13}C dimensions, respectively, 8 k × 256 points in F1 and F2 dimensions, a relaxation delay of 2 s, a delay for the evolution of long-range couplings of 50 ms and 76 scans. In order to quantify the assigned compounds, the integral of the corresponding selected ^{1}H resonances were measured with respect to the integral of TSP methyl group signal normalized to 100. Quantitative results were expressed in μg/mg of dry weight.

2.5. Enzyme Inhibition Assays

Inhibition assays of different enzymes associated with the MS were performed as described below. The selected fruit extract (*P. domestica* subsp. syriaca fruit pulp), dissolved in 1% DMSO (SERVA Electrophoresis GmbH, Aurogene, Rome, Italy) and respective positive controls were tested using different concentrations to obtain half minimal inhibitory concentration (IC_{50}) for each enzyme, by nonlinear regression analysis. The absorbance of the sample blank (buffer in place of enzyme solution) and control (buffer in place of extract) was recorded as well. The inhibition of enzyme activity was calculated using following equation:

$$\% \text{ inhibition} = [(A_{\text{control}} - A_{\text{extract}})/A_{\text{control}}] \times 100$$

2.5.1. α-Amylase Inhibition Assay

The α-amylase from porcine pancreas inhibition assay was performed according to the protocol set up by Cicolari et al. with slight modifications [33]. The reaction mixture contained 20 μL fruit extract solution (concentration range: 0.0625–25 mg/mL) or acarbose (concentration range: 15.56–400 μg/mL), and 20 μL enzyme solution (0.5 mg/mL) in 0.02 M sodium phosphate buffer (pH 6.9 with 0.006 M NaCl), which was preincubated for 10 min at 25 °C. Then, 20 μL of 1% starch solution was added to each tube at timed intervals and allowed to incubate for 10 min at 25 °C. The reaction was stopped by the addition of 40 μL color reagent (DNSA). The test tubes were incubated in a boiling water bath for 10 min, and then cooled to room temperature. Finally, 600 μL of bidistilled water was added to dilute the reaction mixture and the absorbance was read at 540 nm using a microplate reader.

2.5.2. α-Glucosidase Inhibition Assay

The α-glucosidase from *Saccharomyces cerevisiae* inhibition assay was performed according to the protocols set by Cicolari et al. with slight modifications [33]. The reaction mixture containing 50 μL fruit extract solution (concentration range: 0.0313–25 mg/mL) or acarbose (concentration range: 15.56–900 μg/mL), and 100 μL enzyme solution (1 unit/mL) in 0.1 M phosphate buffer (pH 6.9), was incubated in a 96-well plate for 10 min at 25 °C.

After preincubation, 50 μL of 0.1 M phosphate buffer (pH 6.9) solution containing 5 mM p-nitrophenyl-α-D-glucopyranoside was added to each well at timed intervals, and was incubated for 5 min at 25 °C. The absorbance was read at 405 nm using a microplate reader.

2.5.3. HMG-CoA Reductase Inhibition Assay

The assay was conducted according to the manufacturer's protocol (Sigma-Aldrich). The assay was conducted by placing 910 μL phosphate buffer with 5 μL fruit extract (concentration range: 0.0625–30 mg/mL) or 5 μL pravastatin (concentration range: 18.75–300 μM) into microtubes; 20 μL of NADPH and 60 μL of HMG-CoA reductase substrate were then added. The analysis was initiated (time 0) by the addition of 5 μL of HMG-CoA reductase, and incubated at 37 °C. The rate of NADPH consumed was monitored every 15 s for up to 5 min by reading the decrease in absorbance at 340 nm, using the microplate reader.

2.5.4. Pancreatic Lipase Inhibition Assay

Porcine Pancreatic Lipase (PPL) inhibition assay was conducted according to the protocols reported by Nwakiban et al. (2019) [34]. The assay was conducted by mixing 30 μL PPL (2.5 mg/mL in 10 mM MOPS and 1 mM EDTA, pH 6.8) with 850 μL Tris buffer (100 mM Tris-HCl and 5 mM CaCl2, pH 7.0). Then, either 100 μL of fruit extract (concentration range: 1.30–12.5 mg/mL) or orlistat (concentration range: 1–500 μg/mL) was added to the mixture and incubated for at 37 °C for 15 min, followed by the addition of 10 μL substrate (10 mM p-NPB in dimethyl formamide). The mixtures were incubated again at 37 °C for 30 min. The absorbance was read at 405 nm using a microplate reader, to determine the lipase activity by quantifying the hydrolysis of p-NPB to p-nitrophenol.

2.6. Cell Culture

Murine monocyte/macrophage J774 cell line was obtained from the American Type Culture Collection (ATTC TIB 67). The cell line was grown in adhesion in Dulbecco's modified Eagles medium (DMEM) supplemented with glutamine (2 mM, Aurogene Rome, Italy) Hepes (25 mM, Aurogene Rome, Italy) penicillin (100 U/mL, Aurogene Rome, Italy), streptomycin (100 μg/mL, Aurogene Rome, Italy), fetal bovine serum (FBS, 10%, Aurogene Rome, Italy) and sodium pyruvate (1.2%, Aurogene Rome, Italy) (DMEM completed). The cells were plated at a density of ~1 × 10^6 cells in 75 cm^2 culture flasks and maintained at 37 °C under 5% CO_2 in a humidified incubator until 90% confluence. The culture medium was changed every 2 days. Before a confluent monolayer appeared, sub-culturing cell process was carried out. *P. domestica* subsp. syriaca fruit pulp extract was solubilized in DMSO at the concentration of 200 mg/mL (stock solution). Then, it was diluted in DMSO to obtain solutions at the concentrations of 150 mg/mL, 100 mg/mL, 20 mg/mL and 2 mg/mL. Cells were plated to a seeding density of 5.0 × 10^5 in 24 multiwell plates. After 2 h of adhesion, cells were pretreated (for 2 h) with increasing concentration of *P. domestica* subsp. syriaca fruit pulp extract (5 μL of 2, 20, 100, 150 and 200 mg/mL, which correspond to a final concentration in the well (1 mL) of 0.01, 0.1, 0.5, 0.75 and 1 mg/mL). After the preincubation, macrophages were stimulated with or without LPS from *Escherichia coli*, Serotype 0111:B4, (10 μg/mL; 100 μL of solution 100 μg/mL in DMEM completed with FBS, Sigma Aldrich, Milan, Italy) for 24 h [35].

2.6.1. Nitrite, IL-1β and PGE$_2$ Assay

After 24 h of incubation, the supernatants were collected for the nitrite, IL-1β and PGE$_2$ measurement. The nitrite concentration in the samples was measured by the Griess reaction, by adding 100 μL of Griess reagent (0.1% naphthylethylenediamide dihydrochloride in H_2O and 1% sulphanilamide in 5% concentrated H_2PO_4; vol. 1:1; Sigma Aldrich, Milan, Italy) to 100 μL samples. The optical density at 540 nm (OD_{540}) was measured immediately after Griess reagent addition, using ELISA microplate reader (Thermo Scientific, Multiskan GO, Milan Italy). Nitrite concentration was calculated by comparison with OD_{540} of standard solutions of sodium nitrite prepared in culture medium. IL-1β (R&D

Systems, Aurogene, Rome, Italy) and PGE_2 (Cayman Chemical, BertinPharma, Montigny Le Bretonneux, France) levels were measured with commercially available ELISA kits according to the manufacturer's instructions.

2.6.2. Cell Viability

Cell respiration, an indicator of cell viability, was assessed by the mitochondrial-dependent reduction of 3-(4,5-dimethylthiazol-2-yl)-2,5-diphenyltetrazolium bromide (MTT; Sigma Aldrich, Milan, Italy) to formazan. Cells were plated to a seeding density of 1.0×10^5 in 96 multiwell plates. After stimulation with LPS in the absence or presence of test compounds for 24 h, cells were incubated in 96-well plates with MTT (0.2 mg/mL), for 1 h. Culture medium was removed by aspiration and the cells were lysed in DMSO (0.1 mL). The extent of reduction of MTT to formazan within cells was quantified by the measurement of OD_{550}.

2.7. Statistical Analysis

Results from at least three independent experiments carried out in triplicate for TPC and antioxidant assay, and two independent experiments carried out in duplicate for enzyme inhibitory activities of the *P. domestica* fruit extract, were expressed as mean ($\pm$SD) values. Student's t-test was used to determine the level of significance and statistical differences among variables using GraphPad prism. For the in vitro anti-inflammatory studies, the results were expressed as mean $\pm$ standard error (SEM) of the mean of n observations, where n represents the number of experiments performed in different days. Triplicate wells were used for the various treatment conditions. The results were analyzed by one-way ANOVA followed by a Bonferroni post hoc test for multiple comparisons. A p-value less than 0.05 was considered significant. All graphs were generated using GraphPad prism, version 5 (GraphPad, San Diego, CA, USA).

3. Results

3.1. Description of P. domestica Extracts

Two different varieties of *Prunus domestica* L. fruits (*P. domestica* subsp. domestica and *P. domestica* subsp. syriaca), which were separated into pulp and skin, were submitted to three different hydroethanolic extractions to evaluate the effects of ethanol percentage in the extraction solvent on the total polyphenol content and antioxidant activity. The extraction yield calculated for the freeze-dried fruit skin ranged from 37 to 43% regardless of the *Prunus* variety. On the contrary, the extraction yield calculated for the freeze-dried fruit pulp ranged from 59.5 to 67% and from 45 to 49%, for subsp. domestica and subsp. syriaca, respectively (Table 1).

Table 1. Fruit samples and yield of extracts.

Prunus Variety	Common Name	Skin Color	Fruit Part Extracted	Ethanol (%)	Dry Extract (g/g) [1]	Extraction Yield (%)
P. domestica subsp. domestica	Common plum	Purple	Skin	99	0.42	42.0
				70	0.39	39.0
				50	0.38	38.0
			Pulp	99	0.60	59.5
				70	0.63	63.0
				50	0.67	67.0
P. domestica subsp. syriaca	Mirabelle plum	Yellow	Skin	99	0.37	37.0
				70	0.41	41.0
				50	0.43	43.0
			Pulp	99	0.45	45.0
				70	0.46	46.0
				50	0.49	49.0

[1] The weight of dry extract obtained in grams per gram of sample used for extraction.

3.2. Total Phenolic Contents and In Vitro Antioxidant Activity

A total of 12 hydroethanolic extracts were evaluated for their total phenolic content (TPC) and in vitro antioxidant activity (Table 2). The pulp of *P. domestica* subsp. syriaca extracted with 50% ethanol showed the highest TPC (12.9 mg GAE/g on dry weight basis), followed by fruit skin of *P. domestica* L. subsp. domestica extracted with 70% ethanol containing 12.8 mg GAE/g on dry weight basis. Overall, the fruit extracts extracted with 99% of ethanol exhibited relatively less phenolics, with fruit pulp of *P. domestica* subsp. syriaca showing the lowest content of total phenolics (6.5 mg GAE/g on dry weight basis). In general, fruit skin showed the highest antioxidant activity with fruit skin of *P. domestica* L. subsp. domestica (extracted with 50% ethanol) showing a Trolox equivalent concentration of 1944.1 μM/g, while fruit pulp of *P. domestica* subsp. syriaca (extracted with 99% ethanol) showed the lowest Trolox equivalent concentration of 585.5 μM/g. On the basis of the higher polyphenol content, to which the inhibitory activity of the enzymes involved in MS is generally ascribed [29], the good antioxidant activity, the higher yield (49%), and the lower percentage of ethanol used as extraction solvent (50%), the *P. domestica* subsp. syriaca fruit pulp extract obtained with 50% hydroethanolic solution was selected for the subsequent chemical characterization.

Table 2. Total phenolic content and Trolox equivalent concentration of the extracts obtained from the two varieties of *P. domestica*.

Prunus Variety	Fruit Part Extracted (Ethanol %)	TPC (GAE/g on Dry Weight Basis)	Trolox Equivalent Concentration (μM/g on Dry Weight Basis)
P. domestica subsp. domestica	skin (99%)	9.1 ± 1.0	1282.4 ± 84.1 [a]
	skin (70%)	12.8 ± 0.9 [a]	1826.2 ± 216.4
	skin (50%)	11.0 ± 0.6 [b]	1944.1 ± 138.1 [b]
	pulp (99%)	7.2 ± 1.0	630.5 ± 44.1 [a]
	pulp (70%)	11.3 ± 0.2 [a]	1611.9 ± 289.5
	pulp (50%)	9.7 ± 0.2 [b]	1290.7 ± 155.5 [b]
P. domestica subsp. syriaca	skin (99%)	7.0 ± 0.2	708.0 ± 25.1
	skin (70%)	11.2 ± 1.4	1597.4 ± 88.2 [c]
	skin (50%)	7.9 ± 0.8 [c]	1602.1 ± 368.1
	pulp (99%)	6.5 ± 0.4	578.5 ± 53.5
	pulp (70%)	10.0 ± 0.9	727.7 ± 43.9 [c]
	pulp (50%)	12.9 ± 1.7 [c]	1119.4 ± 93.1

Data are expressed as mean ± SD ($n = 3$). The assigned values of different letters in a column show significant difference among the mean values ($p < 0.05$); TPC, Total phenolic content.

Based on the assessment of total polyphenol content and antioxidant activity, on the higher yield (49%), and the lower percentage of ethanol used as extraction solvent (50%), the fruit pulp extract of *P. domestica* subsp. syriaca obtained with 50% hydroethanolic solution was selected, and in view of the high content of glucose and sucrose determined via NMR, it was subjected to the chemical precipitation of sugars by treatment with absolute ethanol, followed by ultra-freezing temperature. The organic solvent was removed under reduced pressure by a rotary evaporator, and the dry extract obtained from the fruit pulp extract of *P. domestica* subsp. syriaca without sugars was kept at −20 °C for subsequent biological assays.

3.3. UHPLC-HRMS Profile

The list of the metabolites occurring in the hydroethanolic (50%) extract obtained from *P. domestica* subsp. syriaca fruit pulp, with tentative identification based on accurate mass and fragmentation pattern compared against reference MS/MS spectra reported in silico and in previous literature, is reported in Table 3. In particular, 23 compounds belonging to different classes (organic and hydroxycinnamic acids and flavonoids, both aglycone and glycosylated) were identified in the extract (Figure 2). Hydroxycinnamic and quinic acid derivatives were the most abundant compounds, retaining the 46.7% of total peak area, followed by procyanidins, in particular dimer (17%), monomers (13.10%) and trimers (6.9%). Lastly, flavonol glycosides represented the remaining 7.9%. (Table 4)

Table 3. Identified compounds in *P. domestica* subsp. syriaca fruit pulp extract according to the retention time (RT), compound, *m/z* and MS/MS, molecular formula, and mass accuracy, reported as part per million (ppm) error.

Peak	r_t	Compound	[M-H]⁻	MS/MS	Molecular Formula	Error (ppm)
1	0.60	Citric acid	191.0227	111.0103; 173.0103	$C_6H_8O_7$	1.57
2	3.12	Chlorogenic acid	353.0874	173.0489; 191.0576	$C_{16}H_{18}O_9$	−1.13
3	4.68	Coumaroylquinic acid Isomer	337.0945	163.0417; 119.0558	$C_{16}H_{18}O_8$	4.75
4	5.45	Catechin	289.0729	245.0816	$C_{15}H_{14}O_6$	3.81
5	6.08	(+) Epicatechin dimer B type	577.1328	407.0787; 289.0728	$C_{30}H_{26}O_{12}$	−4.16
6	6.50	Feruloylquinic acid	367.1053	193.0531; 134.0390	$C_{17}H_{20}O_9$	4.90
7	6.70	Coumaroylquinic acid isomer	337.0928	163.0447; 191.0594	$C_{16}H_{18}O_8$	1.19
8	7.20	Coumaroylquinic acid isomer	337.0952	173.0458; 163.0418	$C_{16}H_{18}O_8$	2.30
9	8.12	(+) Epicatechin	289.0735	245.0816	$C_{15}H_{14}O_6$	5.88
10	8.48	(+) Epicatechin trimer B type	865.1979	407.0790; 287.0569; 577.1344	$C_{45}H_{38}O_{18}$	3.40
11	8.86	(+) Epicatechin dimer B type isomer	577.1344	407.0790; 289.0732	$C_{30}H_{26}O_{12}$	−1.39
12	9.70	Quinic acid derivative	393.1777	149.0465; 191.0561	$C_{17}H_{30}O_{10}$	2.80
13	10.50	Feruloyl-coumaroylquinic acid derivative	559.1665	337.0947; 193.0514	$C_{24}H_{32}O_{15}$	−0.54
14	11.29	Feruloyl-coumaroylquinic acid derivative	559.1670	337.0949; 193.0510	$C_{24}H_{32}O_{15}$	−0.50
15	12.19	Feruloyl-coumaroylquinic acid derivative	559.1677	337.0946; 193.0514	$C_{24}H_{32}O_{15}$	1.61
16	12.32	(+) Epicatechin dimer B type isomer	577.1358	407.0831; 289.0742	$C_{30}H_{26}O_{12}$	1.04
17	12.74	(+) Epicatechin B type trimer isomer	865.2015	407.0778; 287.0569; 577.1344; 543.0905	$C_{45}H_{38}O_{18}$	3.47
18	13.20	Quercetin-rutinoside	609.1477	301.0351; 271.0254; 255.0320	$C_{27}H_{30}O_{16}$	3.47
19	13.48	(+) Epicatechin A type trimer	863.1823	575.1180; 423.0711; 285.0393	$C_{45}H_{36}O_{18}$	−0.20
20	14.04	(+) Epicatechin A type trimer isomer	863.1828	575.1180; 423.0711; 285.0393	$C_{45}H_{36}O_{18}$	−0.12
21	14.82	(+) Epicatechin A type dimer	575.1197	423.0746; 285.0395	$C_{30}H_{24}O_{12}$	1.22
22	15.75	Quercetin-rhamnoside	447.0924	301.0371; 255.	$C_{21}H_{20}O_{11}$	−0.9
23	16.52	(+) Epicatechin A type dimer isomer	575.1187	423.0716; 285.0398	$C_{30}H_{24}O_{12}$	−1.39

Table 4. Retention time (min) and peak area, expressed as percentage of total area of the identified compounds in *P. domestica* subsp. syriaca fruit pulp extract.

Peak	Compound	Retention Time	Area %
1	Citric acid	0.6	7.92
2	Chlorogenic acid	3.12	15.43
3	Coumaroylquinic acid Isomer	4.68	0.74
4	Catechin	5.45	8.61
5	(+) Epicatechin dimer B type	6.08	5.74
6	Feruloylquinic acid	6.5	0.21
7	Coumaroylquinic acid isomer	6.7	0.39
8	Coumaroylquinic acid isomer	7.2	2.42
9	(+) Epicatechin	8.12	4.54

Table 4. *Cont.*

Peak	Compound	Retention Time	Area %
10	(+) Epicatechin trimer B type	8.48	0.24
11	(+) Epicatechin dimer B type isomer	8.86	5.47
12	Quinic acid derivative	9.7	1.98
13	Feruloyl-coumaroylquinic acid derivative	10.5	3.01
14	Feruloyl-coumaroylquinic acid derivative	11.29	19.55
15	Feruloyl-coumaroylquinic acid derivative	12.19	1.49
16	Feruloyl-coumaroylquinic acid derivative	12.32	1.47
17	(+) Epicatechin dimer B type isomer	12.74	6.27
18	(+) Epicatechin B type trimer isomer	13.2	0.79
19	Quercetin-rutinoside	13.48	3.20
20	(+) Epicatechin A type trimer	14.04	4.05
21	(+) Epicatechin A type trimer isomer	14.82	1.81
22	(+) Epicatechin A type dimer	15.75	0.16
23	Quercetin-rhamnoside	16.52	4.50

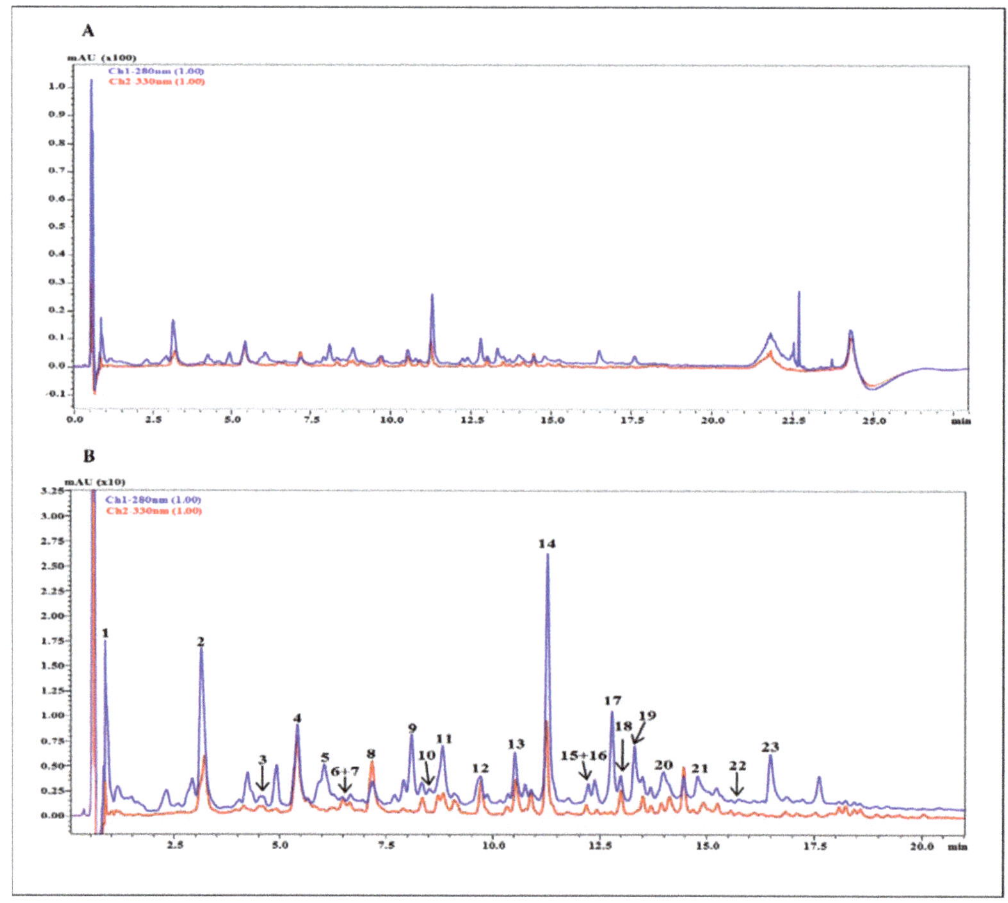

Figure 2. RP-UHPLC chromatograms of *P. domestica* subsp. syriaca fruit pulp extract with UV detection registered at λ 280 nm and 330 nm (**A**), and chromatogram expansion with corresponding peak HRMS assignment (**B**).

3.4. NMR Analysis and Quantification of Sugar and Organic Acid Contents

The ^{1}H spectrum of the *P. domestica* subsp. syriaca fruit pulp extract dissolved in phosphate buffer/D2O shows the presence of glucose, sucrose, xylose and citric, malic and quinic acids. The ^{1}H spectral assignment was obtained by literature data regarding other fruits [36,37] and 2D NMR experiments [38]. The integrals of selected signals due to sugars, namely xylose, glucose, and sucrose at 5.20 ppm, 5.25 ppm, and 5.42 ppm, respectively, and to organic acids, namely citric, malic and quinic acids at 1.88 ppm, 2.54 ppm, and 4.30 ppm, respectively, were used for compound quantification (Table 5). The ^{1}H NMR spectrum with the selected signals used for the quantification of metabolites (Figure 3) and the compound assignments table (Table S1) was also reported. In the case of some compounds, only a partial assignment was obtained due to low concentration of the compound. However, the partial assignment included diagnostic signals that allowed the identification of the reported compounds. Glucose and malic acid turned out to be the sugar and the organic acid, respectively, present in major amounts.

Table 5. Compounds identified in the ^{1}H NMR spectrum of *P. domestica* subsp. syriaca fruit pulp extract dissolved in phosphate buffer/D2O and the corresponding chemical shift signals (ppm) used in the integration process. The compound amounts in μg/mg of dry weight are also reported.

Compound	Chemical Shift (ppm) of Selected Resonances Used for Quantification	μg/mg Dry Weight
Quinic acid	1.88 (CH2-1)	7.50
Citric acid	2.54 (α,γ-CH)	0.84
Malic acid	4.30 (α-CH)	38.49
Xylose	5.20 (CH-1)	0.56
Glucose	5.25 (CH-1)	106.59
Sucrose	5.42 (CH-1)	31.59

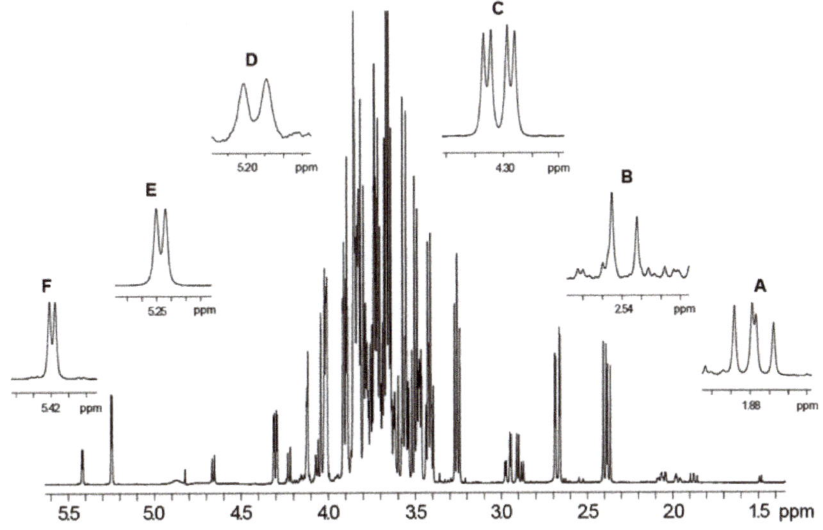

Figure 3. ^{1}H NMR spectrum of *P. domestica* subsp. syriaca fruit pulp extract. Quantified selected NMR signals are reported in expanded regions. (**A**) CH2-1 protons of quinic acid (1.88 ppm), (**B**) α,γ-CH protons of citric acid (2.54 ppm), (**C**) α-CH proton of malic acid (4.30 ppm), (**D**) CH-1 proton of in α-xylose (5.20 ppm), (**E**) CH-1 proton of α-glucose (5.25 ppm), (**F**) CH-1 proton of sucrose (5.42 ppm).

3.5. Preparation of P. domestica Fruit Extract without Sugar

In view of the high sugar content (about 14% of the whole extract) of *P. domestica* subsp. syriaca fruit pulp crude extract, it was subjected to chemical precipitation of sugar contents by treatment with absolute ethanol, supported by ultra-freezing temperature. The percent extraction yield following sugar precipitation was 41.8%.

3.6. Effect of P. domestica Subsp. Syriaca Fruit Pulp Extract on Enzyme Activities

P. domestica subsp. syriaca fruit pulp extract with reduced content of sugars was used for enzyme inhibition activities, with the aim of reducing the substances that may interfere with the enzyme inhibition activities of the vegetable extract. The inhibition of the enzyme activities performed by the extract at increasing concentration and the IC50 values for each enzyme, calculated with the nonlinear regression analysis, have been illustrated in Figure 4.

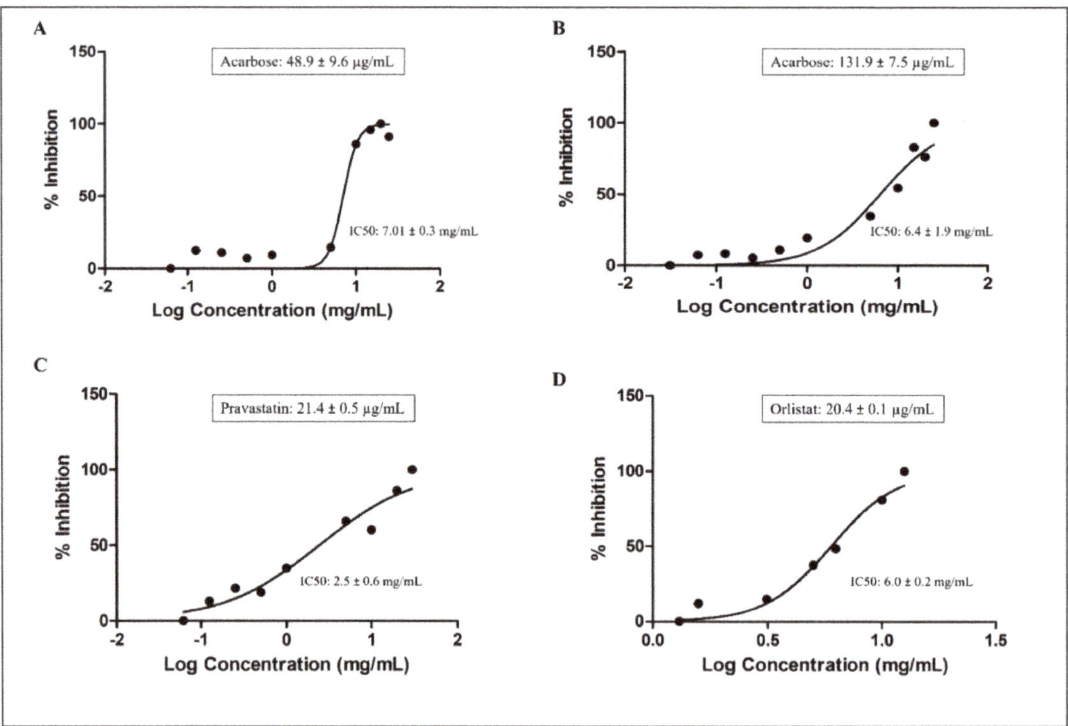

Figure 4. Enzyme inhibition activities and calculated IC50 values of *P. domestica* subsp. syriaca fruit pulp extract. α-amylases inhibition (**A**), α-glucosidase inhibition (**B**), HMG CoA reductase inhibition (**C**), Pancreatic lipase inhibition (**D**). Data expressed as mean ± SD. The IC50 values of the fruit extract against each enzyme were calculated using nonlinear regression analysis.

P. domestica subsp. syriaca fruit pulp extract inhibited α-amylase and α-glucosidase enzymes in a concentration-dependent manner, with an IC50 value of 7.01 mg/mL compared to acarbose, used as positive control (IC50: 48.9 µg/mL) (Figure 4A), and IC50 value of 6.4 mg/mL compared to acarbose (IC50: 131.9 µg/mL) (Figure 4B), respectively. As far as HMG-CoA reductase is concerned, *P. domestica* subsp. syriaca fruit extract inhibited HMG-CoA reductase enzyme in a concentration-dependent manner with an IC50 value of 2.5 mg/mL, while the reference inhibitor pravastatin inhibited HMG-CoA reductase with an IC50 value of 21.4 µg/mL (Figure 4C). Finally, regarding pancreatic lipase, the results suggest that *P. domestica* subsp. syriaca fruit pulp extract inhibited this enzyme in a concentration-dependent manner with an IC50 value of 6.0 mg/mL compared to reference inhibitor orlistat (IC50 value: 20.4 μg/mL) (Figure 4D). Overall, the mean IC50 values of the fruit extract were found to be significantly different from the values obtained from the positive controls ($p < 0.05$).

3.7. In Vitro Anti-Inflammatory Effects of P. domestica Subsp. Syriaca Fruit Pulp Extract

In order to assess the anti-inflammatory proprieties of *P. domestica* subsp. syriaca fruit pulp extract, murine macrophage cell line J774 stimulated with LPS (10 μg/mL, 24 h), a well-known proinflammatory stimulus, was used. The anti-inflammatory activities were assessed by measuring the levels of proinflammatory mediators such as nitrites, PGE$_2$ and IL-1β. Preincubation of J774 macrophages with *P. domestica* subsp. syriaca fruit pulp extract (2 h before LPS treatment) inhibited significantly and in a concentration-dependent manner (0.01, 0.1, 0.5, 0.75 and 1 mg/mL) the production of nitrite (IC50 0.46 mg/mL, Figure 5A), PGE$_2$ (IC50 0.56 mg/mL, Figure 5B) and IL-1β (IC50 0.18 mg/mL, Figure 5C) induced by LPS, starting from the concentration of 0.1 mg/mL. No effects of P. domestica subsp. syriaca fruit pulp extract on proinflammatory mediator production were observed in unstimulated cells (without LPS) (Figure 5D–F). To rule out any alteration of cell viability, an MTT assay was performed and did not show any statistical reduction in cell viability after treatment with extract (Figure 5G).

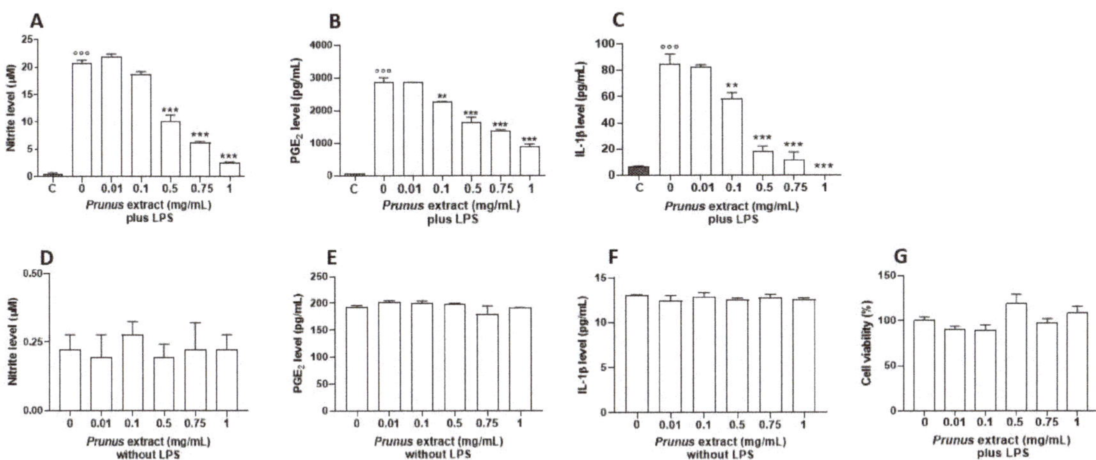

Figure 5. Effect of the *P. domestica* subsp. syriaca fruit pulp extract on LPS-induced nitrite, IL-1β and PGE$_2$ production. J774 cells were pretreated for 2 h with increasing concentrations of the extract ('0.01'. '0.1'. '0.5'. '0.75'. and '1' mg/mL) and then stimulated with LPS ('10' μg/mL) for 24 h (plus LPS). Effects of extract were also evaluated in absence of LPS (without LPS). Unstimulated J774 cells acted as a negative control (**C**). Nitrites (**A,D**), stable end products of NO, were measured in the supernatants by the Griess reaction, whereas IL-1β (**B,E**) and PGE$_2$ (**C,F**) were measured by ELISA. Cell viability was evaluated by the mitochondrial-dependent reduction of MTT to formazan (**G**). ooo $p < 0.001$ vs. unstimulated cells (**C**), $p < 0.001$ vs. unstimulated cells (**C**), *** $p < 0.001$ and ** $p < 0.01$ vs. LPS alone.

4. Discussion

The World Health Organization (WHO) reports that approximately 650 million people live today with obesity, 422 million with T2DM and 1.13 billion people with hypertension [39–41]. Thus, metabolic disorders remain a prevalent and urgent concern in the healthcare field, which today are remedied through pharmacological approaches prescribed to target some specific pathogenetic mechanisms. This, in turn, is often accompanied by adverse effects and poor compliance by patients. [42]. Therefore, the study of new food supplement ingredients for the reduction of risk factors of MS is essential and finding new agents able to modulate the enzyme activities associated with carbohydrate/lipid digestion and cardiometabolic diseases is one of the essential targets in the prevention and treatment of cardiometabolic disorders. As evident by preclinical and clinical trials, *Prunus* species can improve energy homeostasis involving glucose and lipid metabolism, decrease inflammatory mediators, reduce lipid deposition, and modulate gut microbiota, and thus can reverse metabolic dysregulation states [43]. In this study, we explored the ability of a hydroethanolic extract of *P. domestica* subsp. syriaca (Mirabelle plum) fruit pulp to inhibit the activity of various enzymes associated with cardiometabolic disorders. Initially, fruit skin and pulp of two different *P. domestica* varieties (Common plum and Mirabelle plum), extracted with different concentrations of ethanolic solution, were evaluated for TPC, and the Mirabelle plum extract showing the highest TPC was selected for enzyme inhibition assays and evaluation of in vitro anti-inflammatory activity. Chemical profiling of *P. domestica* subsp. syriaca fruit pulp extract was evaluated using a multimethodological approach based on the application of new technologies, consisting of untargeted NMR spectroscopy and untargeted UHPLC-HRMS, which favors a holistic approach as opposed to the traditional reductionist methods, allowing us to overcome the concept of identifying one compound responsible for the obtained biological effect, and to ascribe the bioactivity to the whole phytocomplex [44]. The results suggested the presence of hydroxycinnamic acids (p-coumaroylquinic acid isomers and feruloylquinic acid derivatives), which resulted to be the most represented polyphenols followed by flavanols (catechin, epicatechin, procyanidins), flavonols (rutinoside and rhamnoside derivatives of quercetin), organic acids (quinic acid, citric acid, and malic acid), and carbohydrates (xylose, glucose, and sucrose). The carbohydrate quantitative analysis showed that glucose (106.6 mg/g dry weight of extract) is the main saccharide found in the fruit extract, in agreement with USDA Food Composition Databases, followed by sucrose (31.6 mg/g dry weight of extract) and xylose (0.6 mg/g dry weight of extract). Sugars of the fruit extract were at least in part chemically precipitated with ethanol before proceeding with the enzyme inhibition and the anti-inflammatory assays to remove the constituents that may interfere with the biological activities of the vegetable extract. The inhibition of α-amylase and α-glucosidase enzymes are important to reduce the digestion of complex carbohydrates and in turn the absorption of glucose, with the aim to normalize the blood glucose level both in subjects with mild hyperglycemia and in diabetic patients, to support glucose-lowering medication [19]. Acarbose is a pharmacologic drug currently employed in the treatment of subjects with diabetes due to its potential to inhibit α-amylase and α-glucosidase enzymes, thus reducing carbohydrate digestion, slowing down the absorption of carbohydrates, and decreasing postprandial insulin secretion, in addition to stimulating glucagon-like peptide (GLP-1) release [45]. The extract of *P. domestica* subsp. syriaca fruit pulp showed inhibitory activity against α-amylase and α-glucosidase enzymes with IC50 values of 7.01 mg/mL and 6.4 mg/mL, respectively. While evaluating antidiabetic activity of novel smoothies from selected *Prunus* fruits, Nowicka et al. demonstrated that anthocyanin and flavonol content have the highest impact on α-glucosidase enzyme, whereas flavanols may have the potential to inhibit α-amylase [46]. Some studies also reported that inhibition of α-glucosidase may be associated with the content of hydroxycinnamic acid derivatives such as ferulic acid or *p*-coumaric acids [47]. The researchers indicated that flavonols can interact with hydroxycinnamic acids or anthocyanins, which may increase the inhibition of α-glucosidase [48]. It has also been suggested that procyanidins-rich fruits are effective α-amylase inhibitors, possibly by the formation of enzyme-tannin complexes resulting in the prevention of the enzyme from interacting with starch [48]. The literature shows an inhibition of α-amylase and α-glucosidase enzymes with numerous botanical extracts, including *Elateriospermum tapos* Blume [49], *Xylopia parviflora* Spruce, *Monodora myristica* (Gaertn.) Dunal, *Tetrapleura tetraptera* (Schum. & Thonn.) Taub., *Dichrostachys glomerata* (Forssk.) Chiov., *Aframomum melegueta* K.Schum., *Aframomum citratum* (C.Pereira) K.Schum [34] and *Adansonia digitata* L. [33]. Different *Prunus* fruits (including Common European plum, 'vlaškača', damson plum, white damson, purple-leaf cherry plum, white cherry plum, red cherry plum, sweet cherry, sweet cherry-wild type, sour cherry, steppe cherry, mahaleb cheery, blackthorn, and peach) extracted with 50% ethanol exhibited inhibition of α-amylase (IC50 value range: 1.11–136.23 mg/mL) and α-glucosidase (IC50 value range: 0.41–28.44 mg/mL) [50]. *Prunus* species in general showed a greater affinity towards α-glucosidase enzyme compared to α-amylase [50]. Altogether, these data suggest that inhibition of the enzymes involved in the digestion of carbohydrates by vegetable extracts may have promising potential in the management of glucose metabolism disorders. Statins are effective lipid-lowering agents, widely used as a first-line therapy in the atherosclerotic CVDs, which are known to competitively inhibit HMG-CoA reductase enzyme (rate-limiting enzyme of cholesterol synthesis) [51]. Considering the same approach of decreasing cholesterol synthesis, *P. domestica* subsp. syriaca fruit pulp extract was tested against the HMG-CoA reductase activity, showing an IC50 value of 2.5 mg/mL. The mean IC50 value calculated for the fruit extract was significantly higher from the reference statin (pravastatin). Susilowati et al. (2020) performed in silico analysis while evaluating antihyperlipidemic effects of apple peel extract, which showed the highest HMG CoA reductase inhibition by catechin, epicatechin, quercetin (aglycosidal form) and chlorogenic acid [52]. Several other vegetable extracts have already shown HMG-CoA reductase inhibitory activities, including, but not limited to *Basella alba* L. [53], *Syzygium polyanthum* (Wight) Walp. [54], *Ficus palmata* Forssk. [55], and *Amaranthus viridis* L. [56]. Inhibition of pancreatic lipase is a clinically validated approach in the treatment of obesity, as it reduces the hydrolysis of fats and decreases their absorption [57]. The Food and Drug Administration (FDA) approved orlistat in 1999 for the pharmacological management of obesity in conjunction with a reduced caloric diet, while in 2007 it was approved as an over-the-counter (OTC) agent for weight loss in overweight adults (18 years or older) [58]. Later, the FDA revised the label for orlistat by adding a new warning about severe liver injury, which has been reported rarely with this drug [59]. Following concerns about the possible cause of severe hepatic toxicities with orlistat, the European Medicines Agency completed a review for this medicine, where the Agency's Committee for Medicinal Products for Human Use (CHMP) concluded that the benefits of orlistat continue to outweighs the risks, but also recommended that marketing authorizations should ensure that the safety information on rarely occurring liver injuries be provided on the product information of all orlistat-containing medicines [60]. As reported in the present study, *P. domestica* subsp. syriaca fruit pulp extract inhibited the pancreatic lipase enzyme with an IC50 value of 6.0 mg/mL, although this value was significantly higher than that determined for orlistat. Hydroxycinnamic acids and proanthocyanidins have proven efficacy against lipase activity [61,62]. The presence of hydroxyl groups in the molecule (more potent), methoxy groups (less potent) and position of hydroxyl groups in the phenolic ring could influence the activity of polyphenols in inhibiting lipase enzyme [61]. Moreover, flavan-3-ol esters showed a stronger lipase inhibition compared to non-esterified flavanols such as catechin and epicatechin [61]. Plant species of different families (*Vitis vinifera* L., *Rhus coriaria* L., *Origanum dayi* Post, *Quercus infectoria* G.Olivier, *Eucalyptus galbie*, *Rosa damascene*, and *Levisticum officinale* W.D.J.Koch) showed considerable inhibition of pancreatic lipase enzyme [63,64]. Nowicka et al. observed an inhibition of pancreatic lipase enzyme with *Prunus persica* L. Batsch fruits (different cultivars), with an IC50 value ranging from 0.07 to 2.06 mg/mL [65]. On the whole, although the values of IC50 are much higher than those found for drugs, it must be considered that the extract, being derived from a food commonly consumed in diet and having been deprived of sugars, should not show adverse effects unlike medicines, and therefore it can be taken in larger quantities and for very long periods. In vivo studies are needed to confirm inhibitory activity against the enzymes considered. Lifestyle modifications (including diet and exercise) and pharmacological agents (such as peroxisome proliferator-activated receptor (PPAR)-α agonists, angiotensin-converting enzyme (ACE) inhibitors and angiotensin receptor blockers (ARBs)) all target inflammation in various ways, and thus can reduce MS-associated complications. Ongoing research studies are uncovering inflammatory pathways (related to obesity, T2DM, and MS), which may be potential targets for novel preventive and treatment strategies with the aim of improving overall patient quality of life and reducing mortality, preferentially by preventing the adverse sequelae from MS [66]. *P. domestica* subsp. syriaca fruit pulp extract showed promising results against nitrite, IL-1β, and PGE$_2$ levels in LPS-stimulated macrophages in a concentration-dependent manner, with an IC50 of 0.46, 0.18 and 0.56 mg/mL, respectively. These results are in agreement with those previously published [67,68], although our extract was more active. In particular, dried plum polyphenols significantly suppressed the production of NO and COX-2 in LPS-stimulated RAW 264.7 macrophages at the concentration of 100 and 100 mg/mL, respectively [68]. The higher anti-inflammatory effects of *P. domestica* subsp. syriaca fruit pulp extract could be due to the presence of polyphenolic compounds in different concentrations, mainly flavonoids (catechin and (+) epicatechin) and hydroxy cinnamic acids (chlorogenic acid and feruloyl-coumaroylquinic acid derivative). Each compound inhibited the production of proinflammatory mediators at different concentrations, and whether the anti-inflammatory effects of this prunus extract are due to additively or synergistically polyphenolic action is not known. For example, inhibitory effects of catechin [69] and chlorogenic acid [70] on various inflammatory mediators using LPS-stimulated RAW 264.7 macrophages were reported. In particular, chlorogenic acid completely inhibited NO production in LPS-stimulated macrophage RAW 264.7 cells at the concentration of 40 ug/mL [70]. A medicinal plant *Inonotus (I.) sanghuang* (rich in rutin, chlorogenic acid, isorhamnetin, quercetin, and quercitrin) improved insulin resistance and MS by reducing inflammation via modulation of the crosstalk between macrophages and adipocytes [71]. Grape powder extract (rich in quercetin-3-glucoside, catechin, epicatechin, rutin, gallic acid and resveratrol) showed to decrease LPS-stimulated inflammation in macrophages by affecting the gene expression of IL-6, IL-8, IL-1β and TNF-α, and in turn decreased insulin resistance [72]. *Sambucus nigra* L. fruit extract alleviated insulin resistance by suppressing the enhanced production of NO, TNF-α, IL-6, and PGE$_2$, where the presence of cyanidin-based anthocyanins, flavan-3-ols, flavonols, and hydroxycinnamic acids were detected [73].

The main strength of this study is represented by the comprehensive characterization of the *P. domestica* extract by different and complementary technologies and the evaluation of some in vitro functional activities of this extract on mechanisms related to the pathophysiology of cardiometabolic diseases. A limitation of this study is that we did not evaluate additional specific aspects, such as modulation of triglyceride content or glucose uptake in appropriate cell models. The assessment of these and other related parameters will be the object of future research, in order to complete the functional characterization of *P. domestica* extracts and lay the scientific foundations for the placing on the market of new food supplements based on *P. domestica* extracts.

5. Conclusions

Nutraceuticals and functional foods are known to play an important role in the maintenance of human health and wellbeing through the prevention of chronic diseases. A substantial increase in the worldwide usage of vegetable extracts has been observed in recent decades, probably due to the increasing trend of consumer propensity towards preventive care through natural substances. Currently, more than 80% of world population is relying on the use of vegetable products for their primary health concerns [74]. The present investigation studied the potential nutraceutical benefits of *P. domestica* subsp. syriaca fruit pulp extract and found that it inhibited key enzymes involved in the metabolism of carbohydrates and lipids as well as in cholesterol synthesis, and attenuated LPS-stimulated release of proinflammatory mediators. The chemical characterization showed the presence of polyphenolic contents, which potentially justifies these biological properties. Taken together, these findings point towards a potential to modulate some molecular mechanisms involved in the pathophysiology of cardiometabolic diseases. Further studies using in vitro and in vivo models are required to better characterize these properties and their potential applications for human health.

Supplementary Materials: The following are available online at https://www.mdpi.com/article/10.3390/nu14020340/s1, Table S1: Sugars and organic acids identified in the 600.17 MHz ^{1}H spectra of P. domestica fruit pulp extracts. Asterisks indicate signals selected for integration.

Author Contributions: Conceptualization, M.D.(Maria Daglia), P.M. and A.D.M.; methodology, H.U., E.S., C.S., D.D. and G.D.M.; software, H.U., A.R., M.D.(Marco Dacrema); validation, H.U., E.S., C.S. and A.R.; formal analysis, H.U., A.R., L.M. and P.C.; investigation, H.U., E.S., A.R. and P.C.; resources, H.U., A.R., M.D.(Marco Dacrema) and P.C.; data cura1tion, H.U. and C.S.; writing—original draft preparation, H.U., E.S., C.S., and A.R.; writing—review and editing, H.U., E.S., C.S., A.R., P.M. and M.D.(Maria Daglia); visualization, H.U. and M.D.(Maria Daglia); supervision, M.D.(Maria Daglia). All authors have read and agreed to the published version of the manuscript.

Funding: This research received no external funding.

Informed Consent Statement: Not applicable.

Conflicts of Interest: The authors declare no conflict of interest.

References

1. Belwal, T.; Bisht, A.; Devkota, H.P.; Ullah, H.; Khan, H.; Pandey, A.; Bhatt, I.D.; Echeverría, J. Phytopharmacology and Clinical Updates of Berberis Species against Diabetes and Other Metabolic Diseases. *Front. Pharmacol.* **2020**, *11*, 41. [CrossRef] [PubMed]
2. Costa, L.A.; Canani, L.H.; Lisbôa, H.R.K.; Tres, G.S.; Gross, J.L. Aggregation of Features of the Metabolic Syndrome Is Associated with Increased Prevalence of Chronic Complications in Type 2 Diabetes. *Diabet Med.* **2004**, *21*, 252–255. [CrossRef]
3. McCracken, E.; Monaghan, M.; Sreenivasan, S. Pathophysiology of the Metabolic Syndrome. *Clin. Dermatol.* **2018**, *36*, 14–20. [CrossRef] [PubMed]
4. Åberg, F.; Helenius-Hietala, J.; Puukka, P.; Färkkilä, M.; Jula, A. Interaction between Alcohol Consumption and Metabolic Syndrome in Predicting Severe Liver Disease in the General Population. *Hepatology* **2018**, *67*, 2141–2149. [CrossRef]
5. Mongraw-Chaffin, M.; Foster, M.C.; Anderson, C.A.M.; Burke, G.L.; Haq, N.; Kalyani, R.R.; Ouyang, P.; Sibley, C.T.; Tracy, R.; Woodward, M.; et al. Metabolically Healthy Obesity, Transition to Metabolic Syndrome, and Cardiovascular Risk. *J. Am. Coll. Cardiol.* **2018**, *71*, 1857–1865. [CrossRef]
6. Scarpellini, E.; Tack, J. Obesity and Metabolic Syndrome: An Inflammatory Condition. *Dig. Dis.* **2012**, *30*, 148–153. [CrossRef]
7. Magni, P.; Liuzzi, A.; Ruscica, M.; Dozio, E.; Ferrario, S.; Bussi, I.; Minocci, A.; Castagna, A.; Motta, M.; Savia, G. Free and Bound Plasma Leptin in Normal Weight and Obese Men and Women: Relationship with Body Composition, Resting Energy Expenditure, Insulin-sensitivity, Lipid Profile and Macronutrient Preference. *Clin. Endocrinol.* **2005**, *62*, 189–196. [CrossRef]
8. Popko, K.; Gorska, E.; Stelmaszczyk-Emmel, A.; Plywaczewski, R.; Stoklosa, A.; Gorecka, D.; Pyrzak, B.; Demkow, U. Proinflammatory Cytokines Il-6 and TNF-α and the Development of Inflammation in Obese Subjects. *Eur. J. Med. Res.* **2010**, *15*, 120–122. [CrossRef]
9. Tangvarasittichai, S.; Pongthaisong, S.; Tangvarasittichai, O. Tumor Necrosis Factor-A, Interleukin-6, C-Reactive Protein Levels and Insulin Resistance Associated with Type 2 Diabetes in Abdominal Obesity Women. *Indian J. Clin. Biochem.* **2016**, *31*, 68–74. [CrossRef] [PubMed]
10. Chakraborty, S.; Roy, S.; Rahaman, M. Epidemiological Predictors of Metabolic Syndrome in Urban West Bengal, India. *J. Family Med. Prim. Care* **2015**, *4*, 535. [CrossRef]
11. de Filippis, A.; Ullah, H.; Baldi, A.; Dacrema, M.; Esposito, C.; Garzarella, E.U.; Santarcangelo, C.; Tantipongpiradet, A.; Daglia, M. Gastrointestinal Disorders and Metabolic Syndrome: Dysbiosis as a Key Link and Common Bioactive Dietary Components Useful for Their Treatment. *Int. J. Mol. Sci.* **2020**, *21*, 4929. [CrossRef] [PubMed]
12. Prasad, H.; Ryan, D.A.; Celzo, M.F.; Stapleton, D. Metabolic Syndrome: Definition and Therapeutic Implications. *Postgrad. Med.* **2012**, *124*, 21–30. [CrossRef] [PubMed]
13. Rask Larsen, J.; Dima, L.; Correll, C.U.; Manu, P. The Pharmacological Management of Metabolic Syndrome. *Expert Rev. Clin. Pharmacol.* **2018**, *11*, 397–410. [CrossRef] [PubMed]
14. Matfin, G. Developing Therapies for the Metabolic Syndrome: Challenges, Opportunities, and the Unknown. *Ther. Adv. Endocrinol. Metab.* **2010**, *1*, 89–94. [CrossRef] [PubMed]
15. Russo, R.; Gallelli, L.; Cannataro, R.; Perri, M.; Calignano, A.; Citraro, R.; Russo, E.; Gareri, P.; Corsonello, A.; de Sarro, G. When Nutraceuticals Reinforce Drugs Side Effects: A Case Report. *Curr. Drug Saf.* **2016**, *11*, 264–266. [CrossRef]
16. Li, X.T.; Liao, W.; Yu, H.J.; Liu, M.W.; Yuan, S.; Tang, B.W.; Yang, X.H.; Song, Y.; Huang, Y.; le Cheng, S.; et al. Combined Effects of Fruit and Vegetables Intake and Physical Activity on the Risk of Metabolic Syndrome among Chinese Adults. *PLoS ONE* **2017**, *12*, e0188533. [CrossRef]
17. EFSA Panel on Food Additives and Nutrient Sources added to Food (ANS); Younes, M.; Aggett, P.; Aguilar, F.; Crebelli, R.; Dusemund, B.; Filipič, M.; Frutos, M.J.; Galtier, P.; Gott, D.; et al. Scientific Opinion on the Safety of Monacolins in Red Yeast Rice. *EFSA J.* **2018**, *16*, e05368.
18. Birari, R.B.; Bhutani, K.K. Pancreatic Lipase Inhibitors from Natural Sources: Unexplored Potential. *Drug Discov.* **2007**, *12*, 879–889. [CrossRef] [PubMed]

19. Tundis, R.; Loizzo, M.R.; Menichini, F. Natural Products as α-Amylase and α Glucosidase Inhibitors and Their Hypoglycaemic Potential in the Treatment of Diabetes: An Update. *Mini Rev. Med. Chem.* **2010**, *10*, 315–331. [CrossRef]

20. Lin, S.-H.; Huang, K.-J.; Weng, C.-F.; Shiuan, D. Exploration of Natural Product Ingredients as Inhibitors of Human HMG-CoA Reductase through Structure-Based Virtual Screening. *Drug Des. Devel. Ther.* **2015**, *9*, 3313. [PubMed]

21. Young Shin, J.; Young Kim, J.; Tak Kang, H.; Hwa Han, K.; Yong Shim, J. Effect of Fruits and Vegetables on Metabolic Syndrome: A Systematic Review and Meta-Analysis of Randomized Controlled Trials. *Int. J. Food Sci. Nutr.* **2015**, *66*, 416–425. [CrossRef] [PubMed]

22. Lee, M.; Lim, M.; Kim, J. Fruit and Vegetable Consumption and the Metabolic Syndrome: A Systematic Review and Dose-Response Meta-Analysis. *Br. J. Nutr.* **2019**, *122*, 723–733. [CrossRef] [PubMed]

23. El-Beltagi, H.S.; El-Ansary, A.E.; Mostafa, M.A.; Kamel, T.A.; Safwat, G. Evaluation of the Phytochemical, Antioxidant, Antibacterial and Anticancer Activity of Prunus Domestica Fruit. *Not. Bot. Horti Agrobot. Cluj Napoca* **2019**, *47*, 395–404. [CrossRef]

24. Igwe, E.O.; Charlton, K.E. A Systematic Review on the Health Effects of Plums (*Prunus domestica* and *Prunus salicina*). *Phytother. Res.* **2016**, *30*, 701–731. [CrossRef]

25. Furchner-Evanson, A.; Petrisko, Y.; Howarth, L.; Nemoseck, T.; Kern, M. Type of Snack Influences Satiety Responses in Adult Women. *Appetite* **2010**, *54*, 564–569. [CrossRef]

26. Nilsson, A.; Ostman, E.; Holst, J.; Björck, I. Including Indigestible Carbohydrates in the Evening Meal of Healthy Subjects Improves Glucose Tolerance, Lowers Inflammatory Markers, and Increases Satiety after a Subsequent Standardized Breakfast. *J. Nutr.* **2008**, *138*, 732–739. [CrossRef]

27. di Matteo, G.; Spano, M.; Esposito, C.; Santarcangelo, C.; Baldi, A.; Daglia, M.; Mannina, L.; Ingallina, C.; Sobolev, A.P. NMR Characterization of Ten Apple Cultivars from the Piedmont Region. *Foods* **2021**, *10*, 289. [CrossRef]

28. Ingallina, C.; Spano, M.; Sobolev, A.P.; Esposito, C.; Santarcangelo, C.; Baldi, A.; Daglia, M.; Mannina, L. Characterization of Local Products for Their Industrial Use: The Case of Italian Potato Cultivars Analyzed by Untargeted and Targeted Methodologies. *Foods* **2020**, *9*, 1216. [CrossRef]

29. Singleton, V.L.; Rossi, J.A. Colorimetry of Total Phenolics with Phosphomolybdic-Phosphotungstic Acid Reagents. *Am. J. Enol. Vitic.* **1965**, *16*, 144–158.

30. Kok, J.M.L.; Jee, J.M.; Chew, L.Y.; Wong, C.L. The Potential of the Brown Seaweed Sargassum Polycystum against Acne Vulgaris. *J. Appl. Phycol.* **2016**, *28*, 3127–3133. [CrossRef]

31. Mass Bank of North America. Available online: https://mona.fiehnlab.ucdavis.edu/ (accessed on 18 July 2021).

32. Sirius. Available online: https://bio.informatik.uni-jena.de/software/sirius/ (accessed on 18 July 2021).

33. Cicolari, S.; Dacrema, M.; Tsetegho Sokeng, A.J.; Xiao, J.; Atchan Nwakiban, A.P.; di Giovanni, C.; Santarcangelo, C.; Magni, P.; Daglia, M. Hydromethanolic Extracts from *Adansonia digitata* L. Edible Parts Positively Modulate Pathophysiological Mechanisms Related to the Metabolic Syndrome. *Molecules* **2020**, *25*, 2858. [CrossRef] [PubMed]

34. Nwakiban, A.P.A.; Sokeng, A.J.; Dell'Agli, M.; Bossi, L.; Beretta, G.; Gelmini, F.; Tchamgoue, A.D.; Agbor, G.A.; Kuiaté, J.R.; Daglia, M.; et al. Hydroethanolic Plant Extracts from Cameroon Positively Modulate Enzymes Relevant to Carbohydrate/Lipid Digestion and Cardio-Metabolic Diseases. *Food Funct.* **2019**, *10*, 6533–6542. [CrossRef]

35. Galuppo, M.; Rossi, A.; Giacoppo, S.; Pace, S.; Bramanti, P.; Sautebin, L.; Mazzon, E. Use of Mometasone Furoate in Prolonged Treatment of Experimental Spinal Cord Injury in Mice: A Comparative Study of Three Different Glucocorticoids. *Pharmacol. Res.* **2015**, *99*, 316–328. [CrossRef] [PubMed]

36. Capitani, D.; Sobolev, A.P.; Delfini, M.; Vista, S.; Antiochia, R.; Proietti, N.; Bubici, S.; Ferrante, G.; Carradori, S.; Salvador, F.R.D.; et al. NMR Methodologies in the Analysis of Blueberries. *Electrophoresis* **2014**, *35*, 1615–1626. [CrossRef]

37. Mannina, L.; Sobolev, A.P.; Viel, S. Liquid State 1H High Field NMR in Food Analysis. *Prog. Nucl. Magn. Reson. Spectrosc.* **2012**, *66*, 1–39. [CrossRef] [PubMed]

38. Capitani, D.; Mannina, L.; Proietti, N.; Sobolev, A.P.; Tomassini, A.; Miccheli, A.; di Cocco, M.E.; Capuani, G.; de Salvador, R.; Delfini, M. Monitoring of Metabolic Profiling and Water Status of Hayward Kiwifruits by Nuclear Magnetic Resonance. *Talanta* **2010**, *82*, 1826–1838. [CrossRef]

39. WHO Hypertension. Available online: https://www.who.int/ne%0Aws-room/fact-sheets/detail/hypertension (accessed on 23 June 2021).

40. WHO Diabetes. Available online: https://www.who.int/news-roo%0Am/fact-sheets/detail/diabetes (accessed on 23 June 2021).

41. WHO Obesity and Overweight. Available online: https://www.who.int/news-room/fact-sheets/detail/obesity-and-overweight (accessed on 23 June 2021).

42. Mandlekar, S.; Hong, J.L.; Tony Kong, A.N. Modulation of Metabolic Enzymes by Dietary Phytochemicals: A Review of Mechanisms Underlying Beneficial versus Unfavorable Effects. *Curr. Drug Metab.* **2006**, *7*, 661–675. [CrossRef] [PubMed]

43. Ullah, H.; de Filippis, A.; Khan, H.; Xiao, J.; Daglia, M. An Overview of the Health Benefits of Prunus Species with Special Reference to Metabolic Syndrome Risk Factors. *Food Chem. Toxicol.* **2020**, *144*, 111574. [CrossRef] [PubMed]

44. Moracci, L.; Traldi, P.; Agostini, M. Mass Spectrometry for a Holistic View of Natural Extracts of Phytotherapeutic Interest. *Mass Spectrom. Rev.* **2020**, *39*, 553–573. [CrossRef]

45. Kumar Singla, R.; Singh, R.; Kumar Dubey, A. Important Aspects of Post-Prandial Antidiabetic Drug, Acarbose. *Curr. Top. Med. Chem.* **2016**, *16*, 2625–2633. [CrossRef]

46. Nowicka, P.; Wojdyło, A.; Samoticha, J. Evaluation of Phytochemicals, Antioxidant Capacity, and Antidiabetic Activity of Novel Smoothies from Selected Prunus Fruits. *J. Funct. Foods* **2016**, *25*, 397–407. [CrossRef]
47. Wang, T.; Li, X.; Zhou, B.; Li, H.; Zeng, J.; Gao, W. Anti-Diabetic Activity in Type 2 Diabetic Mice and α-Glucosidase Inhibitory, Antioxidant and Anti-Inflammatory Potential of Chemically Profiled Pear Peel and Pulp Extracts (*Pyrus* spp.). *J. Funct. Foods* **2015**, *13*, 276–288. [CrossRef]
48. Boath, A.S.; Stewart, D.; McDougall, G.J. Berry Components Inhibit α-Glucosidase in Vitro: Synergies between Acarbose and Polyphenols from Black Currant and Rowanberry. *Food Chem.* **2012**, *135*, 929–936. [CrossRef] [PubMed]
49. Nor-Liyana, J.; Siroshini, K.T.; Nurul-Syahirah, M.B.; Chang, W.L.; Nurul-Husna, S.; Daryl, J.A.; Khairul-Kamilah, A.K.; Hasnah, B. Phytochemical Analysis of Elateriospermum Tapos and Its Inhibitory Effects on Alpha-Amylase, Alpha-Glucosidase and Pancreatic Lipase. *J. Trop. For. Sci.* **2019**, *31*, 240–248.
50. Popović, B.M.; Blagojević, B.; Kucharska, A.Z.; Agić, D.; Magazin, N.; Milović, M.; Serra, A.T. Exploring Fruits from Genus Prunus as a Source of Potential Pharmaceutical Agents–In Vitro and in Silico Study. *Food Chem.* **2021**, *358*, 129812. [CrossRef]
51. Catapano, A.L.; Graham, I.; de Backer, G.; Wiklund, O.; Chapman, M.J.; Drexel, H.; Hoes, A.W.; Jennings, C.S.; Landmesser, U.; Pedersen, T.R.; et al. 2016 ESC/EAS Guidelines for the Management of Dyslipidaemias: The Task Force for the Management of Dyslipidaemias of the European Society of Cardiology (ESC) and European Atherosclerosis Society (EAS) Developed with the Special Contribution of the Europea. *Atherosclerosis* **2016**, *253*, 281–344. [CrossRef]
52. Susilowati, R.; Jannah, J.; Maghfuroh, Z.; Kusuma, M.T. Antihyperlipidemic Effects of Apple Peel Extract in High-Fat Diet-Induced Hyperlipidemic Rats. *J. Adv. Pharm. Technol. Res.* **2020**, *11*, 128. [CrossRef]
53. Baskaran, G.; Salvamani, S.; Ahmad, S.A.; Shaharuddin, N.A.; Pattiram, P.D.; Shukor, M.Y. HMG-CoA Reductase Inhibitory Activity and Phytocomponent Investigation of Basella Alba Leaf Extract as a Treatment for Hypercholesterolemia. *Drug Des. Devel. Ther.* **2015**, *9*, 509. [CrossRef]
54. Hartanti, L.; Yonas, S.M.K.; Mustamu, J.J.; Wijaya, S.; Setiawan, H.K.; Soegianto, L. Influence of Extraction Methods of Bay Leaves (Syzygium Polyanthum) on Antioxidant and HMG-CoA Reductase Inhibitory Activity. *Heliyon* **2019**, *5*, e01485. [CrossRef] [PubMed]
55. Iqbal, D.; Khan, M.S.; Khan, A.; Khan, M.; Ahmad, S.; Srivastava, A.K.; Bagga, P. In Vitro Screening for β-Hydroxy-β-Methylglutaryl-Coa Reductase Inhibitory and Antioxidant Activity of Sequentially Extracted Fractions of Ficus Palmata Forsk. *Biomed Res. Int.* **2014**, *2014*, 1–10. [CrossRef]
56. Salvamani, S.; Gunasekaran, B.; Shukor, M.Y.; Shaharuddin, N.A.; Sabullah, M.K.; Ahmad, S.A. Anti-HMG-CoA Reductase, Antioxidant, and Anti-Inflammatory Activities of Amaranthus Viridis Leaf Extract as a Potential Treatment for Hypercholesterolemia. *Evid. Based Complement. Alternat. Med.* **2016**, *2016*, 1–10. [CrossRef]
57. Pilitsi, E.; Farr, O.M.; Polyzos, S.A.; Perakakis, N.; Nolen-Doerr, E.; Papathanasiou, A.E.; Mantzoros, C.S. Pharmacotherapy of Obesity: Available Medications and Drugs under Investigation. *Metabolism* **2019**, *92*, 170–192. [CrossRef]
58. FDA Orlistat (Marketed as Alli and Xenical) Information. Available online: https://www.fda.gov/drugs/postmarket-drug-safety-information-patients-and-providers/orlistat-marketed-alli-and-xenical-information (accessed on 2 August 2021).
59. FDA FDA Drug Safety Communication: Completed Safety Review of Xenical/Alli (Orlistat) and Severe Liver Injury. Available online: https://www.fda.gov/drugs/postmarket-drug-safety-information-patients-and-providers/fda-drug-safety-communication-completed-safety-review-xenicalalli-orlistat-and-severe-liver-injury (accessed on 2 August 2021).
60. European Medicines Agency European Medicines Agency Starts Review of Orlistat-Containing Medicines. Available online: https://www.ema.europa.eu/en/news/european-medicines-agency-starts-review-orlistat-containing-medicines (accessed on 2 August 2021).
61. Buchholz, T.; Melzig, M.F. Polyphenolic Compounds as Pancreatic Lipase Inhibitors. *Planta Med.* **2015**, *81*, 771–783. [CrossRef] [PubMed]
62. McDougall, G.J.; Stewart, D. The Inhibitory Effects of Berry Polyphenols on Digestive Enzymes. *Biofactors* **2005**, *23*, 189–195. [CrossRef] [PubMed]
63. Gholamhoseinian, A.; Shahouzehi, B.; Sharifi-Far, F. Inhibitory Effect of Some Plant Extracts on Pancreatic Lipase. *Int. J. Pharmacol.* **2010**, *6*, 18–24. [CrossRef]
64. Jaradat, N.; Zaid, A.N.; Hussein, F.; Zaqzouq, M.; Aljammal, H.; Ayesh, O. Anti-Lipase Potential of the Organic and Aqueous Extracts of Ten Traditional Edible and Medicinal Plants in Palestine; a Comparison Study with Orlistat. *Medicines* **2017**, *4*, 89. [CrossRef]
65. Nowicka, P.; Wojdyło, A.; Laskowski, P. Inhibitory Potential against Digestive Enzymes Linked to Obesity and Type 2 Diabetes and Content of Bioactive Compounds in 20 Cultivars of the Peach Fruit Grown in Poland. *Plant Foods Hum. Nutr.* **2018**, *73*, 314–320. [CrossRef]
66. Welty, F.K.; Alfaddagh, A.; Elajami, T.K. Targeting Inflammation in Metabolic Syndrome. *Transl. Res.* **2016**, *167*, 257–280. [CrossRef] [PubMed]
67. Silvan, J.M.; Michalska-Ciechanowska, A.; Martinez-Rodriguez, A.J. Modulation of Antibacterial, Antioxidant, and Anti-Inflammatory Properties by Drying of *Prunus domestica* L. Plum Juice Extracts. *Microorganisms* **2020**, *8*, 119. [CrossRef]
68. Hooshmand, S.; Kumar, A.; Zhang, J.Y.; Johnson, S.A.; Chai, S.C.; Arjmandi, B.H. Evidence for Anti-Inflammatory and Antioxidative Properties of Dried Plum Polyphenols in Macrophage RAW 264.7 Cells. *Food Funct.* **2015**, *6*, 1719–1725. [CrossRef]

69. Sunil, M.A.; Sunitha, V.S.; Santhakumaran, P.; Mohan, M.C.; Jose, M.S.; Radhakrishnan, E.K.; Mathew, J. Protective Effect of (+)–Catechin against Lipopolysaccharide-Induced Inflammatory Response in RAW 264.7 Cells through Downregulation of NF-KB and P38 MAPK. *Inflammopharmacology* **2021**, *29*, 1139–1155. [CrossRef]
70. Lee, S.H.; Lee, S.Y.; Son, D.J.; Lee, H.; Yoo, H.S.; Song, S.; Oh, K.W.; Han, D.C.; Kwon, B.M.; Hong, J.T. Inhibitory Effect of 2′-Hydroxycinnamaldehyde on Nitric Oxide Production through Inhibition of NF-KB Activation in RAW 264.7 Cells. *Biochem. Pharmacol.* **2005**, *69*, 791–799. [CrossRef]
71. Zhang, M.; Xie, Y.; Su, X.; Liu, K.; Zhang, Y.; Pang, W.; Wang, J. Inonotus Sanghuang Polyphenols Attenuate Inflammatory Response via Modulating the Crosstalk between Macrophages and Adipocytes. *Front. Immunol.* **2019**, *10*, 286. [CrossRef] [PubMed]
72. Overman, A.; Bumrungpert, A.; Kennedy, A.; Martinez, K.; Chuang, C.C.; West, T.; Dawson, B.; Jia, W.; McIntosh, M. Polyphenol-Rich Grape Powder Extract (GPE) Attenuates Inflammation in Human Macrophages and in Human Adipocytes Exposed to Macrophage-Conditioned Media. *Int. J. Obes.* **2010**, *34*, 800–808. [CrossRef] [PubMed]
73. Zielińska-Wasielica, J.; Olejnik, A.; Kowalska, K.; Olkowicz, M.; Dembczyński, R. Elderberry (*Sambucus nigra* L.) Fruit Extract Alleviates Oxidative Stress, Insulin Resistance, and Inflammation in Hypertrophied 3T3-L1 Adipocytes and Activated RAW 264.7 Macrophages. *Foods* **2019**, *8*, 326. [CrossRef] [PubMed]
74. Ekor, M. The Growing Use of Herbal Medicines: Issues Relating to Adverse Reactions and Challenges in Monitoring Safety. *Front. Pharmacol.* **2014**, *4*, 177. [CrossRef]

Article

Cameroonian Spice Extracts Modulate Molecular Mechanisms Relevant to Cardiometabolic Diseases in SW 872 Human Liposarcoma Cells

Achille Parfait Atchan Nwakiban [1,†], Anna Passarelli [2,†], Lorenzo Da Dalt [2], Chiara Olivieri [2], Tugba Nur Demirci [2], Stefano Piazza [2], Enrico Sangiovanni [2], Eugénie Carpentier-Maguire [3], Giulia Martinelli [2], Shilpa Talkad Shivashankara [4], Uma Venkateswaran Manjappara [4], Armelle Deutou Tchamgoue [5], Gabriel Agbor Agbor [5], Jules-Roger Kuiate [1], Maria Daglia [6,7], Mario Dell'Agli [2,*] and Paolo Magni [2,8,*]

[1] Department of Biochemistry, Faculty of Science, University of Dschang, Dschang P.O. Box 96, Cameroon; achilestyle@yahoo.fr (A.P.A.N.); jrkuiate@yahoo.com (J.-R.K.)

[2] Department of Pharmacological and Biomolecular Sciences, Università degli Studi di Milano, 20133 Milan, Italy; anna.passarelli2393@gmail.com (A.P.); lorenzo.dadalt@unimi.it (L.D.D.); chia.olivieri@gmail.com (C.O.); tugba.demirci@studenti.unimi.it (T.N.D.); stefano.piazza@unimi.it (S.P.); enrico.sangiovanni@unimi.it (E.S.); giulia.martinelli@unimi.it (G.M.)

[3] Department of Science and Technology, University of Lille, Rue de Lille, 59160 Lille, France; eugenie.carpentiermaguire@gmail.com

[4] Department of Lipid Science, CSIR-Central Food Technological Research Institute (CFTRI), Mysore 570 020, India; shilpa25188@gmail.com (S.T.S.); umamanjappara@cftri.res.in (U.V.M.)

[5] Institute of Medical Research and Medicinal Plants Studies (IMPM), Yaoundé 4123, Cameroon; armelle_d2002@yahoo.fr (A.D.T.); agogae@yahoo.fr (G.A.A.)

[6] Department of Pharmacy, University of Naples Federico II, 80131 Naples, Italy; maria.daglia@unina.it

[7] International Research Center for Food Nutrition and Safety, Jiangsu University, Zhenjiang 212013, China

[8] IRCCS MultiMedica, Sesto San Giovanni, 20099 Milan, Italy

[*] Correspondence: mario.dellagli@unimi.it (M.D.); paolo.magni@unimi.it (P.M.); Tel.: +39-0250318398 (M.D.); +39-0250318229 (P.M.)

[†] These authors contributed equally to this work.

Citation: Atchan Nwakiban, A.P.; Passarelli, A.; Da Dalt, L.; Olivieri, C.; Demirci, T.N.; Piazza, S.; Sangiovanni, E.; Carpentier-Maguire, E.; Martinelli, G.; Shivashankara, S.T.; et al. Cameroonian Spice Extracts Modulate Molecular Mechanisms Relevant to Cardiometabolic Diseases in SW 872 Human Liposarcoma Cells. *Nutrients* **2021**, *13*, 4271. https://doi.org/10.3390/nu13124271

Academic Editor: Flávio Reis

Received: 9 November 2021
Accepted: 25 November 2021
Published: 26 November 2021

Publisher's Note: MDPI stays neutral with regard to jurisdictional claims in published maps and institutional affiliations.

Abstract: The molecular pathophysiology of cardiometabolic diseases is known to be influenced by dysfunctional ectopic adipose tissue. In addition to lifestyle improvements, these conditions may be managed by novel nutraceutical products. This study evaluated the effects of 11 Cameroonian medicinal spice extracts on triglyceride accumulation, glucose uptake, reactive oxygen species (ROS) production and interleukin secretion in SW 872 human adipocytes after differentiation with 100 μM oleic acid. Triglyceride content was significantly reduced by all spice extracts. Glucose uptake was significantly increased by *Tetrapleura tetraptera*, *Aframomum melegueta* and *Zanthoxylum leprieurii*. Moreover, *Xylopia parviflora*, *Echinops giganteus* and *Dichrostachys glomerata* significantly reduced the production of ROS. Concerning pro-inflammatory cytokine secretion, we observed that *Tetrapleura tetraptera*, *Echinops giganteus*, *Dichrostachys glomerata* and *Aframomum melegueta* reduced IL-6 secretion. In addition, *Xylopia parviflora*, *Monodora myristica*, *Zanthoxylum leprieurii*, and *Xylopia aethiopica* reduced IL-8 secretion, while *Dichrostachys glomerata* and *Aframomum citratum* increased it. These findings highlight some interesting properties of these Cameroonian spice extracts in the modulation of cellular parameters relevant to cardiometabolic diseases, which may be further exploited, aiming to develop novel treatment options for these conditions based on nutraceutical products.

Keywords: Cameroonian spice extracts; SW 872 adipocytes; triglyceride accumulation; glucose uptake; oxidative stress; pro-inflammatory cytokines

1. Introduction

The use of well-characterized nutraceutical products, in addition to lifestyle changes, is an interesting option for prevention and management of chronic cardiovascular and metabolic diseases, including obesity, type 2 diabetes mellitus (T2DM), atherosclerosis and the metabolic syndrome [1], especially in milder pathological conditions [2,3]. In this context, dysfunction of the adipose tissue and derangement of adipocyte biology play a pivotal role involving several pathophysiological mechanisms including insulin resistance, oxidative stress, and low-grade chronic inflammation. Interestingly, many plant extracts used in traditional medicine were found to mitigate obesity and related dysmetabolic conditions in in-vitro experimental models as well as in clinical studies [4]. Our research group, along with other researchers, has previously investigated the composition and the molecular effects of several spice extracts from Cameroonian plants commonly used in traditional medicine. In particular, we explored the ability of these extracts to modulate the activity of some enzymes relevant to the control of cardiometabolic functions [5], the antioxidant and anti-inflammatory activities by gastric epithelial cells [6] and glucose uptake and antioxidant activity in human HepG2 cells [7].

Based on this knowledge, in this study we extended the evaluation of some potentially useful effects of these Cameroonian spice extracts in the context of adipocyte biology and pathophysiology. To address this issue, we took advantage of human SW 872 human liposarcoma cells differentiated to adipocytes by oleic acid (OA) as the model system, which previously proved suitable for testing the in-vitro metabolic activities of plant-derived extracts [8]. More specifically, the present study aimed at assessing the effects of these spice extracts on triglyceride (TG) accumulation, glucose uptake modulation, antioxidant activity and regulation of pro-inflammatory cytokine release.

2. Materials and Methods

2.1. Chemicals

Bovine insulin (Cat. No. I6634), oleic acid (OA) (Cat. No. O1257), ($\pm$)-6-Hydroxy 2,5,7,8 tetramethylchromane-2-carboxylic acid (Trolox) (Cat. No. 93510), dimethylsulfoxide (DMSO) (Cat. No. D8418), metformin hydrochloride (Cat. No. M0605000), epigallocatechin gallate (EGCG) (Cat. No. E4143), 2-deoxy-2-[(7-nitro-2,1,3-benzoxadiazol-4-yl) amino]-D-glucose (2-NBDG) (Cat. No. N13195) and resveratrol (Cat. No. R5010) were obtained from Merck Life Science (Readington, NJ, USA).

5-(and-6)-chloromethyl-2′,7′-dichlorodihydrofluorescein diacetate (CM-H2DCFDA) (Cat. No. C6827) was obtained from Thermo Fisher Scientific (Waltham, MA, USA). Fetal Bovine Serum (FBS) (Cat. No. ECS0180L) (Euroclone, Milan, Italy).

2.2. Preparation of Plant Extracts

Plant material comprised 11 spices (*Xylopia aethiopica* (Dunal) A.Rich., *Xylopia parviflora* (A. Rich) Benth, *Scorodophloeus zenkeri* Harms, *Echinops giganteus* A.Rich., *Monodora myristica* (Gaertn.) Dunal, *Tetrapleura tetraptera* (Schum. & Thonn.) Taub., *Dichrostachys glomerata* (Forssk.) Chiov., *Afrostyrax lepidophyllus* Mildbr., *Aframomum melegueta* K.Schum., *Aframomum citratum* (C.Pereira) K.Schum., and *Zanthoxylum leprieurii* Guill. & Perr.) harvested in different sites of the region of West Cameroon. Samples included fruits, seeds and roots, as identified in the National Herbarium of Cameroon (https://irad.cm/index.php/fr/, accessed on 14 October 2021) in Yaoundé (Cameroon) compared to the deposited specimens. The preparation of the hydroethanolic extracts was conducted as previously reported [5]. Briefly, air-dried and powdered samples (100 g) were extracted using a 70% hydroethanolic mixture at room temperature and in dark conditions. The mixture was filtered, concentrated under reduced pressure, and frozen to give crude extracts. They were then lyophilized and stored at $-20\,^{\circ}$C. Spice extract stock solutions (100 mg/mL) dissolved in DMSO were prepared, aliquoted, and kept at $-80\,^{\circ}$C. The stock solution of spice extracts in DMSO was diluted in culture medium to the appropriate concentrations for cell treatments. For more details, please check Table 1 in [7]. The same concentration of DMSO (in any case never exceeding 0.1%) was also added to all control cultures.

Table 1. Triglyceride accumulation in differentiated SW 872 adipocytes: effect of spice extracts.

	Oleic Acid (100 µM)	Triglyceride (% of Oleic Acid-Treated Cells)	
		+24 h	+48 h
undifferentiated control	−	49.6 ± 4.1 ***	64 ± 0.7 ***
differentiated control	+	100	100
resveratrol (10 µM)	+	94.9 ± 5.9	69.6 ± 3.0 ***
Xylopia aethiopica	+	96.5 ± 7.6	85.5 ± 2.3 *
Xylopia parviflora	+	88.2 ± 1.0	86.2 ± 1.6 *
Scorodophloeus zenkeri	+	97.9 ± 9.6	81.5 ± 4.9 *
Monodora myristica	+	90.4 ± 9.0	84.7 ± 3.2 *
Tetrapleura tetraptera	+	90.9 ± 4.5	86.2 ± 2.2 *
Echinops giganteus	+	104.2 ± 5.9	88.7 ± 0.9 *
Afrostyrax lepidophyllus	+	102.5 ± 6.4	83.5 ± 10.1 *
Dichrostachys glomerata	+	102.6 ± 2.5	82.6 ± 6.8 *
Aframomum melegueta	+	104 ± 5.2	87.0 ± 4.7 *
Aframomum citratum	+	98.3 ± 4.7	84 ± 1.8 *
Zanthoxylum leprieurii	+	98.5 ± 8.8	86.6 ± 5.9 *

All spice extracts were used at 10 µg/mL. Data are expressed as % of oleic acid-treated cells taken as 100; mean ± SD, $n = 3$, ($p < 0.05$) (one-way ANOVA multiple comparison); * $p < 0.05$, *** $p < 0.001$ vs. the differentiated control group.

2.3. Cell Cultures and Differentation

The SW 872 human liposarcoma (ATCC® HTB-92 ™) cell line was from the American Type Culture Collection (ATCC®, Manassas, VA, USA) and grown as recommended. Cells were cultured in DMEM-F12 culture medium with L-Glutamine and HEPES 25 mM supplemented with 10% fetal bovine serum (FBS), 1% penicillin (100 U/mL), and streptomycin (100 µg/mL). They were kept in culture in 100-mm diameter Petri dishes at 37 °C in a humidified atmosphere containing 5% CO_2. Once they reached 80100% confluence, SW 872 cells were treated with 100 µM OA to initiate cellular differentiation into adipocytes [9,10].

2.4. Oil-Red-O Staining

To assess intracellular lipid accumulation (a marker of adipocyte differentiation) of SW 872 cells, we used the Oil-red-O (ORO) staining. SW 872 cells were seeded in 24-well plates, were allowed to adhere until 90–100%, confluence and then were treated with 100 µM OA for 7 days. After medium removal, cells were washed with phosphate-buffered saline (PBS) and fixed with 4% formaldehyde in PBS for 1 h (room temperature). ORO working solution (0.2% ORO in 40% isopropanol) was added to the culture dish and incubated at room temperature for 20 min. After a wash with sterile distilled water, cell images were collected with a light microscope (ZEISS, Milan, Italy).

2.5. MTS (3-(4,5-Dimethylthiazol-2-yl)-5-(3-carboxymethoxyphenyl)-2-(4-sulfophenyl)-2H-tetrazolium) Cell Viability Assay

Cell viability assay [11] was performed using the Cell Titer 96 aqueous non-radioactive cell proliferation assay (Promega, Madison, WI, USA) following the method previously described [12] with minor modifications. Briefly, SW 872 cells were seeded in a sterile, flat-bottom 96-well plate at a density of 2×10^5 cells/well and incubated at 37 °C for 24 h in a humidified incubator containing 5% CO_2. Extracts at different concentrations (1, 10, 25, 50, and 100 µg/mL) were prepared in fresh serum-free DMEM-F12 medium and 100 µL of each treatment was added to each well and then incubated for 24 and 48 h. Subsequently, 20 µL of the MTS reagent in combination with the electron coupling agent phenazine methosulfate was added to each well and allowed to react for 1 h at 37 °C. After 2 min of shaking at lowest intensity, the absorbance at 490 nm was measured (EnSpire PerkinElmer Multimode Plate Reader). In the same conditions, controls and blanks, consisting of cells with media containing DMSO ($\leq$0.1%) and wells containing media without cells, respectively, were performed. The cell viability values were determined using the equation:

$$\% \ cell \ viability = \left(\frac{mean \ sample \ absorbance - mean \ blank \ absorbance}{mean \ control \ absorbance - mean \ blank \ absorbance} \right) \times 100 \qquad (1)$$

Three separate experiments run in triplicate replicates were conducted.

2.6. Morphological Analysis

Cells were cultured in sterile, flat-bottom 6-cm^2 dishes for 24 h. Different concentrations of extracts (1, 10, 20 and 100 μg/mL) were prepared in serum-free media and 2 mL of each treatment was added to each dish and incubated for 24 h and 48 h. After treatment, cells were visualized with a light microscope (ZEISS, Milan, Italy) using 10×, 20× and 32× magnifications.

2.7. Interleukin-6 and Interleukin-8 Measurement

Interleukin-6 (IL-6) and interleukin-8 (IL-8) medium content was quantified using human IL-6 and IL-8 ELISA kits, as previously described [13]. Briefly, Corning 96-well EIA/RIA plates from Merck Life Science (Milan, Italy) were coated with the antibody provided in the ELISA Kit (Peprotech, London, UK) overnight at 4 °C. After blocking the reaction, 200 μL of samples in duplicate were transferred into wells and incubated at room temperature for 1 h. The amount of IL-6 and IL-8 in the samples was detected by the EnSpire Plate Reader (signal read: 450 nm, 0.1 s) using biotinylated and streptavidin–HRP conjugate antibodies, evaluating the 3,3′,5,5′ tetramethylbenzidine (TMB) substrate reaction. Quantification of IL-6 and IL-8 was done using an optimized standard curve provided with the ELISA Kit (8.0–1000.0 pg/mL). Resveratrol (10 μM) and EGCG (40 μM) were used as positive control for IL-6 and IL-8, respectively.

2.8. Triglyceride Content Measurement

To determine the accumulation of TG in the cells we used a sensitive assay using a TG quantification reagent (Vaktro Scientific, Patras, Greece). In this assay, TG are converted to free fatty acids and glycerol. Then, glycerol is oxidized to generate a colorimetric (570 nm)/fluorometric (λex = 535 nm/λem = 587 nm) product. The kit can detect 2 pmole–10 nmole (2–10,000 μM range) of TG. All samples and standards were measured in duplicate. The data obtained were interpolated in an appropriate triglycerides standard curve after background correction.

2.9. Measurement of ROS Production

The ntracellular ROS level in SW 872 cells were determined by a fluorometric assay, applying the oxidant-sensitive fluorescence probe CM-H$_2$DCFDA according to the method described by Piazza et al. [14]. Cells were seeded in 96-well black plates at a density of 2×10^5 cells/well, and after reaching 90% confluence, differentiated as previously described for 7 days with 100 μM oleic acid. Further, they were incubated with plant extracts at different concentrations ranging from 0 to 20 μg/mL for 24 h. Cells were then treated with 20 μM CM-H$_2$DCFDA and then incubated for 1 h with 500 μM hydrogen peroxide to induce ROS production. The resulting fluorescence intensities were quantified at an excitation wavelength of 485 nm and an emission wavelength of 535 nm using the EnSpire Plate Reader (Perkin Elmer, Milan, Italy).

2.10. Glucose Uptake FACS Analysis

Differentiated and non-differentiated SW 872 cells were cultured in 12-well plates in DMEM F-12 + 10% FBS. After reaching about 80% confluence, cells were incubated with spice extracts (10–20 μg/mL) or insulin at different timing and concentrations. After incubation, cells were washed with PBS and treated with 20 μM 2-NBDG for 30 min in MEM

Eagle w/Earle's BSS without glucose. Finally, the 2-NBDG uptake was measured detecting the emitted fluorescence with FACS (BD LSRFortessa™ Flow Cytometer, Milan, Italy).

2.11. Western Blotting Analysis

Total protein extracts from SW 872 cells were obtained by lysing cells in 150 μL RIPA buffer containing a mix of protease and phosphatase inhibitors (Roche Diagnostics, Monza, Italy). Forty μg of proteins and a molecular mass marker (Novex® Sharp Protein Standard, InvitrogenTM; Life Technologies, Monza, Italy) were separated on 10% sodium dodecylsulfate-polyacrylamide gel (SDS-PAGE) under denaturing and reducing conditions. They were then transferred to a nitrocellulose membrane by using the iBlotTM Gel Transfer Device (InvitrogenTM; Life Technologies). The membranes were washed with Tris-Buffered Saline-Tween 20 (TBS-T) and non-specific binding sites were blocked in TBS-T containing 5% BSA (Merck Life Science, Milan, Italy) or non-fat dried milk at room temperature for 60 min. The blots were incubated overnight at 4 °C with anti-pAKT(Ser473), (1:150; Millipore, Merck Life Science), anti-AKT (1:1000; Cell Signaling Technology, Milan, Italy), and anti β -Actin (Millipore, diluted 1:5000) (5% BSA or non-fat dried milk). Membranes were washed with TBS-T and then exposed for 30 min at room temperature to a diluted solution (5% non-fat dried milk) of the secondary antibodies. Immunoreactive bands were detected by exposure of the membranes to ClarityTM Western ECL chemiluminescent substrates (Bio-Rad Laboratories, Segrate, Italy) for 5 min and images were acquired with a ChemiDocTM XRS System (Bio-Rad Laboratories, Milan, Italy). Densitometric readings were assessed using the ImageLab software(6.1 version, Bio-Rad Laboratories, Milan, Italy).

2.12. Statistical Analysis

Results from at least three independent experiments carried out in triplicate were expressed as mean ±SD, values or as a mean percentage (%) compared to a control. Graphs and data were analyzed using GraphPad Prism version 8. The statistical analysis was made with one-way ANOVA followed by Tukey test, considering statistically significant $p < 0.05$.

3. Results

3.1. Spice Extracts up to 25 μg/mL Do Not Affect SW 872 Cell Viability and Morphology

Preliminary experiments were conducted to assess the effect of 11 Cameroonian spice extracts on the viability of differentiated SW 872 cells by MTS assay and morphological analysis. The extraction protocol, molecular composition, and potential cytotoxic effects in other cell lines of the spice extracts used in this study have been previously reported [5–7]. The MTS assay was conducted after treatment with different concentrations (1–100 μg/mL) of extracts for 24 h and 48 h. Viability threshold was set at 80% [15]. All spice extracts incubated for 24 h or 48 h proved non toxic up to the concentration of 25 μg/mL (Figure S1). More specifically, after 24 h of incubation, 6/11 extracts proved toxic at 50 μg/mL and after 48 h 2/11 plants were toxic already at 25 μg/mL. Toxicity at 100 μg/mL was observed for 10/11 extracts in SW 872 cells, independently of the incubation time (Figure S1). We also investigated whether spice extracts affected cell morphology in addition to viability. The morphology of differentiated SW 872 cells was not affected by treatment for 24 h and 48 h with 1–25 μg/mL spice extracts. At 100 μg/mL, the maximum concentration tested, all but one (*A. lepidophyllus*) spice extracts produced morphological changes, including cell death, rounding, and shrinking (representative images obtained with *X. aethiopica* and *A. lepidophyllus* extracts are shown in Figure S2). Based upon these results obtained by MTS assay and morphological analysis, all treatments with spice extracts were conducted for 24 h or 48 h at the concentration of 20 μg/mL or lower.

3.2. All Spice Extracts Reduce Triglyceride Accumulation in Differentiated Adipocytes

SW 872 cells were differentiated to adipocytes with 100 μM OA for seven days, according to a previously published protocol [8,9]. This treatment led to a marked intracellular lipids accumulation, as observed by ORO staining analysis (Figure 1) and quantification of TG content (Table 1), indicating the acquisition of a mature adipocyte phenotype. Of interest, differentiation with OA resulted in a doubling of TG accumulation ($p < 0.05$), compared to untreated cells (Table 1). We then assessed whether treatment with spice extracts was able to modulate TG accumulation in differentiated SW 872 cells, which were treated for 24 h and 48 h with spice extracts (all at 10 μg/mL). TG content did not change after a 24 h exposure to extracts. However, after 48 h, all extracts were able to significantly ($p < 0.05$) decrease TG accumulation ($-11.3\%/-18.5\%$). Resveratrol (10 μM), taken as active control, significantly ($p < 0.05$) reduced TG accumulation by 30.4% (Table 1). We then assessed whether treatment with spice extracts was able to modulate TG accumulation in differentiated SW 872 cells, which were treated for 24 h and 48 h with spice extracts (all at 10 μg/mL). TG content did not change after a 24 h exposure to extracts. However, after 48 h, all extracts were able to significantly ($p < 0.05$) decrease TG accumulation ($-11.3\%/-18.5\%$). Resveratrol (10 μM), taken as active control, significantly ($p < 0.05$) reduced TG accumulation by 30.4% (Table 1).

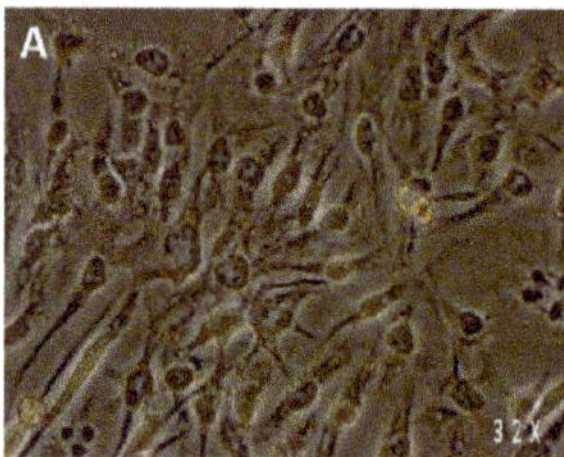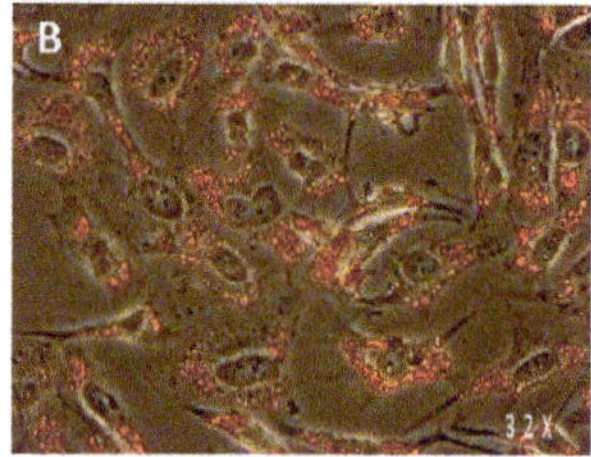

Figure 1. Intracellular lipid content (Oil-red-O staining) of SW 872 cells after 7 days of incubation without any treatment (**A**) or with 100 μM oleic acid (**B**).

3.3. Tetrapleura tetraptera, Aframomum melegueta and Zanthoxylum leprieurii Increase Glucose Uptake in Differentiated Adipocytes

Glucose uptake is a pivotal event of adipocyte biology [16] and was assessed here in both undifferentiated and differentiated SW 872 cells by FACS analysis, measuring the uptake of 2-NBDG. Basal glucose uptake was significantly reduced (-54.8% vs controls; $p < 0.001$) upon OA-differentiation compared to undifferentiated controls (Figure 2A). Insulin-activated glucose uptake was assessed according to time-course (15–60 min) and dose-response (10 nM–1 μM) parameters (Figure S3A,B). A significant increase of glucose uptake was however observed in differentiated cells after 60 min incubation with 10 nM (+16.8%; $p < 0.05$) and 100 nM (+18.8%; $p < 0.01$) insulin (Figure S3B). The uptake of 2-NBDG was significantly ($p < 0.001$) increased by *T. tetraptera* (+40.8%), *A. melegueta* (+41.7%) and *Z. leprieurii* (+56.6%), all at the concentration of 10 μg/mL (Figure 2B,C). This increment was even greater than that elicited by 100 nM insulin (+18.8%; $p < 0.01$). To test whether such increase was dose dependent, we selected the three most effective extracts and added *S. zenkeri* since in preliminary experiments this spice extract was effective at 20 μg/mL (Figure 2D).

We then explored the activation of the Akt signalling pathway, known to be associated with insulin receptors, under non-differentiated and OA-differentiated conditions and observed that 100 nM insulin treatment elicited a rapid and prolonged phosphorylation of Akt in OA-differentiated cells but not in non-differentiated cells (Figure 3A). This insulin-driven activation of Akt was not affected by a 24-h pretreatment of SW 872 cells with the three extracts that were found more effective in promoting glucose uptake (*A. melegueta*, *Z. leprieurii*, and *T. tetraptera*) (Figure 3B).

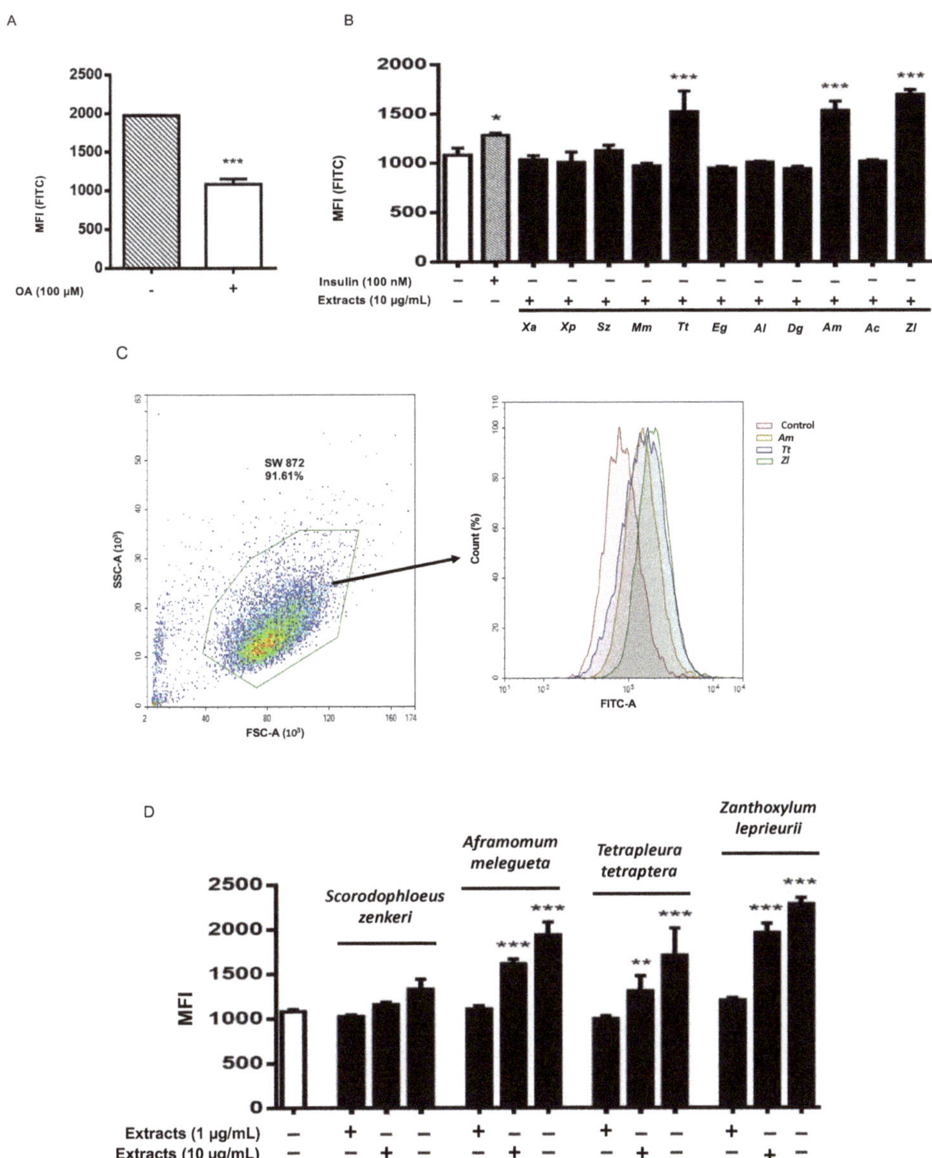

Figure 2. Glucose uptake in SW 872 cells treated with spice extracts. (**A**) Basal glucose uptake in non-differentiated and differentiated cells (100 μM OA); *** $p < 0.001$ (unpaired *t*-test). (**B**) Differentiated cells were incubated for 1 h with 100 nM insulin or treated for 24 h with 10 μg/mL of each spice extract. *Xa, Xylopia aethiopica; Xp, Xylopia parviflora; Sz, Scorodophloeus zenkeri; Mm, Monodora myristica; Tt, Tetrapleura tetraptera; Eg, Echinops giganteus; Al, Afrostyrax lepidophyllus; Dg, Dichrostachys glomerata; Am, Aframomum melegueta; Ac, Aframomum citratum; Zl, Zanthoxylum leprieurii.* * $p < 0.05$, *** $p < 0.001$. (**C**) Mean fluorescence intensity (MFI) of FACS analysis obtained in differentiated SW 872 cells by treatment with *Aframomum melegueta* (*Am*), *Zanthoxylum leprieurii* (*Zl*) and *Tetrapleura tetraptera* (*Tt*). The control (red area) is represented by differentiated SW 872 cells MFI. (**D**) Differentiated cells were treated for 24 h with the indicated extracts at 1–10–20 μg/mL. Glucose uptake was assessed in SW 872 cells by FACS analysis. One experiment (*n* = 3) is shown as representative of 3 separated experiments, each in triplicate. Results are expressed as mean ± SD. ** $p < 0.01$, *** $p < 0.001$ (one-way ANOVA multiple comparison).

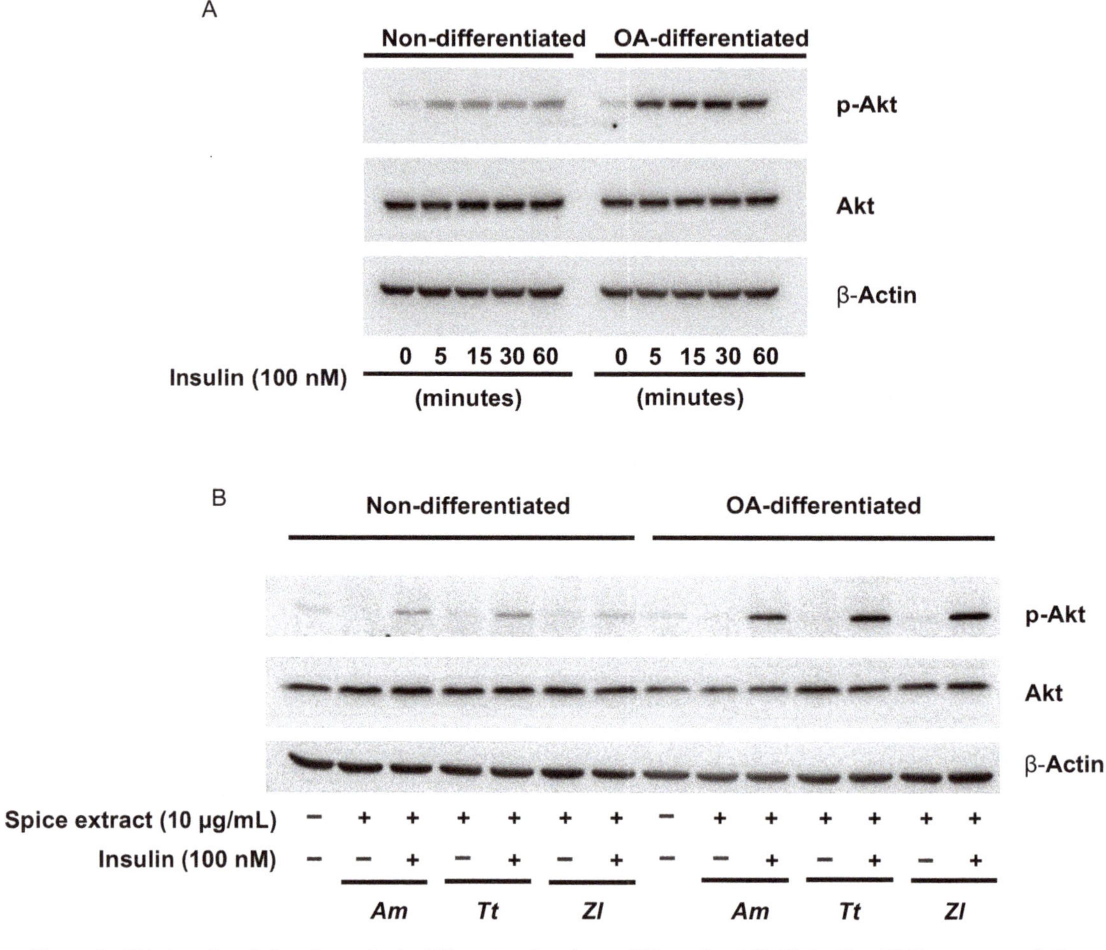

Figure 3. Akt phosphorylation dynamics in differentiated and non-differentiated SW 872 cells. (**A**) Time course of Akt phosphorylation upon treatment with 100 nM insulin. (**B**) Combined effect of 24-h exposure to 10 µg/mL spice extracts (*A. melegueta* (*Am*), *Z. leprieurii* (*Zl*) and *T. tetraptera* (*Tt*)) and 1-h treatment with 100 nM insulin on Akt phosphorylation.

3.4. Spice Extracts Modulate IL-6 and IL-8 Release from Differentiated Adipocytes

The release of pro-inflammatory cytokines, like IL-6 and IL-8, is a central pathophysiological feature of disfunctional adipocytes [17]. OA-differentiation of SW 872 cells resulted in a significant ($p < 0.001$) increase of IL-6 release (Figure 4A), which was significantly reduced by exposure (24 h, 10 µg/mL) to *A. melegueta* (−43.1%; $p < 0.001$), *D. glomerata* (−39.9%; $p < 0.001$), *T. tetraptera* (−29.7%; $p < 0.05$) and *E. giganteus* (−29%; $p < 0.05$) (Figure 4B). Resveratrol, selected as positive control, resulted in a 67.5% ($p < 0.001$) reduction (Figure 4B). Conversely, IL-8 release was significantly reduced ($p < 0.01$) upon OA-differentiation of SW 872 cells (Figure 4C). IL-8 secretion was further decreased by exposure to *X. parviflora*, (−36.8%; $p < 0.001$), *X. aethiopica* (−21.1%; $p < 0.05$), *M. myristica* (−24.3%; $p < 0.01$) and *Z. leprieurii* (−32.7%; $p < 0.01$) (Figure 4D). On the contrary, treatment with *D. glomerata* and *A. citratum* resulted in a marked increase of IL-8 release (+58.6% and +78.7%, respectively; $p < 0.001$).

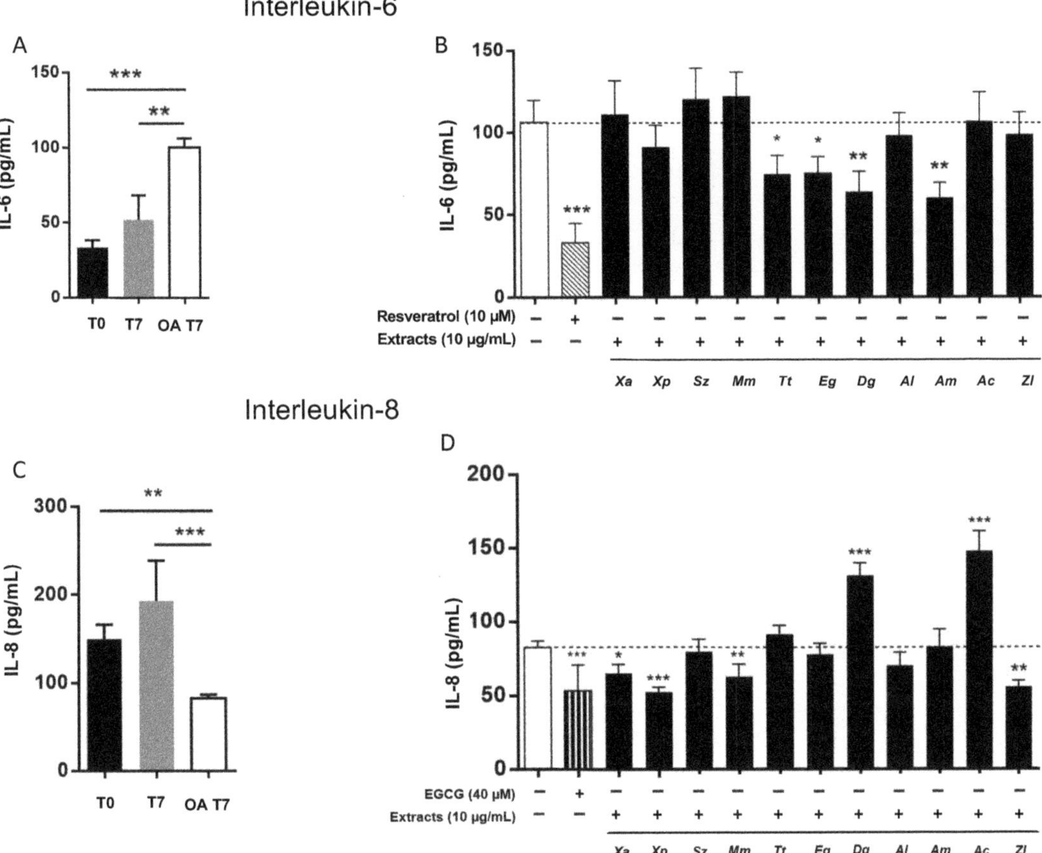

Figure 4. Effects of spice extracts on interleukin-6 (IL-6) and interleukin-8 (IL-8) release by differentiated SW 872 cells. (**A**) Basal IL-6 content was determined at T0 and T7 (1 and 7 days after seeding, respectively) and at OA T7 (after 7-day differentiation with 100 μM oleic acid (OA)). (**B**) OA-differentiated cells were treated with 10 μg/mL spice extracts or the positive controls (10 μM resveratrol or 40 μM epigallocatechin gallate (EGCG)). (**C**) Basal IL-8 content was determined at T0 and T7 (1 and 7 days after seeding, respectively) and at OA T7 (after 7-day differentiation with 100 μM OA). (**D**) OA-differentiated cells were treated with 10 μg/mL spice extracts or the positive controls (10 μM resveratrol or 40 μM EGCG). IL-6 and IL-8 content in the culture medium was determined after 24 h incubation. One experiment ($n = 3$) is shown as representative of 3 separate experiments, each in triplicate. Results are shown as mean ± SD. Data are expressed as pg/mL. * $p < 0.05$, ** $p < 0.01$, *** $p < 0.001$ (one-way ANOVA multiple comparison). *Xa*: *Xylopia aethiopica*; *Xp*: *Xylopia parviflora*; *Sz*: *Scorodophloeus zenkeri*; *Mm*: *Monodora myristica*; *Tt*: *Tetrapleura tetraptera*; *Eg*: *Echinops giganteus*; *Dg*: *Dichrostachys glomerata*; *Al*: *Afrostyrax lepidophyllus*; *Am*: *Aframomum melegueta*; *Ac*: *Aframomum citratum*; *Zl*: *Zanthoxylum leprieurii*.

3.5. Spice Extracts Affect Reactive Oxygen Species Production in Differentiated Adipocytes

The protective effect of antioxidants on cells is related to their ability to reduce the level of intracellular ROS generation. To assess whether spice extracts could reduce intracellular oxidative stress, intracellular ROS generation was quantified in differentiated SW 872 cells using the CM-H$_2$DCFDA assay. Cells were differentiated with OA for seven days and treated or not for another 24 h with spice extracts before being exposed for 1 h to 500 μM H$_2$O$_2$. Treatment with 500 μM H$_2$O$_2$ alone produced in a 2.5-fold increase in intracellular ROS levels (Table 2).

Table 2. Intracellular ROS production in differentiated SW 872 adipocytes: effect of spice extracts.

	H_2O_2 (500 µM)	Relative Intracellular ROS Level (%)
Control	−	35.9 ± 0.2 ***
H_2O_2 (500 µM)	+	100
Trolox (500 µM)	+	61.8 ± 8.6 ***
Xylopia aethiopica	+	255.7 ± 9.3 ***
Xylopia parviflora	+	49.6 ± 3.53 **
Scorodophloeus zenkeri	+	94.7 ± 1.6
Monodora myristica	+	60.0 ± 8.9 ***
Tetrapleura tetraptera	+	72.7 ± 1.5 **
Echinops giganteus	+	56.4 ± 1.2 ***
Afrostyrax lepidophyllus	+	75.4 ± 0.9 *
Dichrostachys glomerata	+	66.6 ± 0.6
Aframomum melegueta	+	122.5 ± 9.9 **
Aframomum citratum	+	112.8 ± 0.4
Zanthoxylum leprieurii	+	89.9 ± 3.3

All plant extracts were used at 10 µg/mL. Data are expressed as % of H_2O_2-treated cells taken as 100; mean ± SD, n = 3. One experiment (n = 3) is shown as representative of 3 separated experiments, each in triplicate. Results are expressed as mean ± SD. * $p < 0.05$, ** $p < 0.01$, *** $p < 0.001$ (one-way ANOVA multiple comparison).

When cells were pre-treated for 24 h with spice extracts (all at 10 µg/mL), ROS production induced by 1 h treatment with H_2O_2 was reduced to a different extent by all spices except *A. citratum*, *Z. leprieurii*, and *S. zenkeri*, which were ineffective. Moreover, *X. aethiopica* and *A. melegueta* extracts significantly increased ROS production (Table 2). In addition, the most effective extracts (*X. parviflora*, *E. giganteus, and D. glomerata*) were found to exert their ROS scavenging activity in a concentration-dependent fashion (Figure 5).

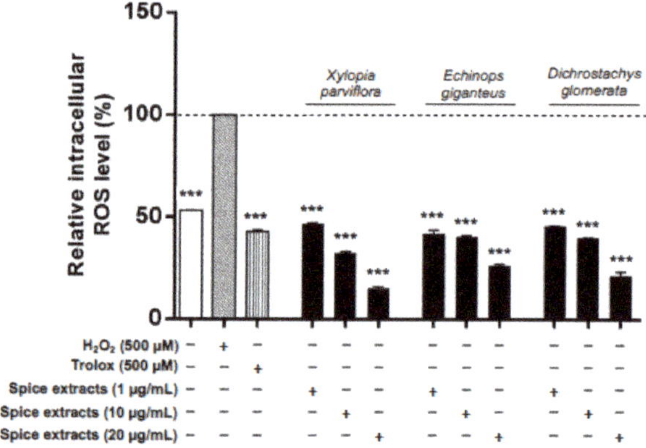

Figure 5. Concentration-dependent modulation of intracellular ROS release in SW 872 cells. The 3 most potent spice extracts were selected to evaluate the dose response. Concentration-dependent reduction of intracellular ROS production in SW 872 cells by selected spice extracts (*Xylopia parviflora, Echinops giganteus*, and *Dichrostachys glomerata*). Data are expressed as % of control taken as 100; mean ±SD, n = 3. One experiment (n = 3) is shown as representative of 3 separated experiments, each in triplicate. Results are expressed as mean ±SD. *** $p < 0.001$ (one-way ANOVA multiple comparison).

4. Discussion

Nutraceutical approaches to prevent and treat cardiometabolic conditions are nowadays a relevant and promising research field [18], in some instances taking advantage of knowledge derived from traditional medicine. In the present study, we evaluated the effects of a set of plant extracts used in Cameroon as nutritional spices and medicinal agents [19,20] on cellular events related to the molecular pathophysiology of cardiometabolic diseases. In consideration that ectopic adipose tissue, especially when dysfunctional, is a well-known important player in cardiovascular and metabolic diseases [21], in the present study, we selected human SW 872 differentiated adipocytes as the model system [22,23]. SW 872 cells show, under basal conditions, an immature phenotype and constitutively express several genes involved in fatty acid metabolism, such as lipoprotein lipase (LPL), cholesterol ester transfer protein (CETP), cluster of differentiation 36 (CD36), peroxisome proliferator-activated receptor (PPAR)α, PPARγ and LDL receptor-related protein (LRP) [22]. For the purpose of this study, SW 872 were differentiated to mature adipocytes using OA, according to established protocols [9]. We observed an overall inhibitory action across all plant extracts regarding TG content of SW 872 cells since the massive TG accumulation induced by OA treatment was reduced by −11/−18% by all extracts after 48 h. Consistent with our data, it has been reported that lipid accumulation was significantly decreased in 3T3-L1 adipocytes by resveratrol [24], which was selected here as a positive control. All the tested spice extracts are rich in a variety of primary metabolites (glycerol, fructofuranose, glucopyranose, etc) and secondary metabolites (chlorogenic acid, catechin, pimaric acid, etc.) [5] that orchestrate the observed molecular responses.

The occurrence of some plant specificity in this effect is suggested by our previous observation that *Adansonia digitata* L. extracts did not affect TG accumulation in differentiated SW 872 adipocytes [8]. The development of obesity is accompanied by adipocyte hypertrophy and hyperplasia [25] leading to excess of lipid accumulation [26], and the observed effects may at least in part explain the actions of some of the tested plants in animal models. For example, extracts from *A. melegueta* seeds were found to reduce adipose tissue in obese mice [19], and high-carbohydrate and high-fat diet-induced obesity and diabetes in rats were attenuated by *T. tetraptera* extract [27]. Dysfunctional adipocytes, as observed in T2DM and obesity, also develop reduced glucose uptake [28–30]. In this study, *T. tetraptera*, *A. melegueta* and *Z. leprieurii* increased glucose uptake in a dose-dependent manner. Interestingly, in another in-vitro model, the HepG2 cells, *T. tetraptera* and *A. melegueta*, but not *Z. leprieurii* increased glucose uptake [7], underlining the relationship between the complex molecular composition of the spice extracts, the observed effect and the specificity linked to the tested cell model. In differentiated SW 872 cells, the tested spice extracts do not appear to modulate the Akt upstream pathway whether or not in the presence of insulin, suggesting that they may increase glucose uptake by promoting other molecular events, like glucose transporter translocation to the membrane. Moreover, insulin resistance, glucose uptake impairment and metabolic disfunction appear to be sustained by the secretion from dysfunctional adipocytes of several pro-inflammatory cytokines [31] as well as by the resultant ROS production [1,32]. Based on these considerations, we evaluated the secretory profile of IL-6 and IL-8, two NF-kB-dependent molecular mediators released in the context of chronic inflammation [7]. Among the 11 tested extracts, four (*A. melegueta, D. glomerata, T. tetraptera* and *E. giganteus*) were effective in significantly reducing (−29/−43%) IL-6 release. *T. tetraptera*, and *D. glomerata* were also found to similarly reduce IL-6 release by GES-1 gastric epithelial cells [6]. In addition, four spice extracts (*X. parviflora, X. aethiopica, M. myristica*, and *Z. leprieurii*) were able to reduce (−21/−36%) IL-8 release, whereas this was markedly increased by *D. glomerata* and *A. citratum*. Notably, *D. glomerata* extract showed a dichotomic effect, reducing IL-6 and increasing IL-8 secretion. It is important to highlight that the release of each cytokine was negatively modulated by fully different sets of plants. Although it is challenging to exactly establish the role of each extract and their metabolites in IL-6 and IL-8 modulation, it is well known that shogaol, gingerol, chlorogenic acid, pimaric acid, and other molecules identified in our extracts [5] are well known inhibitors of NF-κB [14,33–36].

Furthermore, we assessed the effects of treatment with these Cameroonian spice extracts on H_2O_2-induced ROS production in differentiated SW 872 cells. All extracts, except *A. citratum, Z. leprieurii* and *S. zenkeri* were able to reduce ROS production. *X. parviflora, E. giganteus* and *D. glomerata* were the most effective and showed a dose-related activity. These antioxidant effects could be explained by the high content in phenolic compounds and the potent in-vitro antioxidant capacity previously reported by our group [7]. ROS production was however increased by *X. aethiopica* and *A. melegueta* extracts.

Taken together, the findings of the present study, summarized in Table 3, suggest that each of these spice extracts display a rather peculiar profile of activity on differentiated SW 872 adipocytes. This deserves further exploitation, in order to highlight in a more robust way the peculiar utilization of each spice extract to target specific functions (i.e., antioxidant vs. anti-inflammatory or promoting glucose uptake). Nutraceutical products with these activities may find their application in subjects with obesity, metabolic syndrome, T2DM and the related atherosclerotic cardiovascular disease risk, either alone, especially in milder conditions [37], or in combination with selected drugs in order to avoid drug dosage increase, prevent some adverse effects and increase the overall efficacy [4,38]. Interestingly, some pathophysiological mechanisms underlying these conditions, like oxidative stress and chronic low-grade inflammation, do not seem properly managed by the current pharmacology, and thus, well-characterized plant extracts could allow to specifically target them.

Table 3. Summary of the specific effects of the tested Cameroonian spice extracts on differentiated SW 872 adipocytes.

	Triglyceride Reduction	Glucose Uptake Stimulation	ROS Production	IL-6 Reduction	IL-8 Reduction
Xylopia aethiopica	−14.5%		+55.8%		−21.1%
Xylopia parviflora	−13.8%		−50.5%		−36.8%
Scorodophloeus zenkeri	−18.5%				
Monodora myristica	−15.3%		−40%		−24.3%
Tetrapleura tetraptera	−13.8%	+40.8%	−27.4%	−29.7%	
Echinops giganteus	−11.3%		−43.6%	−29%	
Afrostyrax lepidophyllus	−16.5%		−24.6%		
Dichrostachys glomerata	−17.4%			−40%	
Aframomum melegueta	−13%	+41.7%		−43.1%	
Aframomum citratum	−16%				−58.6%
Zanthoxylum leprieurii	−13.4%	+56.6%			−32.7%

This study has some limitations. OA-differentiated SW 872 cells resemble in some respects human adipocytes although this differentiation protocol is associated with peculiar molecular effects related to PPARγ activation, which therefore need to be taken into consideration in the interpretation of the obtained results. It is also important to underline that the extraction protocol utilized in this study [5] may differ from that used by others, making in some cases the comparison with other studies on the same plants difficult. This work, together with the others published in the last years, contributes to define and enrich the study of the biological activities of a set of Cameroonian spices, particularly in the context of dysfunctional adipocyte cell biology.

5. Conclusions

The results of this study show that the tested Cameroonian spice extracts display interesting activities on several molecular features of differentiated SW 872 human adipocytes. All extracts were able to reduce TG accumulation, while the ability to promote glucose uptake, reduce pro-inflammatory cytokines release, and counteract ROS production was limited to panels of three to six plant extracts (Table 3). Such variety of effects may be the basis for the development of nutraceutical products with very specific effects or combining more than one extract to achieve complementary effects. Moreover, *T. tetraptera* stands out as the most versatile plant since it was found to positively modulate most parameters. Since the 11 spice extracts showed rather variable viability profiles, with some extracts being toxic at concentrations just higher than the most effective doses (10–20 µg/mL), a word of caution should be spent regarding the potential toxicity of some plants when used at higher concentrations and/or prolonged treatments.

The findings of the present study, conducted in a human adipocyte in-vitro model, highlight some potential health properties of these Cameroonian spices and suggest the opportunity of further studies in the context of experimental and human cardiometabolic diseases.

Supplementary Materials: The following are available online at https://www.mdpi.com/article/10.3390/nu13124271/s1, Figure S1: SW 872 cells viability (MTS assay) after treatment with all extracts; Figure S2: Intracellular complexity of non-differentiated and differentiated SW 872 cells; Figure S3: Glucose uptake in SW 872 cells. Time course (15–60 min) and dose, response to insulin (10 nM–1 μM) of glucose uptake in non-differentiated (A) and differentiated (B) cells.

Author Contributions: Research design: P.M., A.P.A.N., M.D. (Mario Dell'Agli) and A.P.; conducted experiments: A.P.A.N., A.P., L.D.D., C.O., T.N.D., E.C.-M. and S.P.; contributed reagents, materials, analysis tool: S.P., G.M., T.N.D., L.D.D. and A.P.A.N.; analysis and interpretation of data: A.P.A.N., P.M., L.D.D., E.S. and A.P.; writing manuscript: P.M., A.P.A.N. and A.P.; critically read the manuscript: A.D.T., G.A.A., S.T.S., U.V.M., M.D. (Mario Dell'Agli), J.-R.K. and M.D. (Maria Daglia). All authors have read and agreed to the published version of the manuscript.

Funding: This work was supported by funds to P.M. (Transition grant PSR2015-1720PMAGN_01 from Università degli Studi di Milano, Milan, Italy) andand to M.D.A. (personal funds).

Informed Consent Statement: The data presented in this study are available on request from the corresponding authors.

Data Availability Statement: The data presented in this study are available on request from the corresponding authors.

Acknowledgments: APAN is the recipient of a research fellowship from the Italian Ministry of Foreign Affairs and International Cooperation (MAECI), to which the authors are grateful.

Conflicts of Interest: This study is an original research carried out by the mentioned authors and, thus, the authors declare that there is no conflict of interests regarding the publication of this paper.

References

1. Manna, P.; Jain, S.K. Obesity, Oxidative Stress, Adipose Tissue Dysfunction, and the Associated Health Risks: Causes and Therapeutic Strategies. *Metab. Syndr. Relat. Disord.* **2015**, *13*, 423–444. [CrossRef]
2. Ruscica, M.; Gomaraschi, M.; Mombelli, G.; Macchi, C.; Bosisio, R.; Pazzucconi, F.; Pavanello, C.; Calabresi, L.; Arnoldi, A.; Sirtori, C.R.; et al. Nutraceutical approach to moderate cardiometabolic risk: Results of a randomized, double-blind and crossover study with Armolipid Plus. *J. Clin. Lipidol.* **2014**, *8*, 61–68. [CrossRef]
3. Cicolari, S.; Pavanello, C.; Olmastroni, E.; del Puppo, M.; Bertolotti, M.; Mombelli, G.; Catapano, A.L.; Calabresi, L.; Magni, P. Interactions of Oxysterols with Atherosclerosis Biomarkers in Subjects with Moderate Hypercholesterolemia and Effects of a Nutraceutical Combination (*Bifidobacterium longum* BB536, Red Yeast Rice Extract) (Randomized, Double-Blind, Placebo-Controlled Study). *Nutrients* **2021**, *13*, 427. [CrossRef] [PubMed]
4. Salehi, B.; Ata, A.; Anil Kumar, N.V.; Sharopov, F.; Ramírez-Alarcón, K.; Ruiz-Ortega, A.; Abdulmajid Ayatollahi, S.; Tsouh Fokou, P.V.; Kobarfard, F.; Amiruddin Zakaria, Z.; et al. Antidiabetic Potential of Medicinal Plants and Their Active Components. *Biomolecules* **2019**, *9*, 551. [CrossRef] [PubMed]
5. Atchan Nwakiban, A.P.; Sokeng, A.J.; Dell'Agli, M.; Bossi, L.; Beretta, G.; Gelmini, F.; Deutou Tchamgoue, A.; Agbor Agbor, G.; Kuiate, J.R.; Daglia, M.; et al. Hydroethanolic plant extracts from Cameroon positively modulate enzymes relevant to carbohydrate/lipid digestion and cardio-metabolic diseases. *Food Funct.* **2019**, *10*, 6533–6542. [CrossRef]
6. Nwakiban, A.P.A.; Fumagalli, M.; Piazza, S.; Magnavacca, A.; Martinelli, G.; Beretta, G.; Magni, P.; Tchamgoue, A.D.; Agbor, G.A.; Kuiaté, J.R.; et al. Dietary Cameroonian Plants Exhibit Anti-Inflammatory Activity in Human Gastric Epithelial Cells. *Nutrients* **2020**, *12*, 3787. [CrossRef]
7. Nwakiban, A.P.A.; Cicolari, S.; Piazza, S.; Gelmini, F.; Sangiovanni, E.; Martinelli, G.; Bossi, L.; Carpentier-Maguire, E.; Tchamgoue, A.D.; Agbor, G.; et al. Oxidative Stress Modulation by Cameroonian Spice Extracts in HepG2 Cells: Involvement of Nrf2 and Improvement of Glucose Uptake. *Metabolites* **2020**, *10*, 182. [CrossRef]
8. Cicolari, S.; Dacrema, M.; Tsetegho Sokeng, A.J.; Xiao, J.; Atchan Nwakiban, A.P.; Di Giovanni, C.; Santarcangelo, C.; Magni, P.; Daglia, M. Hydromethanolic Extracts from *Adansonia digitata* L Edible Parts Positively Modulate Pathophysiological. *Molecules* **2020**, *25*, 2858. [CrossRef] [PubMed]
9. Guennoun, A.; Kazantzis, M.; Thomas, R.; Wabitsch, M.; Tews, D.; Seetharama Sastry, K.; Abdelkarim, M.; Zilberfarb, V.; Strosberg, A.D.; Chouchane, L. Comprehensive molecular characterization of human adipocytes reveals a transient brown phenotype. *J. Transl. Med.* **2015**, *13*, 135. [CrossRef]

10. Carmel, J.F.; Tarnus, E.; Cohn, J.S.; Bourdon, E.; Davignon, J.; Bernier, L. High expression of apolipoprotein E impairs lipid storage and promotes cell proliferation in human adipocytes. *J. Cell Biochem.* **2009**, *106*, 608–617. [CrossRef]

11. Butterweck, V.; Nahrstedt, A. What is the best strategy for preclinical testing of botanicals? A critical perspective. *Planta Med.* **2012**, *78*, 747–754. [CrossRef]

12. Kim, Y.M.; Jang, M.S. Anti-obesity effects of Laminaria japonica fermentation on 3T3-L1 adipocytes are mediated by the inhibition of C/EBP-alpha/beta and PPAR-gamma. *Cell Mol. Biol.* **2018**, *64*, 71–77. [CrossRef] [PubMed]

13. Sangiovanni, E.; Di Lorenzo, C.; Colombo, E.; Colombo, F.; Fumagalli, M.; Frigerio, G.; Restani, P.; Dell'Agli, M. The effect of in vitro gastrointestinal digestion on the anti-inflammatory activity of *Vitis vinifera* L. leaves. *Food Funct.* **2015**, *6*, 2453–2463. [CrossRef]

14. Piazza, S.; Pacchetti, B.; Fumagalli, M.; Bonacina, F.; Dell'Agli, M.; Sangiovanni, E. Comparison of Two *Ginkgo biloba* L. Extracts on Oxidative Stress and Inflammation Markers in Human Endothelial Cells. *Mediat. Inflamm.* **2019**, *2019*, 6173893. [CrossRef]

15. Biological Evaluation of Medical Devices—Part 5: Tests for In Vitro Cytotoxicity. Évaluation Biologique des Dispositifs Médicaux—Partie 5: Essais Concernant la Cytotoxicité In Vitro, Third Edition. International Organization for Standardization. Posted June 1. 2009. Available online: https://nhiso.com/wp-content/uploads/2018/05/ISO-10993-5-2009.pdf (accessed on 25 November 2021).

16. Krycer, J.R.; Quek, L.E.; Francis, D.; Zadoorian, A.; Weiss, F.C.; Cooke, K.C.; Nelson, M.E.; Diaz-Vegas, A.; Humphrey, S.J.; Scalzo, R.; et al. Insulin signaling requires glucose to promote lipid anabolism in adipocytes. *J. Biol. Chem.* **2020**, *295*, 13250–13266. [CrossRef]

17. Shoshani, O.; Livne, E.; Armoni, M.; Shupak, A.; Berger, J.; Ramon, Y.; Fodor, L.; Gilhar, A.; Peled, I.J.; Ullmann, Y. The effect of interleukin-8 on the viability of injected adipose tissue in nude mice. *Plast. Reconstr. Surg.* **2005**, *115*, 853–859. [CrossRef] [PubMed]

18. Liu, S.; Chang, X.; Yu, J.; Xu, W. Cherry Polyphenol Reduces High-Fat Diet-Induced Obesity in C57BL/6 Mice by Mitigating Fat Deposition, Inflammation, and Oxidation. *J. Agric. Food Chem.* **2020**, *68*, 4424–4436. [CrossRef] [PubMed]

19. Kuete, V.; Krusche, B.; Youns, M.; Voukeng, I.; Fankam, A.G.; Tankeo, S.; Lacmata, S.; Efferth, T. Cytotoxicity of some Cameroonian spices and selected medicinal plant extracts. *J. Ethnopharmacol.* **2011**, *134*, 803–812. [CrossRef] [PubMed]

20. Hattori, H.; Mori, T.; Shibata, T.; Kita, M.; Mitsunaga, T. 6-Paradol Acts as a Potential Anti-obesity Vanilloid from Grains of Paradise. *Mol. Nutr. Food Res.* **2021**, *65*, e2100185. [CrossRef]

21. Neeland, I.J.; Ross, R.; Després, J.P.; Matsuzawa, Y.; Yamashita, S.; Shai, I.; Seidell, J.; Magni, P.; Santos, R.D.; Arsenault, B.; et al. Visceral and ectopic fat, atherosclerosis, and cardiometabolic disease: A position statement. *Lancet Diabetes Endocrinol.* **2019**, *7*, 715–725. [CrossRef]

22. Wassef, H.; Bernier, L.; Davignon, J.; Cohn, J.S. Synthesis and secretion of apoC-I and apoE during maturation of human SW872 liposarcoma cells. *J. Nutr.* **2004**, *134*, 2935–2941. [CrossRef] [PubMed]

23. Lehrke, M.; Lazar, M.A. The many faces of PPARgamma. *Cell* **2005**, *123*, 993–999. [CrossRef] [PubMed]

24. Rayalam, S.; Yang, J.Y.; Ambati, S.; Della-Fera, M.A.; Baile, C.A. Resveratrol induces apoptosis and inhibits adipogenesis in 3T3-L1 adipocytes. *Phytother. Res.* **2008**, *22*, 1367–1371. [CrossRef]

25. Nishimura, S.; Manabe, I.; Nagasaki, M.; Hosoya, Y.; Yamashita, H.; Fujita, H.; Ohsugi, M.; Tobe, K.; Kadowaki, T.; Nagai, R.; et al. Adipogenesis in obesity requires close interplay between differentiating adipocytes, stromal cells, and blood vessels. *Diabetes* **2007**, *56*, 1517–1526. [CrossRef]

26. Kim, M.R.; Kim, J.W.; Park, J.B.; Hong, Y.K.; Ku, S.K.; Choi, J.S. Anti-obesity effects of yellow catfish protein hydrolysate on mice fed a 45% kcal high-fat diet. *Int. J. Mol. Med.* **2017**, *40*, 784–800. [CrossRef]

27. Kuate, D.; Kengne, A.P.; Biapa, C.P.; Azantsa, B.G.; Abdul Manan Bin Wan Muda, B. Tetrapleura tetraptera spice attenuates high-carbohydrate, high-fat diet-induced obese and type 2 diabetic rats with metabolic syndrome features. *Lipids Health Dis.* **2015**, *14*, 50. [CrossRef]

28. Furukawa, S.; Fujita, T.; Shimabukuro, M.; Iwaki, M.; Yamada, Y.; Nakajima, Y.; Nakayama, O.; Makishima, M.; Matsuda, M.; Shimomura, I. Increased oxidative stress in obesity and its impact on metabolic syndrome. *J. Clin. Investig.* **2004**, *114*, 1752–1761. [CrossRef] [PubMed]

29. Cawthorn, W.P.; Sethi, J.K. TNF-alpha and adipocyte biology. *FEBS Lett.* **2008**, *582*, 117–131. [CrossRef]

30. Zhang, W.Y.; Lee, J.J.; Kim, I.S.; Kim, Y.; Myung, C.S. Stimulation of glucose uptake and improvement of insulin resistance by aromadendrin. *Pharmacology* **2011**, *88*, 266–274. [CrossRef]

31. Kobashi, C.; Asamizu, S.; Ishiki, M.; Iwata, M.; Usui, I.; Yamazaki, K.; Tobe, K.; Kobayashi, M.; Urakaze, M. Inhibitory effect of IL-8 on insulin action in human adipocytes via MAP kinase pathway. *J. Inflamm.* **2009**, *6*, 25. [CrossRef]

32. Issa, N.; Lachance, G.; Bellmann, K.; Laplante, M.; Stadler, K.; Marette, A. Cytokines promote lipolysis in 3T3-L1 adipocytes through induction of NADPH oxidase 3 expression and superoxide production. *J. Lipid Res.* **2018**, *59*, 2321–2328. [CrossRef]

33. Suh, S.J.; Kwak, C.H.; Chung, T.W.; Park, S.J.; Cheeeei, M.; Park, S.S.; Seo, C.S.; Son, J.K.; Chang, Y.C.; Park, Y.G.; et al. Pimaric acid from Aralia cordata has an inhibitory effect on TNF-α-induced MMP-9 production and HASMC migration via down-regulated NF-κB and AP-1. *Chem. Biol. Interact.* **2012**, *199*, 112–119. [CrossRef] [PubMed]

34. Saha, A.; Blando, J.; Silver, E.; Beltran, L.; Sessler, J.; DiGiovanni, J. 6-Shogaol from dried ginger inhibits growth of prostate cancer cells both in vitro and in vivo through inhibition of STAT3 and NF-κB signaling. *Cancer Prev. Res. (Phila)* **2014**, *7*, 627–638. [CrossRef]

35. Liang, N.; Sang, Y.; Liu, W.; Yu, W.; Wang, X. Anti-Inflammatory Effects of Gingerol on Lipopolysaccharide-Stimulated RAW 264.7 Cells by Inhibiting NF-κB Signaling Pathway. *Inflammation* **2018**, *41*, 835–845. [CrossRef] [PubMed]
36. Ruifeng, G.; Yunhe, F.; Zhengkai, W.; Ershun, Z.; Yimeng, L.; Minjun, Y.; Xiaojing, S.; Zhengtao, Y.; Naisheng, Z. Chlorogenic acid attenuates lipopolysaccharide-induced mice mastitis by suppressing TLR4-mediated NF-κB signaling pathway. *Eur. J. Pharm.* **2014**, *729*, 54–58. [CrossRef]
37. Dinda, B.; Dinda, M.; Roy, A.; Dinda, S. Dietary plant flavonoids in prevention of obesity and diabetes. *Adv. Protein Chem. Struct. Biol.* **2020**, *120*, 159–235. [CrossRef]
38. Samson, S.L.; Garber, A.J. Metabolic syndrome. *Endocrinol. Metab. Clin.* **2014**, *43*, 1–23. [CrossRef] [PubMed]

Article

Urinary Tartaric Acid, a Biomarker of Wine Intake, Correlates with Lower Total and LDL Cholesterol

Inés Domínguez-López [1,2], Isabella Parilli-Moser [1,2], Camila Arancibia-Riveros [1], Anna Tresserra-Rimbau [1,2], Miguel Angel Martínez-González [2,3,4], Carolina Ortega-Azorín [2,5], Jordi Salas-Salvadó [2,6,7], Olga Castañer [2,8], José Lapetra [2,9], Fernando Arós [2,10], Miquel Fiol [2,11], Lluis Serra-Majem [2,12], Xavier Pintó [2,13], Enrique Gómez-Gracia [2,14], Emilio Ros [2,15], Rosa M. Lamuela-Raventós [1,2,*] and Ramon Estruch [2,16,*]

[1] Department of Nutrition, Food Science and Gastronomy, XIA, School of Pharmacy and Food Sciences, INSA, University of Barcelona, 08028 Barcelona, Spain; idominguez@ub.edu (I.D.-L.); iparillim@ub.edu (I.P.-M.); carancri77@alumnes.ub.edu (C.A.-R.); annatresserra@ub.edu (A.T.-R.)

[2] Centro de Investigación Biomédica en Red Fisiopatología de la Obesidad y la Nutrición (CIBEROBN), Instituto de Salud Carlos III, 28029 Madrid, Spain; mamartinez@unav.es (M.A.M.-G.); carolina.ortega@uv.es (C.O.-A.); jordi.salas@urv.cat (J.S.-S.); ocastaner@imim.es (O.C.); joselapetra543@gmail.com (J.L.); luisfernando.aros@ehu.eus (F.A.); miguel.fiol@ssib.es (M.F.); lserra@dcc.ulpgc.es (L.S.-M.); xpinto@bellvitgehospital.cat (X.P.); egomezgracia@uma.es (E.G.-G.); eros@clinic.cat (E.R.)

[3] Department of Preventive Medicine and Public Health, School of Medicine, University of Navarra, 31008 Pamplona, Spain

[4] Department of Nutrition, Harvard TH Chan School of Public Health, Boston, MA 02115, USA

[5] Department of Preventive Medicine and Public Health, School of Medicine, University of Valencia, 46010 Valencia, Spain

[6] Unitat de Nutrició Humana, Departament de Bioquímica i Biotecnologia, Hospital Universitari San Joan de Reus, Universitat Rovira i Virgili, 43201 Reus, Spain

[7] Institut d'Investigació Sanitària Pere Virgili (IISPV), 43201 Reus, Spain

[8] Cardiovascular Epidemiology Unit, Municipal Institute for Medical Research (IMIM), 08007 Barcelona, Spain

[9] Research Unit, Department of Family Medicine, Distrito Sanitario Atención Primaria Sevilla, 41010 Sevilla, Spain

[10] Department of Cardiology, Hospital Txangorritxu, 01009 Vitoria, Spain

[11] Institut Universitari d'Investigació en Ciències de la Salut (IUNICS), 07122 Palma de Mallorca, Spain

[12] Department Clinical Sciences, University of Las Palmas de Gran Canaria, 35016 Palmas de Gran Canaria, Spain

[13] Lipid Unit, Department of Internal Medicine, IDIBELL-Hospital Universitari de Bellvitge, L'Hospitalet de Llobregat, FIPEC, 08908 Barcelona, Spain

[14] Department of Epidemiology, School of Medicine, University of Malaga, 29010 Málaga, Spain

[15] Lipid Clinic, Endocrinology and Nutrition Service, Institut d'Investigacions Biomèdiques August Pi Sunyer (IDIBAPS), Hospital Clínic, 08036 Barcelona, Spain

[16] Internal Medicine Department, Institut d'Investigacions Biomèdiques August Pi Sunyer (IDIBAPS), Hospital Clinic, University of Barcelona, 08036 Barcelona, Spain

* Correspondence: lamuela@ub.edu (R.M.L.-R.); restruch@clinic.cat (R.E.)

Citation: Domínguez-López, I.; Parilli-Moser, I.; Arancibia-Riveros, C.; Tresserra-Rimbau, A.; Martínez-González, M.A.; Ortega-Azorín, C.; Salas-Salvadó, J.; Castañer, O.; Lapetra, J.; Arós, F.; et al. Urinary Tartaric Acid, a Biomarker of Wine Intake, Correlates with Lower Total and LDL Cholesterol. *Nutrients* **2021**, *13*, 2883. https://doi.org/10.3390/nu13082883

Academic Editors: J. Mark Brown, Paolo Magni, Andrea Baragetti and Andrea Poli

Received: 1 July 2021
Accepted: 19 August 2021
Published: 22 August 2021

Abstract: Postmenopausal women are at higher risk of developing cardiovascular diseases due to changes in lipid profile and body fat, among others. The aim of this study was to evaluate the association of urinary tartaric acid, a biomarker of wine consumption, with anthropometric (weight, waist circumference, body mass index (BMI), and waist-to-height ratio), blood pressure, and biochemical variables (blood glucose and lipid profile) that may be affected during the menopausal transition. This sub-study of the PREDIMED (Prevención con Dieta Mediterránea) trial included a sample of 230 women aged 60–80 years with high cardiovascular risk at baseline. Urine samples were diluted and filtered, and tartaric acid was analyzed by liquid chromatography coupled to electrospray ionization tandem mass spectrometry (LC-ESI-MS/MS). Correlations between tartaric acid and the study variables were adjusted for age, education level, smoking status, physical activity, BMI, cholesterol-lowering, antihypertensive, and insulin treatment, total energy intake, and consumption of fruits, vegetables, and raisins. A strong association was observed between wine consumption and urinary tartaric acid (0.01 μg/mg (95% confidence interval (CI): 0.01, 0.01), *p*-value < 0.001). Total and low-density lipoprotein (LDL) cholesterol were inversely correlated with urinary tartaric acid (-3.13 µg/mg (-5.54, -0.71), *p*-value = 0.016 and -3.03 µg/mg (-5.62, -0.42), *p*-value = 0.027, respectively), whereas other biochemical and anthropometric variables were unrelated. The results suggest that wine consumption may have a positive effect on cardiovascular health in postmenopausal women, underpinning its nutraceutical properties.

Keywords: PREDIMED; Mediterranean diet; lipid profile; cardiovascular risk; polyphenols; menopause; body fat; biomarkers; tartaric acid

1. Introduction

Cardiovascular disease (CVD) is the leading cause of death worldwide in both sexes. Nevertheless, important sex-specific differences exist. According to the American Heart Association, menopause is listed as a female-specific cardiovascular risk factor (CVRF) [1]. During the menopause transition women experience adverse changes in their lipid profile, body fat distribution, metabolic syndrome risk, and vascular health [2–5]. Previous studies suggested that menopause is associated with increased total and low-density lipoprotein (LDL) cholesterol [6] and changes in body composition such as increased fat mass and loss of lean mass [7]. Changes in blood pressure (BP), waist circumference (WC), body mass index (BMI), and blood glucose and insulin have not been specifically associated with menopause and appear to reflect chronological aging [5,6,8]. Therefore, menopause-induced increases in cholesterol, body fat, and possibly other CVRFs may accelerate the risk of developing CVD.

Diet and lifestyle can also affect the incidence of CVD. Modifiable factors, such as smoking cessation, healthy diet, and regular physical activity, play a crucial role in reducing cardiovascular risk [9]. The Mediterranean diet (MedDiet) has been associated with a better control of several CVRFs [10] through improvements in BP, lipid profile, glucose metabolism, arrhythmia risk, and gut microbiome [11,12]. One of the main characteristics of the MedDiet is the abundant consumption of olive oil, vegetables, fruits, nuts, legumes, fish, and cereals, and moderate wine consumption [13,14]. Epidemiologic studies and randomized clinical trials reported that moderate consumption of wine (1 or 2 glasses/day) during meals has been consistently associated with a lower risk of CVD [15–17]. In the context of a MedDiet, moderate alcohol consumption appears to be synergistic with other cardioprotective components of the MedDiet that increase high-density lipoprotein (HDL) cholesterol, decrease platelet aggregation, promote antioxidant effects, and reduce inflammation [13].

Wine consumption is mainly determined through dietary questionnaires. A biomarker of wine intake reflects its consumption more reliably than a questionnaire, since people may not accurately report the amount of alcohol consumed due to perceived social rejection of excessive consumption [18]. Tartaric acid, the main organic acid in wine and the molecule responsible for wine acidity, is present in high amounts in wine (1.5–4.0 g/L) but is rare in most common foods [19,20]. Urinary tartaric acid has been considered as a sensitive, selective, and robust biomarker of moderate wine intake [21,22]. Therefore, determining tartaric acid stands out as a useful tool to further study the impact of moderate wine drinking on health. The aim of this study was to determine the association between urinary tartaric acid as a biomarker of wine consumption and CVRFs in postmenopausal women at risk of developing CVD.

2. Materials and Methods

2.1. Study Design

This study is a cross-sectional analysis of baseline data from a subsample of participants in the PREDIMED (PREvención con DIeta MEDiterránea) study, a large, parallel-group, multicenter, randomized, controlled, 5-year clinical trial conducted between 2003 and 2009. The objective was to assess the effect of a Mediterranean diet supplemented with extra-virgin olive oil or mixed nuts as compared to a low-fat diet on the primary prevention of CVD in 7447 participants at high cardiovascular risk. Eligible participants were men (55–80 years old) and women (60–80 years old) who had type 2 diabetes mellitus or at least 3 of the following major CVRFs: smoking, hypertension, elevated LDL cholesterol, low HDL cholesterol, overweight or obesity, and/or family history of premature coronary heart disease [23]. All participants provided written informed consent, and the study protocol and procedures complied the ethical standards of the Declaration of Helsinki.

For the present sub-study of the PREDIMED trial, urinary tartaric acid was analyzed in a subsample of women equivalent to 5% of the total female population of the PREDIMED study. The 230 women that were randomly selected had undergone the menopausal transition and their urine samples were available at baseline. Participants who had no available data of total energy intake or reported extreme values (>3500 kcal/day) were excluded from the analysis ($n = 8$).

2.2. Anthropometric, Dietary, and Physical Activity Assessments

Trained personal performed the anthropometric and clinical measurements (height, weight, WC, and BP). BMI was obtained by dividing the body weight in kilograms by the square of height in cm. The waist-to-height ratio (WtHR) was calculated by dividing the WC in centimeters by height in meters. Systolic (SBP) and diastolic blood pressure (DBP) were measured in triplicate with a validated semi-automatic oscillometer (Omron HEM-705CP, Lake Forest, IL, USA). A validated semi-quantitative food frequency questionnaire (FFQ), which included 137 food items [24], and the Minnesota Leisure-Time Physical Activity Questionnaire [25] were used to assess dietary habits over the previous 12 months and physical activity (metabolic equivalent tasks per minute per day, METs min/day) of the participants.

2.3. Clinical Measurements

Medical conditions, family history of disease, and risk factors were collected though a questionnaire during the first screening visit. Biological samples (plasma and urine) were collected at baseline after 12 h overnight fast and stored at $-80\,^\circ$C until assay. Blood glucose, total cholesterol, triglycerides, and HDL cholesterol were determined by standard enzymatic methods, and LDL cholesterol was calculated by the Friedewald equation [26].

2.4. Tartaric Acid Determination

2.4.1. Reagents and Standards

Formic acid (approximately 98%), picric acid (98%, moistened with approximately 33% water), and sodium hydroxide ($\geq$98%) were obtained from Panreac. L-(+)-Tartaric acid and creatinine were purchased from Sigma. The labelled internal standard DL-($\pm$)-tartaric-2,3-d2 acid was obtained from C/D/N Isotopes. Solvents were high-performance liquid chromatography grade, and all other chemicals were analytical reagent grade. Ultrapure water was obtained from a Milli-Q Gradient water purification system (Millipore, Bedford, MA, USA).

Stock solutions of tartaric acid were prepared in water. Working standard solutions that ranged from 0.01 to 5 μg/mL were made by appropriate dilution in 0.5% formic acid in water and then stored in amber glass vials at $-20\,^\circ$C.

2.4.2. Sample Preparation

Determination of urinary tartaric acid was performed following a previously validated stable-isotope dilution LC-ESI-MS/MS method by our research group [27]. Briefly, urine samples (20 μL) were diluted 1:50 (*v:v*) with 0.5% formic acid in water, and 10 μL of a deuterated isotope standard solution in water (DL-($\pm$)-tartaric-2,3-d2 acid, 200 μg/mL) were added. The sample dilution was passed through a 0.20 μm filter and analyzed by LC–ESI-MS/MS. Urinary tartaric acid data were corrected by urine creatinine, measured according to the adapted Jaffé alkaline picrate method for thermo microtiter 96-well plates, according to Medina-Remón et al. [28]. Finally, urinary tartaric acid was expressed as µg of tartaric acid per mg of creatinine. According to previous data, the cut-off of 8.84 µg/mg creatinine was used to discriminate daily consumers and non-consumers of wine [21].

2.4.3. LC–ESI-MS/MS Analysis

After filtration, tartaric acid was analyzed using an Atlantis TE C18, 100 mm × 2.1 mm, 3 µm (Waters, Milford, MA, USA) reversed-phase column coupled for detection to the triple quadrupole mass spectrometer API 3000 (Applied Biosystems, Foster City, CA, USA). The mass spectrometer was operated in negative electrospray ionization mode. The column was maintained at 25 °C throughout the analysis. Mobile phases A and B were 0.5% formic acid in water and 0.5% formic acid in acetonitrile, respectively. The following linear gradient was used: holding at 100%A for 3.5 min, decrease to 10%A over 2 min and holding for 2 min, return to initial conditions for 1.5 min, and re-equilibration for 6 min. The flow rate was set at 350 µL/min and the injection volume was 10 µL. Post-column addition of acetonitrile (250 µL/min) was carried out to improve analyte ionization efficiency. The detection was accomplished in multiple reaction monitoring (MRM) mode, and the following MS/MS transitions were used for quantification and confirmation, respectively: m/z 149/87 and m/z 149/73 for tartaric acid, and m/z 151/88 and m/z 151/74 for the deuterated isotope.

2.5. Statistical Analyses

The baseline characteristics of participants are presented as means and standard deviations (SD) for continuous variables, and frequency (n) and percentage (%) for categorical variables.

The normality of continuous variables was assessed with the Shapiro–Wilk test. The variables without normal distribution were transformed into logarithms. Multiple adjusted linear regression models were used to assess the differences between urinary tartaric acid and wine consumption as well as anthropometric and biochemical measurements. Three different adjustment models were applied. Model 1 was minimally adjusted for age (continuous). Model 2 was additionally adjusted for educational level, smoking status, BMI (except for anthropometric criteria), physical activity, and cholesterol-lowering, antihypertensive, and insulin treatment. Model 3 was further adjusted for total energy intake and consumption of fruits, vegetables, and raisins. We used robust variance estimators to account for the recruitment center in all linear models. To illustrate the relationship between wine consumption and urinary tartaric acid, the mL per month of wine reported in the FFQ were transformed into glasses of wine (with 1 glass equivalent to 100 mL).

Values are shown as 95% confidence interval (CI) and significance for all statistical tests was based on bilateral contrast set at $p < 0.05$. All the statistical analyses were performed using Stata statistical software package version 16.0 (StataCorp LP, College Station, TX, USA).

3. Results

3.1. Study Population

The main characteristics of the PREDIMED participants who were included in this substudy are summarized in Table 1. The mean age of the women was 66.9 + 0.4 years. Their burden of CVRFs was high: 42.1% had been diagnosed with type 2 diabetes, 87.3% with hypertension, and 76.5% with hypercholesterolemia. Among the drug treatments, statins were the most common medication, with 40.72% of them under treatment. Furthermore, 9.1% of the participants were current smokers. Finally, 82% of the participants had a low educational level.

Up to 45.7% of the participants reported wine consumption in the FFQ. The mean concentration of tartaric acid in urine was 28.34 µg/mg creatinine, and 40.4% were considered daily consumers of wine.

Table 1. Baseline characteristics of the women in the study population (*n* = 222).

General Characteristics	
Age, years	66.9 ± 0.4
Type 2 diabetes, *n* (%)	93 (42.08)
Hypertension, *n* (%)	193 (87.34)
Hypercholesterolemia, *n* (%)	169 (76.55)
Medication use, *n* (%)	
ACE inhibitors	64 (28.96)
Diuretics	58 (26.24)
Statins	90 (40.72)
Other lipid-lowering agents	14 (6.33)
Insulin	10 (4.52)
Oral hypoglycemic agents	53 (23.98)
Antiplatelet therapy	46 (20.81)
Current smoker, *n* (%)	20 (9.05)
Leisure-time physical activity, MET·min/week	186.5 ± 10.9
Educational level, *n* (%)	
Low	182 (82.35)
Medium	24 (10.86)
High	15 (6.79)
Wine consumption, *n* (%)	101 (45.70)
Urinary tartaric acid, µg/mg creatinine	28.47 + 4.03
Daily wine consumers, *n* (%)	81 (40.4)
Anthropometric measurements, mean + SD	
Weight, kg	72.9 ± 0.7
BMI, kg/m^2	30.3 ± 0.28
WC, cm	97.5 ± 0.7
WtHR	63.0 ± 0.5
Biochemical measurements, mean + SD	
Total cholesterol, mg/dL	221.0 ± 2.7
LDL cholesterol, mg/dL	136.5 ± 2.4
HDL cholesterol, mg/dL	57.7 ± 1.0
Triglycerides, mg/dL	134.0 ± 5.2
Glucose, mg/dL	118.1 ± 2.4
Blood pressure, mean + SD	
Systolic, mm Hg	148.7 ± 1.2
Diastolic, mm Hg	83.8 ± 0.6
Dietary intake, mean + SD	
Total energy, kcal/day	2161 ± 33
Carbohydrate, % of energy	42.0 ± 0.5
Protein, % of energy	16.8 ± 0.2
Fat, % of energy	39.8 ± 0.5

ACE: angiotensin-converting enzyme; MET: metabolic equivalent task; BMI: body mass index; WC: waist circumference; WtHR: waist-to-height ratio; LDL: low-density lipoprotein cholesterol; HDL: high-density lipoprotein cholesterol. Data are expressed as the mean ± standard deviations (SD) for continuous variables and frequency (*n*) and percentage (%) for categorical variables.

The mean values of anthropometric measurements revealed that most participants were obese, as defined by their BMI, WC, and WtHR data [29] according to the International Diabetes Federation and the American Heart Association [30]. Regarding biochemical measurements, triglycerides and HDL cholesterol were at desirables levels, while total cholesterol, LDL cholesterol and glucose were borderline high [31,32].

The mean energy intake was 2161 kcal/day, of which carbohydrates accounted for 42.0% of the energy consumed, protein intake 16.8%, and fat intake 39.8%.

3.2. Tartaric Acid as a Biomarker of Wine Consumption

After adjustments for several covariates (age, education level, smoking status, physical activity, BMI, cholesterol-lowering, antihypertensive, and insulin treatment, total energy intake, and consumption of fruits, vegetables, and raisins), women who consumed more wine presented higher concentrations of tartaric acid in urine (0.01 μg/mg (95% CI: 0.01, 0.01), *p*-value < 0.001). Figure 1 illustrates the linear relationship between urinary tartaric acid concentrations and wine consumption expressed as glasses of wine, excluding those who reported not consuming wine.

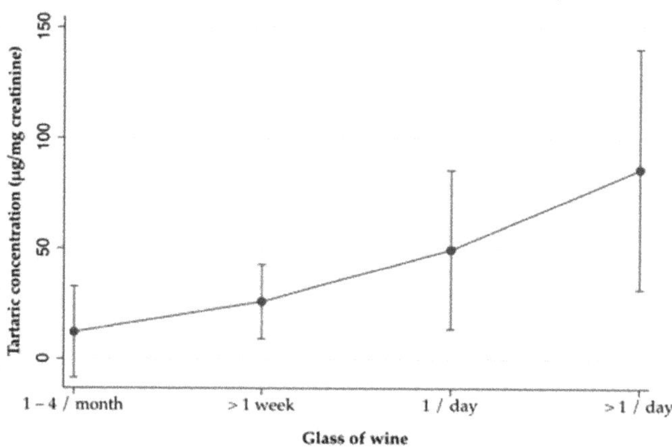

Figure 1. Relationship between urinary tartaric acid concentrations in urine and wine consumption expressed as glasses of wine.

3.3. Anthropometric Measurements and Urinary Tartaric Acid

After adjustment for several covariates, we observed no association between urinary tartaric acid and BMI, WC, weight, WtHR, and systolic or diastolic BP (Table 2).

Table 2. Association between anthropometric variables and urinary tartaric acid (μg/mg creatinine).

		β (95% IC)	*p*-Value
BMI, kg/m^2	Model 1	<−0.01 (−0.01, 0.01)	0.519
	Model 2	<−0.01 (−0.01, 0.01)	0.426
	Model 3	<−0.01 (−0.01, 0.01)	0.973
WC, cm	Model 1	0.56 (−0.25, 1.38)	0.173
	Model 2	0.61 (−0.04, 1.26)	0.064
	Model 3	0.70 (−0.12, 1.51)	0.087
Weight, kg	Model 1	<−0.01 (−0.01, 0.01)	0.770
	Model 2	<−0.01 (−0.01, 0.01)	0.766
	Model 3	<0.01 (−0.01, 0.01)	0.886
WtHR	Model 1	0.29 (−0.25, 0.82)	0.291
	Model 2	0.32 (−0.17, 0.81)	0.175
	Model 3	0.41 (−0.14, 0.96)	0.128

BMI: body mass index; WC: waist circumference; WtHR: waist-to-height ratio; CI: confidence interval. Regression coefficients (95%CI) were obtained from multivariable adjusted linear regression models. β: Non-standardized coefficient. Model 1: adjusted for age. Model 2: adjusted for age, educational level, smoking status, physical activity, and cholesterol-lowering, antihypertensive, and insulin treatment. Model 3: adjusted for age, educational level, smoking status, physical activity, cholesterol-lowering, antihypertensive, and insulin treatment, total energy intake, and consumption of fruits, vegetables, and raisins. We used robust standard errors to account for recruitment center. *p*-values < 0.05 were considered significant.

3.4. Biochemical and Clinical Measurements and Urinary Tartaric Acid

A negative association was observed between urine tartaric acid and total and LDL cholesterol after full adjustment (-3.13 µg/mg (-5.54, -0.71), p-value = 0.016 and -3.03 µg/mg (-5.62, -0.42), p-value = 0.027, respectively). By contrast, no differences were observed for HDL with different concentrations of tartaric acid. Finally, no association was found between triglycerides and glucose and tartaric acid concentrations (Table 3).

Table 3. Association between biochemical variables and blood pressure and urinary tartaric acid (µg/mg creatinine).

		β (95% CI)	*p*-Value
	Model 1	-3.32 (-6.53, -0.10)	0.043
Total cholesterol, mg/dL	Model 2	-3.24 (-5.78, -0.72)	0.017
	Model 3	-3.13 (-5.54, -0.71)	0.016
	Model 1	-3.44 (-6.34, -0.53)	0.021
LDL cholesterol, mg/dL	Model 2	-3.43 (-5.86, -1.00)	0.010
	Model 3	-3.03 (-5.62, -0.42)	0.027
	Model 1	<-0.01 (-0.02, 0.02)	0.689
HDL cholesterol, mg/dL	Model 2	-0.01 (-0.03, 0.01)	0.220
	Model 3	<-0.01 (-0.02, 0.01)	0.633
	Model 1	0.01 (-0.04, 0.05)	0.739
Triglycerides, mg/dL	Model 2	0.02 (-0.03, 0.07)	0.422
	Model 3	0.02 (-0.04, 0.07)	0.525
	Model 1	0.02 (<-0.01, 0.04)	0.092
Glucose, mg/dL	Model 2	0.03 (-0.01, 0.07)	0.119
	Model 3	0.02 (-0.01, 0.06)	0.180
	Model 1	-0.34 (-1.79, 1.11)	0.647
Systolic BP, mmHg	Model 2	-0.09 (-1.53, 1.71)	0.904
	Model 3	0.36 (-1.27, 1.99)	0.633
	Model 1	0.05 (-0.71, 0.81)	0.893
Diastolic BP, mmHg	Model 2	0.24 (-0.47, 0.95)	0.472
	Model 3	0.21 (-0.47, 0.89)	0.502

BP: blood pressure; LDL: low-density lipoprotein cholesterol; HDL: high-density lipoprotein cholesterol; CI: confidence interval. Regression coefficients (95%CI) were obtained from multivariable adjusted linear regression models. β: Non-standardized coefficient. Model 1: adjusted for age. Model 2: adjusted for age, educational level, smoking status, physical activity, BMI, and cholesterol-lowering, antihypertensive, and insulin treatment. Model 3: adjusted for age, educational level, smoking status, physical activity, cholesterol-lowering, antihypertensive, and insulin treatment, total energy intake, and consumption of fruits, vegetables, and raisins. We used robust standard errors to account for recruitment center. *p*-values < 0.05 were considered significant.

4. Discussion

In this sub-analysis of a subset of postmenopausal women participating in the PRED-IMED trial, urinary tartaric acid concentrations as an objective biomarker of wine intake were significantly associated with lower concentrations of total and LDL cholesterol. No associations with anthropometric variables or blood pressure were observed. To the best of our knowledge, the current study is the first to evaluate wine consumption based on a biomarker in a postmenopausal population at increased risk of developing CVD.

Wine consumption has been widely studied due to its beneficial effects on cardiovascular and metabolic health [10]. However, most studies have evaluated wine intake using FFQs or self-questionnaires instead of biological biomarkers, a more reliable and objective way of assessing dietary habits [33]. It has been previously demonstrated that urinary tartaric acid is a specific and sensitive biomarker, as its major sources in the diet are grapes and wine [21,34]. Accordingly, we observed a positive association between wine consumption reported in the FFQs and the concentrations of tartaric acid present in urine. Other phenolic compounds, such as resveratrol and its metabolites, have been proposed as wine biomarkers. However, the resveratrol content in wine is subject to a high variability and its metabolism shows interindividual differences [35]. Thus, selectivity and high correlation with reported intakes make tartaric acid a reliable dietary biomarker of wine consumption.

Different studies have evaluated how alcohol intake affects different parameters of body composition. A cross-sectional study in French adults suggested an inverse association in women of wine intake 100 g/day with BMI and WtHR [36]. Tresserra-Rimbau et al. analyzed the effects of red wine consumption on the prevalence of metabolic syndrome and its components, and found a negative association between moderate red wine consumption and BMI [16]. Tolstrup et al. also described inverse associations between alcohol consumption and WC in women [37], while other studies found no relationship between alcohol consumption and body weight in women [38,39]. The mentioned literature indicates that moderate consumption of wine, an alcoholic beverage that contributes to energy intake, is not related to weight gain or detrimental changes in body composition. Our study supports this notion, as we did not observe any differences in BMI, weight, WC, and WtHR with increasing wine consumption.

Evaluating the effect of alcohol, and specifically wine, on the risk of developing CVD in women is important due to the increase in cardiovascular risk after menopause. Among CVRFs, a recent metanalysis reported that triglycerides, total cholesterol, LDL cholesterol, and the total cholesterol-to-HDL-cholesterol ratio were significantly higher in postmenopausal women compared to premenopausal women, and suggested that age was partly responsible for the differences in lipid levels [40]. We found that women with higher concentrations of tartaric acid presented lower total and LDL cholesterol. Similarly to our results, Rifler et al. reported that after 2 weeks of drinking 250 mL of red wine daily, patients post myocardial infarction presented a 5% decrease in total and LDL cholesterol [41]. Furthermore, Taborsky et al. evaluated the effect of 1 year of wine consumption, and observed a reduction in total and LDL cholesterol [42]. In another clinical trial, authors reported a similar beneficial effect on the lipid profile after consumption of red wine in asymptomatic hypercholesterolemic individuals [43]. The data are almost consistent in showing that wine consumption reduces LDL cholesterol while increasing HDL cholesterol [44,45]. Moderate consumption of alcohol has been associated with higher concentrations of HDL cholesterol and diminished lipid oxidation stress [46]. Resveratrol metabolites in urine, as biomarkers of wine consumption, were significantly associated with lower triglycerides and higher HDL-cholesterol [47]. However, we were unable to confirm that higher urinary tartaric acid as a biomarker of wine consumption was associated with raised HDL cholesterol levels. A probable reason is that the women studied had rather high baseline HDL cholesterol levels, making it more difficult to further increase these with interventions.

Many clinical studies support that light to moderate alcohol consumption, in particular of red wine, is associated with lower CVD rates and an improved lipid profile and inflammatory system [17,48]. However, it remains unknown whether this effect of wine is due to alcohol per se, the phytochemicals of wine, their combined effect, or even the time of drinking, since postprandial oxidative stress after a meal appears to be counteracted by the ingestion of red wine [49]. In this sense, it has been found that wine micro-constituents modulate inflammatory mediators and therefore may be responsible for attenuating postprandial inflammation [50]. In addition, they protect against the effect of ethanol on cytokine secretion, which are involved in inflammatory processes [51]. In support of this view, a randomized clinical trial reported that wine bioactive compounds, such as resveratrol, can decrease total cholesterol by reducing mRNA expression of hepatic 3-hydroxy-3-methyl-glutaryl-CoA (HMG-CoA) reductase, in addition to the increased activation of the sirtuin system in all tissues [52]. Experimental work in cell cultures and animal models has shown that the enhancement of Sirtuin 1 can lead to better metabolic profiles and anti-inflammatory activities, as well as increased reverse cholesterol transport [53]. Overall, evidence supports that wine micro-constituents play a crucial role in the protective effect of wine on cardiovascular health by exerting anti-inflammatory actions.

The main strength of this study is that it used a biological biomarker, tartaric acid, to evaluate wine consumption, instead of less reliable methods such as FFQs or self-reported questionnaires. Moreover, it involved baseline data of participants in the PREDIMED trial; therefore, the results reflect real-life conditions. The main limitations were the modest sample size and the impossibility of determining causality due to the cross-sectional design.

5. Conclusions

The findings from the current cross-sectional study support the notion that wine intake has beneficial nutraceutical effects on the cardiovascular health of postmenopausal women, as its biomarker tartaric acid was associated with lower total and LDL cholesterol concentrations. Randomized trials are needed to confirm these results and determine the impact of wine consumption on cardiovascular health in a sensitive population such as that of postmenopausal women.

Author Contributions: Conceptualization, R.E.; Formal analysis and data curation, I.D.-L.; writing, I.D.-L., I.P.-M., and C.A.-R.; review and editing, A.T.-R., M.A.M.-G., C.O.-A., J.S.-S., O.C., J.L., F.A., M.F., L.S.-M., X.P., E.G.-G., E.R., R.M.L.-R., and R.E. All authors have read and agreed to the published version of the manuscript.

Funding: This research was funded by the Interprofesional del vino by PID2020-114022RB-I00, CICYT [AGL2016- 75329-R], CIBEROBN from the Instituto de Salud Carlos III (ISCIII) from the Ministerio de Ciencia, Innovación y Universidades (AEI/FEDER, UE), and the Generalitat de Catalunya (GC) [2017SGR 196].

Institutional Review Board Statement: The study was conducted according to the guidelines of the Declaration of Helsinki and was approved by the Institutional Review Board of the 11 participating centers. The study was registered with the International Standard Randomized Controlled Trial Number (ISRCTN) 35739639.

Informed Consent Statement: Informed consent was obtained from all subjects involved in the study. Written informed consent was obtained from the patients.

Data Availability Statement: There are restrictions on the availability of data for the PREDIMED trial due to the signed consent agreements around data sharing, which only allow access to external researchers for studies following project purposes. Requestors wishing to access the PREDIMED-Plus trial data used in this study can make a request to the PREDIMED trial Steering Committee chair: restruch@clinic.cat. The request will then be passed to members of the PREDIMED Steering Committee for deliberation.

Acknowledgments: The PREDIMED trial was supported by the official funding agency for biomedical research of the Spanish government (Instituto de Salud Carlos III) through grants provided to research networks specifically developed for the trial: RTIC G03/140 (Coordinator: R.E.) and RTIC RD 06/0045 (Coordinator: M.A.M.-G.). All investigators of the PREDIMED trial belong to Centro de Investigación Biomédica en Red (CIBER), an initiative of Instituto de Salud Carlos III. This sub-study of the PREDIMED trial was specifically funded by the Interprofesional del Vino de España. I.D.-L. thanks the FI-AGAUR Research Fellowship Program, Generalitat de Catalunya [FI_B 00256]. I.P.-M. is thankful for the FI-SDUR (EMC/2703/2019) fellowship from the Generalitat de Catalunya.

Conflicts of Interest: E.R. reports grants, personal fees, non-financial support, and other from the California Walnut Commission during the conduct of the study; grants, personal fees, non-financial support and other from Alexion; and personal fees and other from Amarin, outside the submitted work. R.M.L.-R. reports personal fees from Cerveceros de España, personal fees, and other from Adventia, Wine in Moderation, Ecoveritas S.A., outside the submitted work. R.E. reports grants from the Fundación Dieta Mediterránea (Spain), and Cerveza y Salud (Spain), and personal fees for given lectures from Brewers of Europe (Belgium), the Fundación Cerveza y Salud (Spain), Pernaud-Ricard (Mexico), Instituto Cervantes (Alburquerque, USA), Instituto Cervantes (Milan, Italy), Instituto Cervantes (Tokyo, Japan), Lilly Laboratories (Spain), and the Wine and Culinary International Forum (Spain), as well as non-financial support for the organization of a National Congress on Nutrition and feeding trials with products from Grand Fountain and Uriach Laboratories (Spain).

Abbreviations

Angiotensin-converting enzyme (ACE); blood pressure (BP); body mass index (BMI); confidence interval (CI); cardiovascular disease (CVD); cardiovascular risk factors (CVRFs); food frequency questionnaire (FFQ); high-density lipoprotein (HDL); liquid chromatography with electrospray ionization and tandem mass spectrometry (LC-ESI-MS/MS); low-density lipoprotein (LDL); Mediterranean diet (MedDiet); Prevención con Dieta Mediterránea (PREDIMED); standard deviation (SD); waist circumference (WC); waist to height ratio (WtHR).

References

1. Benjamin, E.J.; Muntner, P.; Alonso, A.; Bittencourt, M.S.; Callaway, C.W.; Carson, A.P.; Chamberlain, A.M.; Chang, A.R.; Cheng, S.; Das, S.R.; et al. Heart Disease and Stroke Statistics-2019 Update: A Report from the American Heart Association. *Circulation* **2019**, *139*, e56–e528. [CrossRef]
2. El Khoudary, S.R.; Greendale, G.; Crawford, S.L.; Avis, N.E.; Brooks, M.M.; Thurston, R.C.; Karvonen-Gutierrez, C.; Waetjen, L.E.; Matthews, K. The menopause transition and women's health at midlife: A progress report from the Study of Women's Health across the Nation (SWAN). *Menopause* **2019**, *26*, 1213–1227. [CrossRef]
3. Matthews, K.A.; Meilahn, E.; Kuller, L.H.; Kelsey, S.F.; Cagguila, A.W.; Wing, R.R. Menopause and the Risk Factors of Coronary Heart Disease. *N. Engl. J. Med.* **1974**, *306*, 802–805.
4. Matthews, K.A.; El Khoudary, S.R.; Brooks, M.M.; Derby, C.A.; Harlow, S.D.; Barinas-Mitchell, E.J.; Thurston, R.C. Lipid Changes around the Final Menstrual Period Predict Carotid Subclinical Disease in Postmenopausal Women. *Stroke* **2017**, *1*, 70–76. [CrossRef]
5. Bonithon-kopp, C.; Scarabin, P.Y.; Darne, B.; Malmejac, A.; Guize, L. Menopause-related changes in lipoproteins and some other cardiovascular risk factors. *Int. J. Epidemiol.* **1990**, *19*, 42–48. [CrossRef] [PubMed]
6. Matthews, K.A.; Crawford, S.L.; Chae, C.U.; Everson-Rose, S.A.; Sowers, M.F.; Sternfeld, B.; Sutton-Tyrrell, K. Are Changes in Cardiovascular Disease Risk Factors in Midlife Women Due to Chronological Aging or to the Menopausal Transition? *J. Am. Coll. Cardiol.* **2009**, *54*, 2366–2373. [CrossRef] [PubMed]
7. Greendale, G.A.; Sternfeld, B.; Huang, M.H.; Han, W.; Karvonen-Gutierrez, C.; Ruppert, K.; Cauley, J.A.; Finkelstein, J.S.; Jiang, S.F.; Karlamangla, A.S. Changes in body composition and weight during the menopause transition. *JCI Insight* **2019**, *4*, 1–14. [CrossRef] [PubMed]
8. Stampfer, M.J.; Hu, F.B.; Manson, J.E.; Rimm, E.B.; Willett, W.C. Primary Prevention of Coronary Heart Disease in Women Through Diet and Lifestyle. *N. Engl. J. Med.* **2000**, *343*, 16. [CrossRef] [PubMed]
9. Hodge, A.M.; English, D.R.; Itsiopoulos, C.; O'Dea, K.; Giles, G.G. Does a Mediterranean diet reduce the mortality risk associated with diabetes: Evidence from the Melbourne Collaborative Cohort Study. *Nutr. Metab. Cardiovasc. Dis.* **2011**, *21*, 733–739. [CrossRef]
10. Marcos, A.; Serra-Majem, L.; Pérez-Jiménez, F.J.; Pascual, V.; Tinahones, F.J.; Estruch, R. Moderate consumption of beer and its effects on cardiovascular and metabolic health: An updated review of recent scientific evidence. *Nutrients* **2021**, *13*, 879. [CrossRef] [PubMed]
11. Mozaffarian, D. Dietary and Policy Prioritites for CVD, diabetes and obesity—a comprehensive review. *Circulation* **2016**, *133*, 187–225. [CrossRef]
12. Casas, R.; Castro-Barquero, S.; Estruch, R.; Sacanella, E. Nutrition and Cardiovascular Health. *Int. J. Mol. Sci.* **2018**, *19*, 3988. [CrossRef] [PubMed]
13. Martínez-González, M.A.; Gea, A.; Ruiz-Canela, M. The Mediterranean Diet and Cardiovascular Health: A Critical Review. *Circ. Res.* **2019**, *124*, 779–798. [CrossRef] [PubMed]
14. Bach-Faig, A.; Berry, E.M.; Lairon, D.; Reguant, J.; Trichopoulou, A.; Dernini, S.; Medina, F.X.; Battino, M.; Belahsen, R.; Miranda, G.; et al. Mediterranean diet pyramid today. Science and cultural updates. *Public Health Nutr.* **2011**, *14*, 2274. [CrossRef]
15. Chiva-Blanch, G.; Arranz, S.; Lamuela-Raventos, R.M.; Estruch, R. Effects of wine, alcohol and polyphenols on cardiovascular disease risk factors: Evidences from human studies. *Alcohol Alcohol.* **2013**, *48*, 270–277. [CrossRef] [PubMed]
16. Tresserra-Rimbau, A.; Medina-Remón, A.; Lamuela-Raventós, R.M.; Bulló, M.; Salas-Salvadó, J.; Corella, D.; Fitó, M.; Gea, A.; Gómez-Gracia, E.; Lapetra, J.; et al. Moderate red wine consumption is associated with a lower prevalence of the metabolic syndrome in the PREDIMED population. *Br. J. Nutr.* **2015**, *113*, S121–S130. [CrossRef] [PubMed]
17. Costanzo, S.; Di Castelnuovo, A.; Donati, M.B.; Iacoviello, L.; De Gaetano, G. Wine, beer or spirit drinking in relation to fatal and non-fatal cardiovascular events: A meta-analysis. *Eur. J. Epidemiol.* **2011**, *26*, 833–850. [CrossRef] [PubMed]
18. Hedrick, V.E.; Dietrich, A.M.; Estabrooks, P.A.; Savla, J.; Serrano, E.; Davy, B.M. Dietary biomarkers: Advances, limitations and future directions. *Nutr. J.* **2012**, *11*, 1. [CrossRef]
19. Velioglu, Y.S. Food acids: Organic acids, volatile organic acids, and phenolic acids. In *Advances in Food Biochemistry*; CRC Press: Boca Raton, FL, USA, 2009; p. 522.
20. Ribéreau-Gayon, P.; Glories, Y.; Maujean, A.; Dubourdieu, D. Organic Acids in Wine. In *Handbook of Enology*; John Wiley & Sons: Hoboken, NJ, USA, 2006; pp. 1–49.

21. Regueiro, J.; Vallverdú-Queralt, A.; Simal-Gándara, J.; Estruch, R.; Lamuela-Raventós, R.M. Urinary tartaric acid as a potential biomarker for the dietary assessment of moderate wine consumption: A randomised controlled trial. *Br. J. Nutr.* **2014**, *111*, 1680–1685. [CrossRef]

22. Lloyd, A.J.; Willis, N.D.; Wilson, T.; Zubair, H.; Chambers, E.; Garcia-Perez, I.; Xie, L.; Tailliart, K.; Beckmann, M.; Mathers, J.C.; et al. Addressing the pitfalls when designing intervention studies to discover and validate biomarkers of habitual dietary intake. *Metabolomics* **2019**, *15*, 1–12. [CrossRef]

23. Martínez-González, M.Á.; Corella, D.; Salas-salvadó, J.; Ros, E.; Covas, M.I.; Fiol, M.; Wärnberg, J.; Arós, F.; Ruíz-Gutiérrez, V.; Lamuela-Raventós, R.M.; et al. Cohort profile: Design and methods of the PREDIMED study. *Int. J. Epidemiol.* **2012**, *41*, 377–385. [CrossRef]

24. Fernández-Ballart, J.D.; Piñol, J.L.; Zazpe, I.; Corella, D.; Carrasco, P.; Toledo, E.; Perez-Bauer, M.; Martínez-González, M.Á.; Salas-Salvadó, J.; Martn-Moreno, J.M. Relative validity of a semi-quantitative food-frequency questionnaire in an elderly Mediterranean population of Spain. *Br. J. Nutr.* **2010**, *103*, 1808–1816. [CrossRef]

25. Elosua, R.; Garcia, M.; Aguilar, A.; Molina, L.; Covas, M.-I.; Marrugat, J. Validation of the Minnesota Leisure Time Spanish Women. *Med. Sci. Sport. Exerc.* **2000**, *32*, 1431–1437. [CrossRef] [PubMed]

26. Estruch, R.; Martínez-González, M.A.; Corella, D.; Salas-Salvadó, J.; Ruíz-Gutiérrez, V.; Covas, M.-I.; Fiol, M.; Gómez-Gracia, E.; López-Sabater, M.C.; Vinyoles, E.; et al. Effects of a Mediterranean-Style Diet on Cardiovascular Risk FactorsA Randomized Trial. *Ann. Intern. Med.* **2006**, *145*, 1–11. [CrossRef] [PubMed]

27. Regueiro, J.; Vallverdú-Queralt, A.; Simal-Gándara, J.; Estruch, R.; Lamuela-Raventós, R. Development of a LC-ESI-MS/MS approach for the rapid quantification of main wine organic acids in human urine. *J. Agric. Food Chem.* **2013**, *61*, 6763–6768. [CrossRef] [PubMed]

28. Medina-Remón, A.; Barrionuevo-González, A.; Zamora-Ros, R.; Andres-Lacueva, C.; Estruch, R.; Martínez-González, M.Á.; Diez-Espino, J.; Lamuela-Raventos, R.M. Rapid Folin-Ciocalteu method using microtiter 96-well plate cartridges for solid phase extraction to assess urinary total phenolic compounds, as a biomarker of total polyphenols intake. *Anal. Chim. Acta* **2009**, *634*, 54–60. [CrossRef] [PubMed]

29. Ashwell, M.; Hsieh, S.D. Six reasons why the waist-to-height ratio is a rapid and effective global indicator for health risks of obesity and how its use could simplify the international public health message on obesity. *Int. J. Food Sci. Nutr.* **2005**, *56*, 303–307. [CrossRef]

30. Alberti, K.G.M.M.; Eckel, R.H.; Grundy, S.M.; Zimmet, P.Z.; Cleeman, J.I.; Donato, K.A.; Fruchart, J.C.; James, W.P.T.; Loria, C.M.; Smith, S.C. Harmonizing the metabolic syndrome: A joint interim statement of the international diabetes federation task force on epidemiology and prevention; National heart, lung, and blood institute; American heart association; World heart federation; International. *Circulation* **2009**, *120*, 1640–1645. [CrossRef]

31. WHO. *Definition and Diagnosis of Diabetes Mellitus and Intermediate Hyperglycemia*; WHO: Geneva, Switzerland, 2006.

32. Care, D.; Suppl, S.S. Classification and diagnosis of diabetes: Standards of medical care in Diabetesd2018. *Diabetes Care* **2018**, *41*, S13–S27. [CrossRef]

33. Marshall, J.R. Methodologic and statistical considerations regarding use of biomarkers of nutritional exposure in epidemiology. *J. Nutr.* **2003**, *133*, 881–887. [CrossRef]

34. Esteban-Fernández, A.; Ibañez, C.; Simó, C.; Bartolomé, B.; Moreno-Arribas, M.V. An Ultrahigh-Performance Liquid Chromatography-Time-of-Flight Mass Spectrometry Metabolomic Approach to Studying the Impact of Moderate Red-Wine Consumption on Urinary Metabolome. *J. Proteome Res.* **2018**, *17*, 1624–1635. [CrossRef]

35. Romero-Pérez, A.I.; Lamuela-Raventós, R.M.; Waterhouse, A.L.; De La Torre-Boronat, M.C. Levels of cis- and trans-Resveratrol and Their Glucosides in White and Rosé Vitis vinifera Wines from Spain. *J. Agric. Food Chem.* **1996**, *44*, 2124–2128. [CrossRef]

36. Lukasiewicz, E.; Mennen, L.I.; Bertrais, S.; Arnault, N.; Preziosi, P.; Galan, P.; Hercberg, S. Alcohol intake in relation to body mass index and waist-to-hip ratio: The importance of type of alcoholic beverage. *Public Health Nutr.* **2005**, *8*, 315–320. [CrossRef]

37. Tolstrup, J.S.; Halkjær, J.; Heitmann, B.L.; Tjønneland, A.M.; Overvad, K.; Sørensen, T.I.A.; Grønbæk, M.N. Alcohol drinking frequency in relation to subsequent changes in waist circumference. *Am. J. Clin. Nutr.* **2008**, *87*, 957–963. [CrossRef] [PubMed]

38. Alcácera, M.A.; Marques-Lopes, I.; Fajó-Pascual, M.; Puzo, J.; Pérez, J.B.; Bes-Rastrollo, M.; Martínez-González, M.Á. Lifestyle factors associated with BMI in a Spanish graduate population: The SUN study. *Obes. Facts* **2008**, *1*, 80–87. [CrossRef]

39. French, M.T.; Norton, E.C.; Fang, H.; Maclean, J.C. Alcohol consumption and body weight. *Health Econ.* **2010**, *19*, 814–832. [CrossRef] [PubMed]

40. Ambikairajah, A.; Walsh, E.; Cherbuin, N. Lipid profile differences during menopause: A review with meta-analysis. *Menopause* **2019**, *26*, 1327–1333. [CrossRef] [PubMed]

41. Rifler, J.P.; Lorcerie, F.; Durand, P.; Delmas, D.; Ragot, K.; Limagne, E.; Mazué, F.; Riedinger, J.M.; D'Athis, P.; Hudelot, B.; et al. A moderate red wine intake improves blood lipid parameters and erythrocytes membrane fluidity in post myocardial infarct patients. *Mol. Nutr. Food Res.* **2012**, *56*, 345–351. [CrossRef] [PubMed]

42. Taborsky, M.; Ostadal, P.; Adam, T.; Moravec, O.; Gloger, V.; Schee, A.; Skala, T. Red or white wine consumption effect on atherosclerosis in healthy individuals (In Vino Veritas study). Taborsky. *Clin. Study* **2017**, *118*, 292–298. [CrossRef]

43. Apostolidou, C.; Adamopoulos, K.; Lymperaki, E.; Iliadis, S.; Papapreponis, P.; Kourtidou-Papadeli, C. Cardiovascular risk and benefits from antioxidant dietary intervention with red wine in asymptomatic hypercholesterolemics. *Clin. Nutr. ESPEN* **2015**, *10*, e224–e233. [CrossRef]

44. Droste, D.W.; Iliescu, C.; Vaillant, M.; Gantenbein, M.; De Bremaeker, N.; Lieunard, C.; Velez, T.; Meyer, M.; Guth, T.; Kuemmerle, A.; et al. A daily glass of red wine associated with lifestyle changes independently improves blood lipids in patients with carotid arteriosclerosis: Results from a randomized controlled trial. *Nutr. J.* **2013**, *12*, 1–9. [CrossRef]

45. Kechagias, S.; Zanjani, S.; Gjellan, S.; Leinhard, O.D.; Kihlberg, J.; Smedby, Ö.; Johansson, L.; Kullberg, J.; Ahlström, H.; Lindström, T.; et al. Effects of moderate red wine consumption on liver fat and blood lipids: A prospective randomized study. *Ann. Med.* **2011**, *43*, 545–554. [CrossRef]

46. Castaldo, L.; Narváez, A.; Izzo, L.; Graziani, G.; Gaspari, A.; Di Minno, G.; Ritieni, A. Red wine consumption and cardiovascular health. *Molecules* **2019**, *24*, 3626. [CrossRef] [PubMed]

47. Zamora-Ros, R.; Urpi-Sarda, M.; Lamuela-Raventós, R.M.; Martínez-González, M.Á.; Salas-Salvadó, J.; Arós, F.; Fitó, M.; Lapetra, J.; Estruch, R.; Andres-Lacueva, C. High urinary levels of resveratrol metabolites are associated with a reduction in the prevalence of cardiovascular risk factors in high-risk patients. *Pharmacol. Res.* **2012**, *65*, 615–620. [CrossRef] [PubMed]

48. Fragopoulou, E.; Choleva, M.; Antonopoulou, S.; Demopoulos, C.A. Wine and its metabolic effects. A comprehensive review of clinical trials. *Metabolism* **2018**, *83*, 102–119. [CrossRef]

49. Covas, M.I.; Gambert, P.; Fitó, M.; de la Torre, R. Wine and oxidative stress: Up-to-date evidence of the effects of moderate wine consumption on oxidative damage in humans. *Atherosclerosis* **2010**, *208*, 297–304. [CrossRef] [PubMed]

50. Argyrou, C.; Vlachogianni, I.; Stamatakis, G.; Demopoulos, C.A.; Antonopoulou, S.; Fragopoulou, E. Postprandial effects of wine consumption on Platelet Activating Factor metabolic enzymes. *Prostaglandins Other Lipid Mediat.* **2017**, *130*, 23–29. [CrossRef] [PubMed]

51. Fragopoulou, E.; Argyrou, C.; Detopoulou, M.; Tsitsou, S.; Seremeti, S.; Yannakoulia, M.; Antonopoulou, S.; Kolovou, G.; Kalogeropoulos, P. The effect of moderate wine consumption on cytokine secretion by peripheral blood mononuclear cells: A randomized clinical study in coronary heart disease patients. *Cytokine* **2021**, *146*, 155629. [CrossRef] [PubMed]

52. Mansur, A.P.; Roggerio, A.; Goes, M.F.S.; Avakian, S.D.; Leal, D.P.; Maranhão, R.C.; Strunz, C.M.C. Serum concentrations and gene expression of sirtuin 1 in healthy and slightly overweight subjects after caloric restriction or resveratrol supplementation: A randomized trial. *Int. J. Cardiol.* **2017**, *227*, 788–794. [CrossRef] [PubMed]

53. Stünkel, W.; Campbell, R.M. Sirtuin 1 (SIRT1): The misunderstood HDAC. *J. Biomol. Screen.* **2011**, *16*, 1153–1169. [CrossRef]

Review

Nuts: Natural Pleiotropic Nutraceuticals

Emilio Ros [1,2,*], Annapoorna Singh [3] and James H. O'Keefe [3]

[1] Lipid Clinic, Endocrinology and Nutrition Service, Institut d'Investigacions Biomediques August Pi Sunyer, Hospital Clinic, University of Barcelona, 08036 Barcelona, Spain
[2] CIBER Fisiopatología de la Obesidad y Nutrición (CIBEROBN), Instituto de Salud Carlos III, 28029 Madrid, Spain
[3] Saint Luke's Mid America Heart Institute, University of Missouri-Kansas City, Kansas City, MO 64110, USA; singhann@umkc.edu (A.S.); jokeefe@saintlukeskc.org (J.H.O.)
* Correspondence: eros@clinic.cat; Tel.: +34-676326989

Abstract: Common nuts (tree nuts and peanuts) are energy-dense foods that nature has gifted with a complex matrix of beneficial nutrients and bioactives, including monounsaturated and polyunsaturated fatty acids, high-quality protein, fiber, non-sodium minerals, tocopherols, phytosterols, and antioxidant phenolics. These nut components synergize to favorably influence metabolic and vascular physiology pathways, ameliorate cardiovascular risk factors and improve cardiovascular prognosis. There is increasing evidence that nuts positively impact myriad other health outcomes as well. Nut consumption is correlated with lower cancer incidence and cancer mortality, and decreased all-cause mortality. Favorable effects on cognitive function and depression have also been reported. Randomized controlled trials consistently show nuts have a cholesterol-lowering effect. Nut consumption also confers modest improvements on glycemic control, blood pressure (BP), endothelial function, and inflammation. Although nuts are energy-dense foods, they do not predispose to obesity, and in fact may even help in weight loss. Tree nuts and peanuts, but not peanut butter, generally produce similar positive effects on outcomes. First level evidence from the PREDIMED trial shows that, in the context of a Mediterranean diet, consumption of 30 g/d of nuts (walnuts, almonds, and hazelnuts) significantly lowered the risk of a composite endpoint of major adverse cardiovascular events (myocardial infarction, stroke, and death from cardiovascular disease) by ≈30% after intervention for 5 y. Impressively, the nut-supplemented diet reduced stroke risk by 45%. As they are rich in salutary bioactive compounds and beneficially impact various health outcomes, nuts can be considered natural pleiotropic nutraceuticals.

Keywords: tree nuts; peanuts; fatty acids; prospective studies; randomized clinical trials; cardiovascular risk; type-2 diabetes; cancer; hypertension; cognitive function; mortality; body weight; blood lipids; inflammation; PREDIMED

Citation: Ros, E.; Singh, A.; O'Keefe, J.H. Nuts: Natural Pleiotropic Nutraceuticals. *Nutrients* **2021**, *13*, 3269. https://doi.org/10.3390/nu13093269

Academic Editors: Paolo Magni, Andrea Baragetti and Andrea Poli

Received: 19 August 2021
Accepted: 14 September 2021
Published: 19 September 2021

1. Introduction

Common tree nuts include almonds, Brazil nuts, cashews, hazelnuts, macadamias, pecans, pine nuts, pistachios and walnuts. Botanically, the peanut (*Arachis Hypogaea*) is a legume, but it has a nutrient profile that is similar to the tree nuts listed above, which qualifies peanuts to be included in the nut food group [1]. The impact of nut consumption on health outcomes has been extensively investigated since the publication in 1992 of the pioneering Adventist Health Study, in which nut consumption was associated for the first time with a lower risk of coronary heart disease (CHD) [2]. Soon after, a landmark randomized clinical trial (RCT) demonstrated that walnut consumption significantly lowered blood cholesterol [3].

Nuts are nutrient-rich foods that have been a staple of humankind's diet throughout our long evolutionary history [4]. However, during the last century, most people in industrialized nations have markedly reduced their consumption of nuts, so that now nuts comprise only a marginal source of dietary energy, except for vegetarians, health-conscious groups such as Seventh Day Adventists, and individuals following diets based on whole natural foods [5]. Tellingly, in the last two decades nut consumption has increased in Western countries in parallel with the United States (US) Food and Drug Administration's issue of a health claim that nut consumption is associated with a reduced risk of CHD [6], inclusion of nuts in many guidelines for health promotion [7], and wide media advertising of their beneficial health effects.

The scientific evidence behind nuts as health-promoting foods stems from both abundant epidemiological observations suggesting that their regular consumption relates inversely to incidence of and mortality from major non-communicable diseases [8] and from RCTs disclosing a consistent cholesterol-lowering effect of nut diets [9]. The mechanisms for these salutary effects include the optimal nutrient composition of nuts, their satiating effects, and their tendency to displace other less healthy foods.

Contrary to the popular belief that, due to the high energy content of nuts, their consumption has a fattening effect, evidence from both epidemiological studies and RCTs suggests that their regular consumption does not lead to increased body weight and may even promote weight loss [10]. This review summarizes current knowledge on the increasingly important topic of nuts as health-promoting foods and their sizable contribution to the nutritional quality of the diet, while laying out the scientific basis to consider them as natural pleiotropic nutraceuticals.

2. Data Sources and Selection of Studies on Nuts and Human Health

For this narrative review we conducted a comprehensive search of the PubMed®/MEDLINE® (https://www.ncbi.nlm.nih.gov/pubmed/ (accessed up to 31 July 2021)) database through July 2021 for English language articles of epidemiological and clinical studies illustrating the effects of exposure to nuts (tree nuts and peanuts) and their components (mainly peanut butter) on health outcomes, and the latest reviews and meta-analyses of these studies. Meta-analyses pooling data from nuts and seeds were excluded. We also searched the references from original research studies and reviews, as well as articles citing clinical studies, reviews, and meta-analyses, as listed by the publishers of individual articles in their websites. Given that the information of the various meta-analyses on the same outcome tends to be redundant, each successive one synthesizing the results of the same studies plus newly published ones, for each outcome, only the information from the most recent meta-analysis is discussed. However, older systematic reviews may be cited if they contain relevant information (i.e., dose–response analyses) not covered in the subsequent meta-analyses. For completeness, the data from well-designed cohort studies published after each specific meta-analysis are also reviewed.

Data were examined for relevance, quality, consistency and independently extracted by the two senior authors (ER, JHO), who reached an agreement when in doubt about a specific citation. Given that few RCTs on the effects of nut consumption on clinical end points are available, we obtain the core of scientific evidence from epidemiologic studies relating frequency of nut consumption to disease outcomes and RCTs of nut-enriched versus control diets for effects on intermediate end points, with particular attention to meta-analyses of such studies.

3. Historical Aspects

Archeological sites throughout the world have produced proof of consumption of hard-shelled nuts by ancient humans going back to the mid-Pleistocene, one million years ago. The oldest evidence of cultivation of the common tree nuts almonds (*Prunus amigdalis*), hazelnuts (*Corylus avellana*), walnuts (English walnuts, *Juglans regia*), and pistachios (*Pistachia vera*) is from Asia, spanning from China to the Middle-East and the Anatolian peninsula (modern Turkey). These trees were subsequently cultivated in Greece, then in the territories of the Roman Empire and the Iberian peninsula, and were extended to all of Europe during the Middle Ages. In the 16th century, in one of the first and boldest food globalizations, European colonizers introduced these tree nuts to the Americas and brought those native to the Americas back to Europe [11]. In North America, there were native hazelnuts and walnuts—the so-called black walnut (*Juglans nigra*), as well as other indigenous nuts such as pecans (*Carya illinoinensis*), while cashews (*Anacardium occidentale*) and Brazil nuts (*Bertholettia excelsa*) are native to South America. Another popular tree nut, macadamias (*Macadamia integrifolia*), is native to Australia. The common pine nuts (*Pinus pinea*) are often obtained from natural forests, mostly in the Mediterranean region, but they are also native to China and North America.

Peanuts (also called groundnuts in some areas) were first cultivated from wild varieties by the ancient Incas from Peru. European explorers during the 16th century first discovered peanut plants, which were being cultivated in Brazil and Mexico, and transported peanuts back to Spain. From there, traders and explorers exported peanuts to Asia and Africa, and eventually to North America in the 1700s. Peanut butter was developed more than a century ago as a soft protein meal for people with poor dentition [12]. Today, peanuts and peanut butter are popular; Americans per capita eat about 6 pounds per year.

In Europe, nut supply is highest in Mediterranean countries [13]. Indeed, nuts are an integral component and a defining feature of the traditional Mediterranean diet, a dietary pattern characterized by high consumption of vegetables, fruits, nuts, olive oil, cereal grains; moderate consumption of fish and alcohol—mostly wine; and a low consumption of dairy products, red meat and meat products, and sweets [14]. Nuts can be are incorporated into the usual diet in different ways, as snacks, mixed in meals or desserts and may be eaten whole (fresh or roasted), in spreads (i.e., almond paste, peanut butter), as oils or hidden in commercial products, sauces, pastries, cakes, ice creams, and baked goods. Due to their high energy content and tasty nature, nuts have been widely introduced into sports snacks and supplements.

4. Nutrient Content

Nuts are nutrient dense foods, coming only after vegetable oils as the natural plant food richest in fat. Their total fat content as percent of weight ranges from 44% in cashews to 76% in macadamias, and they provide 23 to 30 kJ/g (Table 1) [15,16]. However, the fatty acid composition of nuts is salutary because they have a low saturated fatty acid (SFA) content (range, 4% to 16%) and nearly one-half of their total fat content is formed by unsaturated fat—specifically, monounsaturated fatty acids (MUFA) in most nuts, polyunsaturated fatty acids (PUFA) predominating over MUFA in pine nuts, similar amounts of MUFA and PUFA in Brazil nuts, and mostly PUFA in walnuts (Figure 1). Of note, with around 10 g per 100 g, walnuts are particularly rich in α-linolenic acid (ALA), the plant-derived essential omega-3 fatty acid [1]. The favorable lipid content of nuts is an important contributor to the beneficial health effects conferred by their frequent consumption.

Nuts are also contain other macronutrients and bioactives reputed as beneficial for health outcomes. They are a good source of vegetable protein (between 8% and 25% of energy) and are known to have a sizable content of the amino acid L-arginine, which is the substrate for the synthesis of endothelium-derived nitric oxide (NO), the main endogenous vasodilator and blood pressure (BP) regulator [17]. This explains in part why nut consumption helps improve endothelial function and may lower BP. Additionally, nuts are a good source of dietary fiber, ranging from 3 to 12.5 g per 100 g (Table 1) [15,16]. Indeed, a standard 1-oz (28-g) serving of nuts provides 5–10% of daily fiber requirements [1].

Among other nut components, there are several micronutrients that have salutary effects when taken in at doses beyond those necessary to prevent deficiency states. Nuts contain considerable amounts of the B-vitamin folate, peanuts being richest [15,18]. Almonds and hazelnuts are good sources of the antioxidant vitamins including tocopherols (e.g., vitamin E), while all nuts contain polyphenols, which are powerful antioxidants required to protect the germ from oxidative stress and preserve the reproductive potential of the seed [19]. Due to their protective characteristics, most polyphenols reside in the outer peel of nuts (between the shell and the nut), a good reason to eat raw, unpeeled nuts when possible. Walnuts, pistachios, and pecans have the highest polyphenol content (Table 2).

Table 1. Average nutrient composition of tree nuts and oeanuts (per 100 g) [15,16].

Nuts	Energy (kJ)	Protein (g)	Fiber (g)	Fat (g)	SFA (g)	MUFA (g)	PUFA (g)	LA (g)	ALA (g)	Phytosterols (g)
Almonds	2409	21.1	12.5	49.9	3.9	31.5	12.2	12.2	0.00	162
Brazil nuts (dried)	2743	14.3	7.5	66.4	15.1	24.5	24.4	20.5	0.05	72
Cashews	2401	15.3	3	46.4	9.2	27.3	7.8	7.7	0.15	120
Hazelnuts	2669	15.0	9.7	60.8	4.5	45.7	7.9	7.8	0.09	115
Macadamias	2995	7.9	8.0	76	11.9	58.9	1.4	1.3	0.21	119
Peanuts	2372	26	8.5	49.2	6.2	24.4	15.6	15.6	0.00	126
Pecans	2891	9.2	9.6	72.0	6.2	40.8	21.6	20.6	1.00	113
Pine nuts (dried)	2816	13.7	3.7	68.4	4.9	18.8	34.1	33.2	0.16	120
Pistachios	2430	20.6	10.0	47	5.4	25.0	14.0	13.2	0.25	272
Walnuts	2738	15.2	6.7	65.2	6.1	8.9	47.2	38.1	9.08	143

Data for raw nuts, except when specified. ALA, α–linolenic acid; LA, linoleic acid; MUFA, monounsaturated fatty acids; PUFA; polyunsaturated fatty acids; SFA, saturated fatty acids.

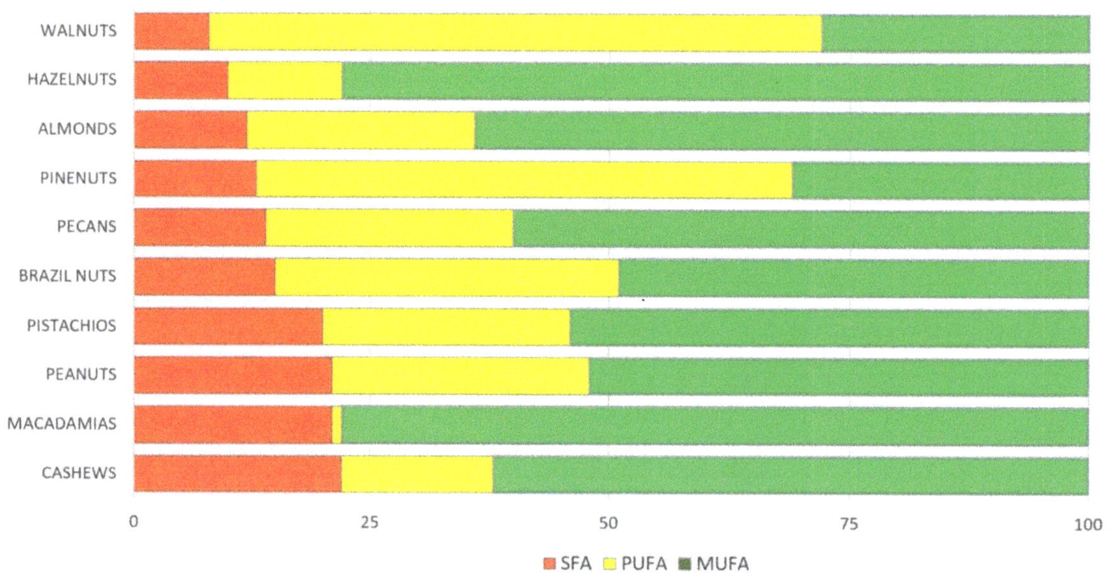

Figure 1. Percent fatty acid profile of common nuts.

Table 2. Average composition of selected micronutrients in tree nuts and peanuts (per 100 g) [15,18].

Nuts	Folate (µg)	Calcium (mg)	Magnesium (mg)	Sodium (mg)	Potassium (mg)	Polyphenols (mg)
Almonds	44	269	270	1	733	287
Brazil nuts	22	160	376	3	659	244
Cashews	69	45	260	16	565	233
Hazelnuts	113	114	163	0	680	671
Macadamias	10	70	118	4	363	126
Peanuts	240	92	168	18	705	406
Pecans	22	70	121	0	410	1284
Pine nuts	34	16	251	2	597	58
Pistachios	49	104	106	6	977	1420
Walnuts	98	98	158	2	441	1579

Nuts are devoid of cholesterol, but their fat fraction contains chemically related non-cholesterol sterols, which belong to a heterogeneous group of molecules known as plant sterols or phytosterols. These compounds are non-nutritive plant components that play a structural role in their cell membranes just as cholesterol does in animal cell membranes [20]. Phytosterols interfere with cholesterol absorption, thus helping lower blood cholesterol concentration when present in doses of 1 g or higher in the intestinal lumen. The mechanism of action of phytosterols depends on their hydrophobic nature, as they have a large hydrocarbon molecule with a higher affinity for micelles than has cholesterol. As phytosterols displace cholesterol from micelles, the amount of sterol available for absorption is reduced. Pistachios and almonds are highest in phytosterols (Table 1). Not unexpectedly, evidence has been provided that phytosterols contribute to the cholesterol-lowering effect of nut consumption [21].

Nuts are also a rich source of beneficial minerals, such as calcium, magnesium, and potassium (Table 2). As in most plant foods, the sodium content of nuts is very low. Low sodium intake coupled with high intake of calcium, magnesium and potassium is associated with protection against hypertension, insulin resistance, and cardiovascular (CV) disease (CVD) [22], besides counteracting bone demineralization. Even lightly salted nuts retain a relatively low sodium content.

In summary, the macronutrients, micronutrients, and phytochemicals of nuts have all been documented to contribute to beneficial health outcomes, particularly a reduced risk of CVD and related metabolic alterations. As shown in Figure 2, bioactive nut components synergize to affect multiple metabolic and vascular physiology pathways leading to decreased cardiometabolic risk. For these reasons, whole unprocessed nuts may be considered as natural pleiotropic nutraceuticals. As such, daily consumption of nuts should be considered an essential feature of a health-promoting diet.

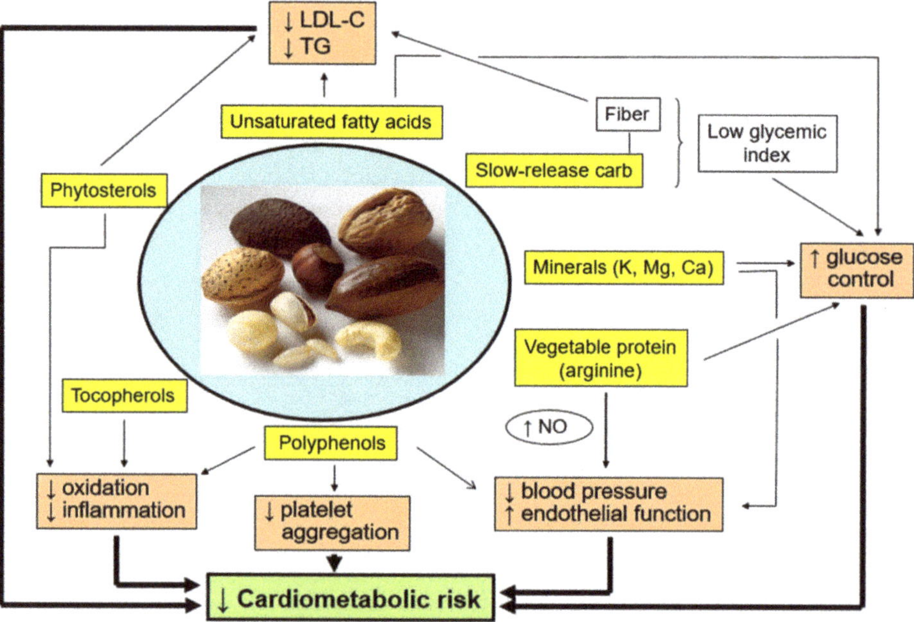

Figure 2. Schematic representation of the effects of nuts on risk of cardiometabolic diseases mediated by their main bioactive nutrients and phytochemicals (yellow boxes), which synergize to positively influence metabolic and vascular physiology pathways (thin arrows and orange boxes). The net effects on intermediate markers of CV risk are lowering of blood cholesterol, improved glycemic control, decreased blood pressure, improved vascular reactivity, and anti-inflammatory actions. Crucially, clinical trials of nuts have demonstrated all such effects. The overall result is reduced cardiometabolic risk (thick arrow connections), as observed in many prospective cohort studies and proven in the PREDIMED trial. Abbreviations: Ca, calcium; K, potassium; LDL-C, LDL-cholesterol; Mg, magnesium; NO, nitric oxide; TG, triglycerides. ↑: increase, ↓: decrease.

5. Nut Consumption and Health Outcomes

The bulk of evidence concerning the effects of nuts on health outcomes stems from prospective studies, the results of which have been summarized in numerous systematic reviews and meta-analyses conducted over the last two decades. Many RCTs have also been conducted examining the effects of nuts on intermediate risk factors, and corresponding meta-analyses have been published. As only one seminal RCT, the PREDIMED study [23], has assessed the effects of a long-term nut-enriched diet on hard CVD outcomes, it will be discussed separately from other RCTs.

5.1. CVD Incidence and Mortality

CVD, mainly CHD and stroke, are the leading causes of death globally. Most CVD could be avoided by addressing and modifying behavioral risk factors, such as incorporating healthy dietary habits [24]. In the last three decades, considerable evidence has accumulated on the effects of frequent nut consumption on CVD outcomes. The most recent systematic review and meta-analysis of 19 prospective cohort studies by Becerra-Tomás et al. [25] found an inverse association between total nut consumption (comparing highest vs. lowest categories) and CVD incidence (Relative Risk [RR] = 0.85; 95% Confidence Interval [CI], 0.80, 0.91; 3 studies), CVD mortality (RR = 0.77; 95%CI, 0.72, 0.82; 14 studies), CHD incidence (RR = 0.82; 95% CI, 0.69, 0.96; 7 studies), CHD mortality (RR = 0.76; 95% CI, 0.67, 0.86; 11 studies), and stroke mortality (RR = 0.83; 95% CI, 0.75, 0.93; 11 studies). No association was ascertained with incident stroke, either ischemic or hemorrhagic, in seven and five studies, respectively. Regarding specific nut types, the reduced risk was noted with tree nuts and peanuts for most CVD outcomes, but not with peanut butter. However, concerning stroke mortality, reduced risk was found for high versus low consumption of peanuts (RR = 0.85; 95% CI, 0.79, 0.92), but not tree nuts. In dose–response analyses, total nut consumption and CVD outcomes showed non-linear inverse associations, with risk reductions up to a consumption of 5 g/day (stroke mortality), 10 g/day (CVD incidence), and 15–20 g/day (CVD and CHD mortality), namely, no further significant reductions were observed above these amounts. The findings of this meta-analysis concur with those of an earlier systematic review by Aune et al. [8].

In proof of the interest of the topic, data from additional large prospective cohort studies relating nut consumption to CVD outcomes, principally CVD mortality, have been released after that meta-analysis [25]. In an analysis of 16,217 men and women with diabetes from the prospective Nurses' Health Study (NHS)-I and -II and the Health Professionals Follow-Up Study (HPFS), highest versus lowest total nut consumption was associated with a lower risk of CVD and CHD incidence and all-cause mortality, with RRs similar to those ascertained in the mentioned meta-analysis, but not with stroke incidence and mortality [26]. In these cohorts, only tree nuts, not peanuts, were associated with reduced CVD outcomes. In a very large (n = 566,398) population-based prospective study in the US with a median follow-up of 15.5 y, data on cause-specific mortality confirmed the inverse association between higher total nut consumption and CVD deaths (Hazard Ratio [HR], 0.70; 95% CI, 0.66, 0.74), while no association for peanut butter consumption was found [27]. The Prospective Urban and Rural Epidemiology (PURE) study, conducted in 16 countries from 5 continents, examined nut consumption in relation to CVD outcomes in 124,329 participants followed for a median of 9.5 y [28]. Overall, CV mortality was lower (RR = 0.72, 95% CI, 0.56, 0.92) in high nut consumers, but no significant effects were detected for CHD or stroke. In the Women's Health Study (n = 39,167) with a mean follow-up of 19 y, higher versus lower nut consumption was associated with lower CVD mortality (HR = 0.73; 95% CI, 0.61, 0.87) [29]. In a recent Iranian population-based prospective cohort study comprised of 6504 participants, those in the highest quartile of nut consumption had a markedly decreased CVD risk (HR = 0.57, 95% CI, 0.47, 0.70) [30]. In a population-based prospective study from Japan (n = 31,552), even though participants consumed very low amounts of nuts (1.6 g/d on average), mostly peanuts, higher versus lower peanut consumption was associated with reduced CVD mortality in women, while only trends towards inverse associations were found in men for peanuts and in both sexes for total nuts [31].

Another report from the large NHS and HPFS prospective cohorts relates nut consumption to CVD risk in a unique way by estimating risk associated with changes in nut consumption, either increases or decreases, during 4-year periods [32]. The researchers found that increasing consumption of total nuts, tree nuts, walnuts and peanuts, but not peanut butter, is associated with reduced risk of total CVD, CHD and stroke, while the converse (increase in risk of CVD and stroke) occurred when participants decreased nut consumption. This is one of few epidemiologic studies supporting a positive effect of nuts on stroke risk, though the PREDIMED RCT showed a marked reduction in stroke in those assigned to the Mediterranean diet with nuts arm [23].

Finally, two very large population surveys in Europe [33] and Latin America [34] analyzed the contribution of dietary factors to CVD mortality and found that one of the most important factors, accounting for the largest number of cardiometabolic deaths, was low nut and seed consumption.

Few prospective studies have examined the relationship of nut consumption with two additional CVD outcomes: atrial fibrillation and heart failure. The cited meta-analysis [25] synthetized data from two cohorts that examined the association of highest vs. lowest nut consumption categories with atrial fibrillation and found a RR of 0.85 (95% CI, 0.73, 0.99). On the other hand, no effect of nuts on heart failure (two studies) was observed [25].

In summary, consistent data from numerous large, well-conducted prospective studies and meta-analyses suggest that nuts are potent cardioprotective foods. The effect of higher nut consumption is strongest on CVD and CHD mortality, with reductions of 25–30%, followed by CVD and CHD incidence and stroke mortality (15–18% reduction), while effects on stroke incidence are less consistent. Total nuts, tree nuts and peanuts, but not peanut butter, generally share the same positive effects on CVD risk. These effects are likely ascribable to nuts' high content of healthy nutrients, such as PUFA, MUFA, non-sodium minerals, vitamins, and polyphenols and their potential to improve intermediate risk factors of CVD, as discussed in the corresponding section.

5.2. Hypertension Incidence and Mortality

The most recent review of epidemiological studies concerning total nut consumption in relation to cardiometabolic outcomes analyzes data from three meta-analyses of prospective studies with outcomes on incident hypertension [35]. An average 15% risk reduction, which was fairly constant in the three meta-analyses, was apparent when comparing high vs. low categories of total nut consumption. Based on data from four prospective studies with 11,962 incident hypertension cases, a 2017 dose–response meta-analysis (included in the review) estimated a 30% attenuation of hypertension risk for each daily serving (1-oz or 28 g) of nuts (RR = 0.70; 95% CI, 0.45, 1.08), with a linear dose–response [36]. No further prospective studies analyzing exposure to nuts in relation to hypertension risk have been published since that review.

In conclusion, in prospective studies nut consumption is associated with a consistent reduction of incident hypertension. As hypertension is the principal risk factor for stroke, this evidence clashes with the generally null epidemiological association of nut consumption with incident stroke, albeit, as discussed, increasing nut consumption is associated with lower stroke mortality. There is, however, sound RCT evidence that nut consumption lowers BP, as discussed in the sections on intermediate markers and health effects of nuts in the PREDIMED trial.

5.3. Diabetes Incidence and Mortality

The effects of nut consumption on risk of type-2 diabetes mellitus (T2D) in epidemiological studies have mostly been inconclusive and controversial [35,37]. A recent meta-analysis of nine studies (six prospective, three cross-sectional) published between 2002 and 2018 reported no association between extremes of consumption of total nuts, tree nuts or peanuts and the risk of T2D [38]. Walnuts, however, appeared to behave differently, as one large prospective study from the NHS and HPFS cohorts included in the meta-analysis found that walnut consumption related inversely to T2D risk (RR = 0.76; 95% CI, 0.62, 0.94) [39], while another large cross-sectional study found an even more beneficial effect of walnuts on T2D (RR = 0.47; 95% CI, 0.31, 0.71) [40].

Surprisingly, peanut butter, assessed in two early prospective studies, was inversely associated with T2D risk in the pooled estimate (RR = 0.87; 95% CI, 0.77, 0.98). Yet, only the results of one of the two analyzed studies favored peanut butter for T2D risk. Notably, risk estimates for total nuts changed substantially from nonsignificant to significant for lower T2D risk when time-updated measurement of body mass index (BMI) obtained during follow-up was excluded from the model (RR = 0.85; 95% CI, 0.75, 0.95), which supports body weight changes as a mediator of the reduction in T2D risk [38]. In fact, there is increasing evidence from prospective studies that regular nut consumption is associated with less long-term weight gain and a lower risk of obesity, while short-term RCTs confirm the lack of fattening effect of nuts [35]. Hence, given that long-term nut consumption is associated with a lower BMI, adjustment for BMI may conceal the true relationship between nuts and T2D, as also highlighted in the review of meta-analyses of nut studies by Kim et al. [35].

The main cause of death in T2D is CVD. Regarding nut consumption in relation to T2D mortality, the analysis of data from participants with T2D in the prospective cohorts of the NHS-I, NHS-II and HPFS showed that highest vs. lowest total nut consumption was associated with a 25% lower reduced CVD mortality and a 27% lower all-cause mortality [26]. In regard to T2D mortality, these data concur with the results of the earlier meta-analysis of Aune et al. [8].

In summary, epidemiological evidence suggests that consumption of total nuts is associated with a reduced T2D risk, which is mediated by nut-associated favorable weight changes that obscure the relationship when BMI is entered as covariate in adjustment models. RCTs also indicate favorable effects of nuts on glycemic control, as discussed in the section of intermediate markers below. Among the tree nuts, to date only the walnut has been associated with lower T2D risk. The unique nutrient composition of walnuts, rich in ALA, highly bioactive polyphenols and melatonin [41], may explain their differential effects on health outcomes, including T2D. A note of caution is necessary when considering the positive effect of peanut butter on T2D risk, as it was only ascertained in one prospective study. Additional large prospective studies are warranted to further elucidate the effects of nuts on T2D.

5.4. Cancer Incidence and Mortality

Cancer is a major cause of death and constitutes a huge public health hazard worldwide, as the global burden of cancer is expected to increase to 29.5 million new cancer cases and 16.4 million cancer-related deaths by 2040 [42]. At least 40% of cancers could be prevented by addressing modifiable risk factors such as tobacco use, dietary carcinogens, sedentary lifestyle, obesity and infectious agents [43]. In particular, adherence to a wholesome eating pattern such as a traditional Mediterranean diet has been shown to reduce the risk of some cancers by 4% to 57% [44]. There is substantial epidemiological evidence suggesting that consumption of nuts, a staple in the Mediterranean diet, is associated with reduced risks for cancer development and cancer-related deaths.

Based on the pooled results of eight prospective studies, the meta-analysis by Aune et al. showed that highest vs. lowest nut consumption was associated with a significant 15% reduction of total cancer incidence [8]. The most recent and comprehensive meta-analysis by Naghshi et al. comprised 51 epidemiological studies and reported that the summary effect size for risk of cancer, comparing extreme categories of total nut consumption, was similar to that described by Aune et al. [8], with an RR of 0.86 (95% CI, 0.81, 0.92) [45]. In the dose–response analysis, each 5-g/d increase of total nut consumption was associated with 3%, 6%, and 25% lower risks of overall, pancreatic, and colon cancers, respectively. Of note, this inverse dose–response relationship between nuts and incident cancer was not significant for peanuts and peanut butter consumption [45]. This could be due to different nutrient composition in peanuts vs. tree nuts [1]. Peanut butter, though comprised predominantly of ground peanuts that have been roasted, generally also contains additives such as sugar, salt and hydrogenated oils that may hamper its health benefits [46], as shown for CVD. Moreover, when peanuts or other nuts are improperly stored, they can be contaminated with aflatoxins, which are potent carcinogens produced by certain molds.

Nuts may also mitigate the increased breast cancer risk associated with alcohol use, as noted in a cohort study that showed reduced risk of benign breast disease, a precursor for breast cancer, with nut consumption, especially among individuals with substantial alcohol use [47].

The Naghshi et al. meta-analysis [45] also evaluated the effects of nuts on deaths due to malignancy and found statistically significant 18%, 8%, and 13% risk reductions in the risk of cancer mortality with the higher intake of tree nuts, peanuts, and total nuts, respectively. Like for the case of incident cancer, no significant association between peanut butter consumption and cancer mortality was ascertained [45]. Many other studies have evaluated the relationship between nut consumption and risk of cancer mortality. An earlier meta-analysis by Zhang et al. in 2020 showed that total nut consumption was associated with a reduced odds ratio (OR) of cancer-related mortality (OR = 0.90; 95% CI, 0.88, 0.92) [48]. In this meta-analysis, statistically significant inverse associations between nut consumption and cancer site occurrence were present for colorectum, stomach, pancreas, and lung.

In summary, a consistent body of evidence, albeit based exclusively on observational studies, suggests that consumption of nuts may modestly reduce cancer incidence and cancer-related deaths. These benefits appear to be most significant for tree nuts, less so for peanuts, and nonexistent for peanut butter. More large prospective studies and RCTs are needed to clarify this important issue.

5.5. Brain Health

An unwanted consequence of increased lifespan and associated population aging in recent decades is a growing number of elderly individuals at risk of neurodegenerative disorders, particularly Alzheimer's disease—the most common type of dementia. Given that no effective disease-modifying pharmacologic treatments for mild cognitive impairment, a common harbinger of dementia, or dementia itself are available [49], there is an increasing interest in preventive strategies to implement in preclinical and early stages. Among them, lifestyle modifications, including dietary changes, are being actively investigated and there is incipient evidence that they may forestall cognitive decline and even prevent dementia, particularly in individuals at higher risk [50,51]. Brain oxidative stress and inflammation are currently considered to be causal factors leading to age-related neurodegeneration, a reason why dietary patterns and foods with anti-inflammatory properties are the most promising for improving brain health [52]. Additionally, a close link exists between Alzheimer's disease and vascular pathology, and there is evidence that treatment of CV risk factors contributes to maintain neuronal integrity and prevent cognitive dysfunction [53].

Nuts are rich in neuroprotective nutrients such as PUFAs and polyphenols and their increased consumption benefits vascular function and is consistently associated with reduced rates of CVD, therefore it can be predicted that they might also beneficially influence cognition and overall brain health. Although data from cohort studies relating nut consumption to dementia outcomes are lacking, evidence is accumulating on the potential of nuts to improve cognitive function. A recent systematic review synthetized data from 14 epidemiological studies and eight RCTs assessing effects of nut-enriched diets on cognitive function [54]. While some epidemiological studies showed a positive association, the quality of the evidence was low because nine studies were cross-sectional or case–control and only five were prospective. Nevertheless, studies targeting populations at higher risk of cognitive decline tended to have favorable outcomes. Notably, studies that specifically addressed the association between walnut consumption and cognitive performance had more homogeneous results, as out of six walnut studies, including two RCTs, only one failed to find a positive association. This may be due to the highly bioactive nutrients of walnuts previously mentioned in reference to T2D risk [41]. Indeed, many studies using walnuts in experimental models of brain aging and neurodegeneration have consistently uncovered beneficial effects [55]. A large population-based prospective study published after that systematic review, the Singapore Chinese Health Study, supports the cognitive benefit of nuts, as nut consumption at midlife was associated with a dose-dependent reduction in risk of cognitive impairment 20 years later [56]. Interestingly, unsaturated fatty acid intake mediated close to 50% of the beneficial effect, pointing to the fatty acid composition of nuts as relevant in improving cognition.

Late-life depression is a common psychiatric disorder that compromises the quality of life of affected individuals. Depression is also a risk factor for cognitive decline, where chronic inflammation contributes to its pathophysiology, as is the case with neurodegenerative disorders [57]. For a similar reason, depression is also a risk factor for CHD, although the association is bidirectional [58]. Consequently, nuts can be postulated to have a salutary effect on depression. The epidemiologic evidence, however, is scanty and of suboptimal quality. Nut consumption was reported to benefit depressive symptoms in a large cross-sectional study of Chinese adults [59]. In another cross-sectional report from the US, nut consumers, and particularly walnut consumers, disclosed lower depression scores than subjects who were not consuming nuts, and this beneficial effect was more pronounced in women [60]. In that study, food consumption was assessed only via 24 h diet recalls, which can provide strong evidence for frequently consumed foods but, unless repeated, are much weaker for sporadically consumed foods such as nuts. In the Invecchiare in Chianti study, an Italian prospective investigation of 1058 adults followed for up to 9 years with repeated measurements of diet and depression scores, no association between consumption of nuts and depressive symptoms was observed [61].

Brain-beneficial nutrients contained in nuts, such as PUFAs and polyphenols, support their potential to delay cognitive decline, as evidenced in a few prospective studies and RCTs. Furthermore, other nut components such as phytomelatonin, phytosterols, antioxidant tocopherols, and folic acid may also support neurological health and cognitive wellness. Clearly, more well-designed prospective studies and RCTs, preferably conducted in individuals at high risk or with early dementia stages, are warranted to uncover the full potential of nuts to counteract cognitive decline.

5.6. All-Cause Mortality

While the major focus of epidemiological research with nuts has been CVD, many large population-based prospective cohort studies conducted globally have examined associations of exposure to nuts or nut components such as peanut butter with all-cause mortality. The latest meta-analysis by Chen et al. [62] synthetized data from 18 prospective studies and obtained a summary RR for high compared with low nut consumption of 0.81 (95% CI, 0.78, 0.84) for all-cause mortality. When data for total nuts and tree nuts and peanuts were analyzed separately, the RR estimates were similar for the three nut categories. Only two studies examined peanut butter separately from peanuts and their combined RR for all-cause mortality was 0.89 (95% CI, 0.80, 0.99). In dose–response analysis, the RR for all-cause mortality per one additional serving of total nuts per week was 0.96 (95% CI, 0.94, 0.97). Interestingly, dose–response analyses revealed nonlinear inverse associations between nut consumption and mortality, with risk reduction leveling off at consumption of approximately 3 servings/week (equivalent to 12 g/d), which suggests that maximum benefit on survival may be achieved with relatively low doses of nuts.

That low nut doses relate to lower overall mortality is underlined in a recent report from a large (n = 31,552) population-based prospective Japanese study, whereby higher compared with lower nut consumption was associated with reduced all-cause mortality (HR = 0.85, 95% CI, 0.75, 0.96) in men (not in women), in spite of an average consumption of only 1.8 g/d, peanuts accounting for 80% of total nuts [31]. Additionally, in spite of a similarly low average nut consumption in Korea, a recent large cross-sectional population survey relating dietary factors to all-cause and cause-specific mortality using a comparative risk assessment analysis found that a sizable proportion of deaths was related to low consumption of nuts [63]. The results of a recent very large (n = 566,398) population-based prospective study in the US with a median follow-up of 15.5 years support the inverse association between higher nut consumption and total mortality (HR = 0.78; 95% CI, 0.76, 0.81); in contrast consumption of peanut butter was not associated with lower risk of mortality [27]. Finally, the recent report from the PURE study described a significant reduction in total mortality (HR = 0.77; 95% CI, 0.69, 0.87) for highest ($\geq$120 g/week) versus lowest (<30 g/month) nut consumption [28]. In PURE, tree nut consumption was associated with a decreased risk of mortality, whereas peanut consumption disclosed a nonsignificant trend towards a lower mortality risk.

Nut consumption in relation to mortality has also been examined in a large prospective cohort of individuals with T2D. In the NHS and HPFS report on nuts and mortality among 16,217 men and women with diabetes, higher vs. lower nut consumption was associated with a significant 31% reduction in all-cause mortality [26]. When assessed separately, consumption of tree nuts and peanuts related to 33% and 20% lower mortality risks, respectively. Overall, the findings from prospective studies consistently point to an inverse association of nut consumption with all-cause mortality, with an average of 1 in 5 deaths prevented or delayed by nut consumption at moderate levels.

5.7. Intermediate Markers: Adiposity, Lipids, Blood Pressure, Glycemic Control, Endothelial Function, and Inflammation

5.7.1. Adiposity

The steady increase in the prevalence of overweight/obesity worldwide is a major public health problem. Due to the high energy density of nuts, increased body weight with long-term consumption has been an underlying concern. Yet, to the contrary, a growing body of epidemiological evidence suggests that daily nut consumption is a potentially effective strategy in the primary prevention of obesity [64].

A 2014 review of epidemiological and RCT data concluded that evidence was lacking on the common assertion that regular consumption of nuts increased adiposity [10]. This was confirmed in a recent network meta-analysis of 105 RCTs comparing the effects of diets enriched in various tree nuts and peanuts vs. control diets on body weight, BMI, waist circumference (WC), and percent body fat [64]. No significant increase was observed in any adiposity measures with any of the nuts, except for hazelnut-rich diets, which raised WC. On the other hand, results of pairwise comparisons between different nuts indicated that almond diets reduced WC compared to control diets; walnuts also reduced WC compared to pistachio, hazelnut and mixed nuts-enriched diets. In subgroup analyses considering only RCTs specifically designed to assess the weight loss effects of nut consumption, almonds were associated with reduced BMI and walnuts with reduced percent body fat. Importantly, among overweight and obese study subjects, those who consumed nut-enriched diets experienced greater weight loss, reduced BMI and lower WC compared with their counterparts who consumed a nut-free isocaloric control diet (Figure 3).

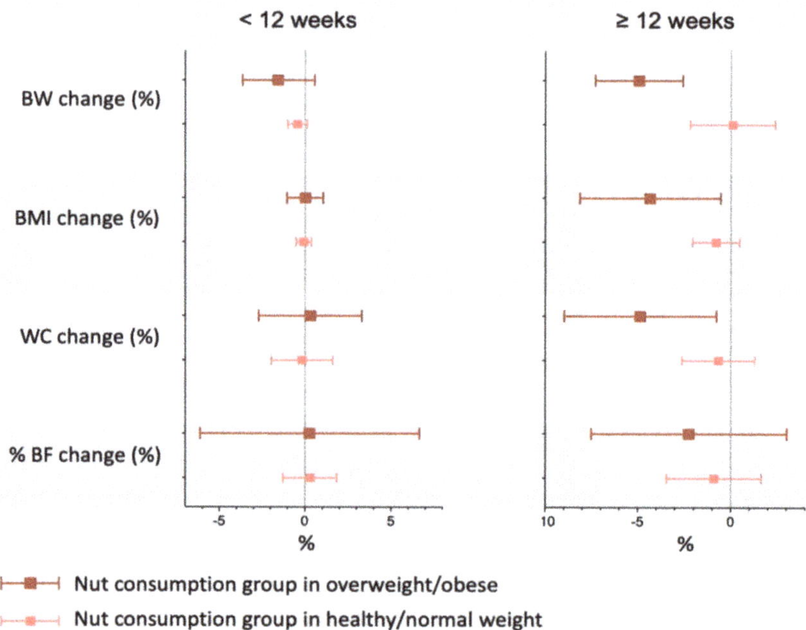

Figure 3. Percentage change for adiposity outcomes in the healthy/normal weight groups vs. the overweight/obesity group with regard to the length of time following the nut interventions in 105 RCTs. Reproduced from reference [64], with permission. BW: body weight; BMI: body mass index; WC: waist circumference; % BF: body fat percentage.

Another recent meta-analysis of nut-feeding trials examined whether providing or not dietary substitution instructions to participants (recommending foods to be replaced by the nuts or just advising to eat the nuts on top of the usual diet) influenced adiposity changes [65]. The results showed the same absence of weight, BMI or WC changes for the two categories of studies.

A very large prospective study involving the three Harvard cohorts of the NHS-I, NHS-II, and HPFS assessed the association between changes in consumption of total and specific nuts per 4 y intervals and weight changes over 20–24 y of follow-up [66]. Increases in nut consumption, per 0.5 servings/d (14 g), were significantly associated with less weight gain per 4 y interval: −0.19 kg (95% CI, −0.21, −0.17) for total nuts, −0.37 kg (95% CI. −0.45, −0.30) for walnuts, −0.36 kg (95% CI, −0.40, −0.31) for other tree nuts, and −0.15 kg (95% CI, −0.19, −0.11) for peanuts. An increase in consumption of total nuts, per 0.5 servings/d, was associated with a modest but significant 3% lower risk of becoming obese, while a similar increase in consumption of walnuts and other tree nuts was associated with a 15% and 11% lower risk of developing obesity, respectively. Increasing peanut consumption, however, was not associated with reduced obesity risk.

Thus, both epidemiological and RCT data point to a slightly beneficial effect of nut consumption on adiposity rather than a harmful effect. That regularly eating a highly energy-dense food does not promote a positive energy balance is of particular interest. Several mechanisms underly the associations between nut consumption and lower risk of weight gain [10]. Nuts require considerable effort at mastication and chewing, and their high fat and fiber content can delay gastric emptying, increase satiety, suppress hunger and promote fullness. The fiber in nuts also increases binding of fatty acids in the gut, leading to greater fecal fat excretion. Similarly, the efficiency of energy absorption from nuts is reduced due to incomplete mastication and encasement of fat within unbroken cell walls in nut particles, hampering the bioaccessibility of fat from nuts in the gastrointestinal tract, with ensuing increases in fecal energy (fat) loss. Finally, there is evidence that the high unsaturated fat levels in nuts enhance fatty acid oxidation and increase thermogenesis and resting energy expenditure, which may also mitigate weight gain.

5.7.2. Blood Lipids

Since the landmark RCT of Sabaté et al. demonstrating the cholesterol-lowering effect of a walnut diet [3], the effects of diets enriched with different nuts on blood lipids and lipoproteins have been examined in many RCTs [35].

To date, the 2015 meta-analysis of Del Gobbo et al. [9] is the most comprehensive. It reviewed 61 intervention trials (42 randomized and 19 non-randomized) lasting from 3 to 26 weeks designed to assess the effects of tree nuts on the blood lipid profile. All trials provided the study nuts to participants rather than simply giving advice to procure the nuts by themselves. Nut consumption (per serving/d) significantly decreased total cholesterol (−4.7 mg/dL), LDL-cholesterol (−4.8 mg/dL), and triglycerides (−2.2 mg/dL), but had no effect on HDL-cholesterol. Walnuts, followed by almonds and pistachios, were the nuts most frequently studied. The LDL-cholesterol lowering effect was dose related in a non-linear fashion, with stronger effects at doses of 60 g/d (approximately 2 servings), while triglyceride lowering had a linear dose–response. There was no heterogeneity by nut type or quality of the control diet. These authors reanalyzed the data as a function of the phytosterol content of nuts in each study and demonstrated that the phytosterol dose was strongly related to the observed LDL-cholesterol reduction, although this association was driven by the total nut dose [21].

An earlier analysis with pooled individual data from 21 RCTs indicated that, for an average consumption of 67 g/d of tree nuts or peanuts (two servings, approximately 20% of energy), the mean estimated reduction of LDL-cholesterol was 10 mg/dL (7%) [67]. Nuts had no significant effect on serum triglycerides, except in participants with triglycerides >150 mg/dL, in whom a significant 10.2 mg/dL reduction was observed. Importantly, there was a clear dose–response in LDL-cholesterol lowering. The statistical power of this pooled analysis allowed detection of differential responses by baseline LDL-cholesterol level (greater response with higher values) and BMI (greater response with lower BMI) (Figure 4). The mean 10% LDL-cholesterol reduction with 2 servings/d of nuts in hypercholesterolemic individuals is similar to that described for functional foods fortified with plant sterols/stanols [68], which epitomizes the nutraceutical properties of nuts as cholesterol-lowering foods. Recently, in a network meta-analysis of 66 RCTs comparing the effects of 10 common food groups (refined grains, whole grains, nuts, legumes, fruits and vegetables, eggs, dairy, fish, red meat, and sugar-sweetened beverages) on cardiometabolic outcomes, nuts were ranked as the best food group at reducing LDL-cholesterol [69].

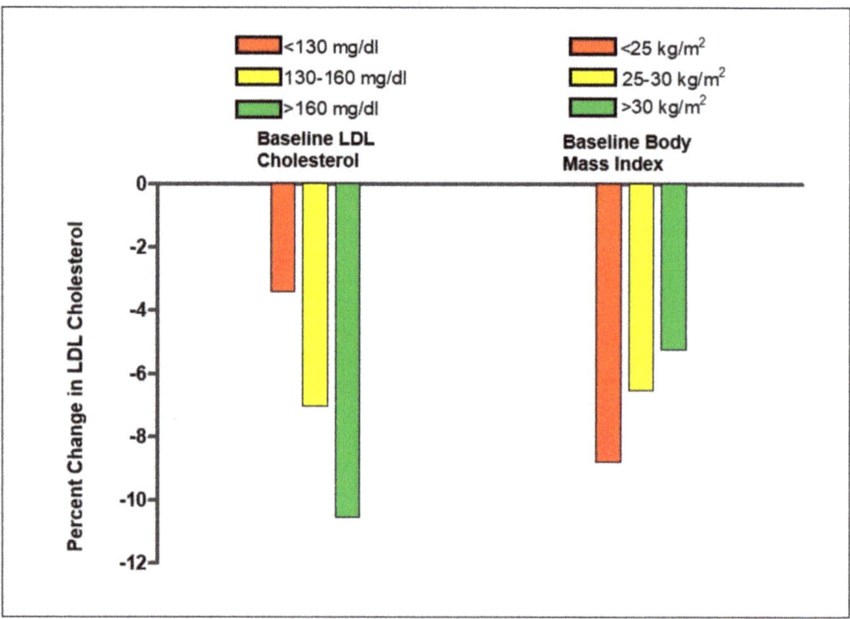

Figure 4. LDL-cholesterol responses to nut diets by baseline LDL-cholesterol and BMI. Data obtained in a pooled study of 25 nut RCTs (adapted from ref. [67] with permission).

The lipid effects for individual nut types have been examined in meta-analyses of RCTs using walnuts (24 studies) [70], almonds (27 studies) [71], pistachios (11 studies) [72], hazelnuts (3 studies) [73], and cashews (3 studies) [74]. All individual nuts except cashews reduced LDL-cholesterol to a similar extent than reported for total nuts in the mentioned systematic reviews [9,67], but cashews had no effect, which may be due to the low number of RCTs analyzed. Finally, a recent network meta-analysis of 34 RCTs of these five nuts for lipid outcomes used analyses based on the surface under the cumulative ranking curves and concluded that diets enriched in pistachios and walnuts were best for lowering LDL-cholesterol and triglycerides compared with the other nut-enriched diets included in the study [75].

5.7.3. Blood Pressure

The effects of nuts on office BP have been reported in many RCTs [9,35]. BP changes were a secondary outcome in the 2015 meta-analysis of Del Gobbo et al. [9], and no effect of nut-enriched diets on either systolic BP (SBP) or diastolic BP (DBP) was found. A 2015 meta-analysis including 21 RCTs of nut diets by Mohammadifard et al. [76] focused on BP changes. Results showed that diets supplemented with nuts had no effect on BP overall, except in individuals without T2D, who disclosed a weighted mean difference (WMD) in SBP of −1.29 mm Hg; (95% CI, −2.35, −0.22). In sub-analyses stratified by nut types, only diets enriched in pistachios resulted in a significant BP reduction, with a WMD of −1.82 mm Hg (95% CI, −2.97, −0.67) for SBP and of −0.80 mm Hg (95% CI, −1.43, −0.17) for DBP, while mixed nuts reduced only DBP, with a WMD of −1.19 mm Hg (95% CI, −2.35, −0.03).

Data on BP changes for specific nut types have also been reported. Thus, the 2018 meta-analysis by Guasch-Ferré et al. [70] of 24 RCTs focused on CV risk factor changes with walnut-enriched diets reported no effect on BP. A recent meta-analysis of 16 RCTs examining the effects of almonds on BP showed no differences for SBP between almond and control diets, but pooled analyses revealed a significant reduction of DBP by almond diets (WMD = −1.30 mm Hg; 95 % CI, −2.31, −0.30) [77]. A meta-analysis of 13 RCTs using pistachios for outcomes of CV risk factors by Asbaghi et al. [78] indicated a significant reduction of SBP (WMD = −2.12 mm Hg; 95 % CI, −3.65, −0.59), which supports the findings of the Mohammadifard et al. meta-analysis [76], although no effect on DBP was found. The meta-analysis of 3 cashew RCTs by Jalali et al. [74] also reported a significant reduction of SBP (WMD = −3.39 mm Hg; 95% CI, −6.13, −0.65), without changes of DBP.

The evidence on the effects of nuts on BP outcomes is inconsistent and, in general, does not support a relevant lowering effect, which contrasts with the epidemiological findings of an association of nut diets with a lower risk of incident hypertension, consistent across different meta-analyses [35,36]. Reasons for the failure of RCTs to detect BP changes with nut-enriched diets may be low statistical power (most RCTs included less than 50 participants), short duration of the intervention, exclusive use of office BP measurements, and the fact that they were usually a secondary outcome of lipid-focused trials, hence were not powered to detect changes in BP. Recently, the 2-year effects of a walnut diet on both office BP and 24-h ambulatory BP (the gold standard of BP measurements) in the Walnuts and Healthy Aging (WAHA) RCT conducted in 236 older individuals were reported [79]. The results showed that, compared with a control diet, a diet supplemented with walnuts at ≈15% of energy resulted in lower office SBP (−4.61 mm Hg) in the whole cohort and reduced 24-h ambulatory SBP (−8.5 mm Hg) in hypertensive participants. No changes in diastolic BP were observed. During the trial, participants in the walnut group required less uptitration of antihypertensive medication and had better overall BP regulation than controls. The WAHA trial overcomes the limitations of prior RCTs concerning BP effects of nut diets and shows a beneficial effect of long-term walnut consumption on SBP.

5.7.4. Glycemic Control

Acute feeding studies have shown that nuts consumed with carbohydrate-rich foods having a high glycemic index reduce postprandial glucose responses in comparison with consumption of the same foods alone in both normoglycemic individuals and those with T2D [80,81], which suggests that nuts may be useful in glycemic control. The evidence from RCTs, however, is mixed. A recent meta-analysis of 40 RCTs with a median duration of 3 months concluded that consumption of tree nuts or peanuts had modest favorable effects on the homeostasis model assessment of insulin resistance (HOMA-IR) (WMD = −0.23) and fasting insulin (WMD = −0.40 μIU/mL), but not on fasting blood glucose or hemoglobin A1c [82]. Subgroup analyses showed similar results whether the study subjects were healthy individuals or those with prediabetes or T2D.

A meta-analysis of 16 RCTs that assessed effects of walnut diets on biomarkers of glycemic control failed to find any benefit [83]. Likewise, a recent in-depth narrative review of almonds and health outcomes based on findings of 64 RCTs and 14 meta-analyses and/or systematic reviews concludes that almonds have inconsistent and/or insignificant beneficial effects on glycemic control [81].

5.7.5. Endothelial Function

The endothelium plays a central role in arterial health and throughout all stages of atherosclerosis. Endothelial function can be viewed as an integrative biomarker of the overall harmful effects of CV risk factors on the arterial wall, a reason why endothelial dysfunction is an independent predictor of future CVD events [84]. Endothelial dysfunction is characterized by a decreased bioavailability of NO and increased expression of pro-inflammatory cytokines and cellular adhesion molecules and can be evaluated non-invasively by several methods; flow-mediated dilation (FMD) measured by brachial artery ultrasound is considered the most sensitive and accurate in assessing endothelial function [85].

Two meta-analyses have summarized results of RCTs testing nut diets for effects on FMD [86,87]. The meta-analysis of Neale et al. [86] examined RCTs of nut diets providing data on inflammatory molecules, but also regarding effects on endothelial function. FMD was explored in nine strata (five testing the effects of walnuts) from eight RCTs, resulting in significant improvements in FMD of the nut versus the control diets (WMD = 0.79%; 95% CI, 0.35, 1.23). When subgroup comparisons were made according to nut type, only the walnut interventions resulted in improved FMD. The meta-analysis of Xiao et al. [87] of 10 RCTs was focused exclusively on effects of nuts on FMD. The pooled estimates showed that nut consumption significantly improved FMD (WMD = 0.41%; 95% CI, 0.18, 0.63). Again, subgroup analyses indicated that only walnut interventions improved FMD. Walnuts are particularly rich in ALA, polyphenols, arginine (the precursor of NO, the endogenous vasodilator) and other bioactives, as reviewed [41], which may explain their differential effects on endothelial function. However, almonds may also improve endothelial function, as shown by a recent 6-wk RCT that tested almond snacks (about 2 servings/day) versus control snacks (muffins) in adults at above-average CV risk for effects on FMD, among other cardiometabolic risk variables [88]. The results showed a noticeable increase in FMD by almonds (WMD = 4.1%; 95% CI, 2.2, 5.9), much higher than that reported in the cited meta-analyses [87,88]. No effects on BP were observed despite the use of 24 h ambulatory BP monitoring.

A recent review analyzed 16 nut intervention trials using noninvasive techniques other than FMD to assess vascular function, such as pulse wave velocity, pulse wave analysis, digital volume pulse, impedance cardiography, and peripheral arterial tonometry [89]. The results were mixed, with only 6 out of 16 studies showing improved vascular function ensuing nut diets.

5.7.6. Inflammation

Chronic non-communicable diseases, such as atherosclerosis with major CV events, obesity, T2D, neurodegenerative disorders, cancer, and auto-immune diseases are characterized by a state of low-grade inflammation, which plays a central role in disease progression and perpetuation. Changes in this inflammatory state can be identified by determination of circulating biomarkers of inflammation, including C-reactive protein (CRP), tumor-necrosis factor-alpha (TNF-α), interleukin-6 (IL-6), E-selectin, and adhesion molecules intercellular adhesion molecule (ICAM)-1 and vascular cell adhesion molecule (VCAM)-1, all of which have garnered much interest in CV risk prediction [90]. The critical role of chronic inflammation in CVD has been substantiated recently by landmark RCTs demonstrating that interventions selectively targeting inflammation can improve clinical outcomes in patients with atherosclerosis [91].

The effects of nut-enriched diets on soluble inflammatory biomarkers have been investigated in many RCTs, usually as secondary outcomes, therefore not powered to detect changes in these outcomes [35]. A 2017 meta-analysis of 32 RCTs by Neale et al. [86] concluded that nut interventions induced no significant changes in inflammatory markers, including CRP, TNF-α, IL-6, ICAM-1 and VCAM-1, or in the anti-inflammatory biomarker adiponectin. A 2018 meta-analysis of 23 RCTs by Xiao et al. [92] showed that nut consumption reduced ICAM-1 (WMD = -0.17; 95% CI, -0.32, -0.03), but had no consistent effects on CRP or other soluble inflammatory molecules. Other meta-analyses focused on the lipid effects of total nuts [9] and walnuts [70] and a recent review on the cardiometabolic effects of almonds [81] concur in reporting no significant changes in CRP levels.

While based on results of generally small and short-term RCTs, it appears that consumption of nuts has a negligible impact on inflammatory markers, a recent report from the large, long-term WAHA trial provides a different view. The walnut intervention at ≈15% of energy for 2 years in 634 older participants recruited in two sites, Barcelona, Spain and Loma Linda, California, resulted in significant mean reductions ranging from 3.5% to 11.5% in several inflammatory biomarkers, including granulocyte-monocyte colony stimulating factor, interferon-γ, IL-1-β, IL-6, TNF-α, and E-selectin, but had no effect on CRP, ICAM-1 or VCAM-1 [93]. Thus, high statistical power and a long duration of the interventions might be necessary to uncover the anti-inflammatory effects of nuts. Regardless, these data provide novel mechanistic insight for the benefit of nut (walnut) consumption on CVD risk beyond that of lipid lowering.

5.8. Other Health Outcomes

Few prospective studies or single RCTs have examined the effects of nuts on alternative health outcomes, such as gallbladder disease, metabolic syndrome (MetS), non-alcoholic fatty liver disease (NAFLD), physical function, healthy aging, bone health, and reproductive health. There is also incipient evidence that nut diets elicit changes in microbiota.

Two reports from the large prospective cohorts of the NHS in women and the HPFS in men examined the association between frequency of nut consumption and risk of gallstone disease. In the NHS, women consuming ≥5 servings of nuts per week had a significantly lower risk of cholecystectomy (RR = 0.75, 95% CI, 0.66, 0.85) than did those who rarely or never consumed nuts [94], while in the HPFS, men consuming ≥5 servings of nuts per week had a significantly lower risk of symptomatic gallstone disease (RR = 0.70, 95% CI, 0.60, 0.86) compared to those who rarely or never consumed nuts [95]. The results of the two studies suggest that regular nut consumption protects men and women equally against gallstone disease. This beneficial effect is attributable to the richness of nuts in bioactive components capable of influencing intestinal bile acid and cholesterol biology, particularly unsaturated fatty acids, fiber, and non-sodium minerals.

An individual meets diagnostic criteria for MetS when harboring at least three of the following risk factors: increased WC, high triglycerides, low HDL-C, elevated BP, and high fasting blood glucose, and this cluster of risk factors increases the risk of CVD and all-cause mortality beyond the risk imparted by each separate factor [96]. The pathophysiological basis of MetS is insulin resistance, generally linked to central fatness, and as such MetS is an epidemic condition worldwide. Lifestyle changes directed to weight loss and cardiometabolic risk factor control are critical for preventing and treating MetS, thus nut consumption might play a role [97]. A 2014 meta-analysis included 49 RCTs of ≥3 weeks duration reporting effects of nut consumption on at least one criterion of the MetS [98]. Pooled analyses showed a beneficial effect of nuts on MetS via modest decreases in triglycerides and fasting blood glucose. However, it should be noted that a recent meta-analysis found no evidence of benefit of nut diets on blood glucose levels [82], which underlines the limitations of present data on a putative beneficial effect of nuts on MetS. Likewise, the large WAHA trial found no effect of a 2-year walnut-enriched diet on MetS [99].

NAFLD, the accumulation of fat (triglycerides) in the liver in the absence of excessive alcohol intake is the hepatic manifestation of MetS, a prevalent condition globally and a public health concern. NAFLD not only increases risk of CVD, but also of liver cirrhosis and hepatocarcinoma [100]. Like in MetS, abdominal obesity and T2D are major drivers of NAFLD, and its primary treatment consists of lifestyle and dietary changes directed at weight loss. The favorable effects of nuts on body weight, glycemic control and CVD risk would predict a beneficial effect in NAFLD, and a few prospective studies have suggested that increased nut consumption is associated with a lower incidence of NAFLD, as recently reviewed [101]. Nuts may contain the carcinogenic agent aflatoxin, a fungal metabolite and mycotoxin that can contaminate improperly stored nuts and other seeds. This has been a reason of concern for patients with NAFLD due to their increased risk of liver cancer. However, in Western countries, where aflatoxin contamination of crops is rare due to strict regulations, health benefits provided by increased nut consumption likely outweigh the risks associated with chronic increases in aflatoxin exposure. This may not be the case in countries known for high rates of aflatoxin contamination of peanuts, like Indonesia [101].

There is evidence from a single prospective study conducted in Spain, where nut consumption is rather high, that it may lower the risk of impaired agility/mobility and increase overall physical function in older individuals [102], while another report from the NHS cohort suggests that consumption of total nuts and, particularly, walnuts is associated with healthy aging, i.e., survival beyond 65 years with no chronic diseases, no memory impairment, no physical disabilities, and intact mental health [103].

Dietary components are important for providing crucial constituents for bone health and regulating cellular metabolism within bone [104]. Nuts might promote bone health because they are rich sources of antioxidant, anti-inflammatory flavonoids, and calcium. Resveratrol is a stilbene-type polyphenol present in some nuts that is a powerful activator of the longevity-linked sirtuin-1 molecule, which regulates processes related to longevity, including apoptosis, DNA repair and energy expenditure [105]. Nevertheless, there are no data from prospective studies relating nut consumption to bone health, while RCTs are limited. A single small RCT assessed the effects of an ALA diet sourced from walnuts and flaxseed oil in comparison with an average American diet and a linoleic acid-rich diet on bone turnover, assessed by serum concentrations of N-telopeptides and bone-specific alkaline phosphatase [106]. N-telopeptide levels were significantly lower following the ALA diet relative to the average American diet, suggesting that plant sources of dietary n-3 PUFA may have a protective effect on bone metabolism via decreased bone resorption. In summary, the evidence on the efficacy of nuts to promote bone health is very limited. Both well-powered prospective studies and RCTs are warranted to examine this important issue.

Concerning reproductive health, in the last decades there has been a steady increase in infertility worldwide, in great part related to declining semen quality. Exposure to environmental toxins, smoking, and unhealthy diets are believed to underlie impaired spermatogenesis [107,108]. Two RCTs have tested nut diets for outcomes of semen quality. Robbins et al. [109] randomized 117 healthy men to consume 75 g of walnuts/d for 12 weeks on top of their usual Western-style diet or usual diet alone and found improved sperm vitality and motility after the walnut diet, but no changes in total sperm count. The FERTINUTS trial [110] was a 14-wk RCT involving 119 healthy men 18–35 y-old that assessed the effects on various sperm parameters of 60 g/day of mixed nuts (30 g walnuts, 15 g almonds, and 15 g hazelnuts) in the context of a Western-style diet vs. the same diet without nuts. Compared to the control group, the nut group showed significant improvements in total sperm count and vitality, motility and morphology. Nuts appear to improve male fertility, but clearly more research is needed.

Another prevailing pathology related to men's sexual heath is erectile dysfunction, a condition in which endothelial dysfunction at the level of penile vasculature is causal in the failure to initiate and/or maintain an erection [111]. Erectile dysfunction is intimately linked to CV risk factors and associated with an increased incidence of CVD [112]. Given that nut consumption is associated with reduced CV risk and improved endothelial function, in part due to their content in arginine, the precursor of the endogenous vasodilator NO, it is plausible that nut diets would benefit erectile function. Indeed, in folk medicine nuts (particularly cashews and walnuts) are promoted as aphrodisiacs and a remedy for impotency. A single RCT, a secondary analysis of the FERTINUTS trial [110], tested the effect of mixed nuts on erectile function in healthy young men [113]. Compared to the control group, small but significant increases in self-reported orgasmic function and sexual desire, but not erectile function, were observed following the nut intervention. There were no between-group differences in changes of peripheral concentrations of NO and E-selectin. Well-powered RCTs conducted in individuals with an objective diagnosis of erectile dysfunction are necessary to reach definitive conclusions on the efficacy of nuts to help men with this prevalent and troublesome pathology.

An expanding area of clinical research is the intestinal microbiome, which is primarily controlled by the nutritional quality of the diet and is believed to play a major role in a vast array of biological functions [114]. Colonic microbiota can be modulated by different lifestyle and dietary factors and impact the risk of developing obesity, T2D and other cardiometabolic diseases, as well as infectious diseases. Nuts have been suggested to have a prebiotic effect (that conferred by a substrate selectively used by the host microorganisms translating into a health benefit) on the gut microbiome [115]. The non-bioaccessible components of nuts (fiber, polymerized polyphenols and fat contained within undigested cell walls in incompletely masticated nut particles) make up a rich supply of nutrients to the intestines for feeding the microbes residing there. The field of nuts and microbiota is still at an early stage, but a recent comprehensive meta-analysis synthesized data from nine RCTs investigating almonds (*n* = 5), walnuts (*n* = 3) and pistachios (*n* = 1) for effects on fecal bacterial diversity [116]. Nut consumption increased the relative abundances of the genera *Clostridium*, *Lachnospira* and *Roseburia*, which are considered beneficial because they produce butyrate, a short-chain fatty acid critical in nourishing the intestinal epithelium and maintaining its integrity. Nut consumption had little overall impact on bacterial diversity, a metric considered as positive for health, except for a marginal enhancement from almond consumption, which could be explained by the particular matrix of almonds and small cell walls limiting fat availability for digestion, but increasing fat delivery to the colon and thus feeding the microbiota [81]. The overall meta-analytical evidence of a modulatory effect of microbiota by nuts is weak because microbial determinations were a secondary outcome in most RCTs, which were not powered to detect changes of this outcome. Nut effects on microbiota is a relevant topic for future research.

5.9. Health Effects of Nuts in the Predimed Trial

The landmark PREDIMED trial targeted both the effects of nut consumption on intermediate cardiometabolic markers and clinical outcomes, such as CVD and T2D, among others. The PREDIMED study [23] was a multicentric, parallel group, nutrition intervention RCT for the primary prevention of CVD. It was conducted in Spain and enrolled 7447 men and women aged 55–80 years at high risk of CVD but no CVD at recruitment. Participants were allocated to three study arms: two Mediterranean diets, supplemented with either extra-virgin olive oil (50 mL or more/day) or mixed nuts (30 g/day: 15 g walnuts, 7.5 g almonds, and 7.5 g hazelnuts), or control diet (advice on a low-fat diet) and followed for 5 y. The supplemental foods (olive oil and raw, unpeeled nuts) were delivered periodically to participants in the corresponding groups. Registered dietitians delivered the interventions at quarterly individual visits and group sessions separate for each group. As PREDIMED intended to assess the effects of the nutrition intervention alone, the diets were energy-unrestricted and increased physical activity was not promoted. The primary end point was a composite of major CVD events (non-fatal myocardial infarction, non-fatal stroke, and CV death). An event adjudication committee, whose members were blinded to group allocation, was responsible for event ascertainment. Attesting to the high CV risk of participants, the mean age was 67 years and the mean BMI was 30 kg/m^2, almost one-half had T2D, two-thirds had dyslipidemia, and 4 out of 5 had hypertension. Since its inception in June 2003, the trial has generated a steady stream of data on the beneficial health effects of Mediterranean diets enriched with either of the supplemental foods, culminating with the publication of results on the primary CVD outcome, demonstrating a ≈30% reduction with the two Mediterranean diets compared with the control diet [23]. The incidence of myocardial infarction was reduced non significantly with the two Mediterranean diets. The main results concerning the Mediterranean diet enriched with nuts are summarized below.

The most striking result was that incident stroke, a component of the PREDIMED main outcome, was significantly reduced by 45% (HR = 0,55; 95% CI, 0.35, 0.86] in the group allocated the Mediterranean with nuts compared to the control group (Figure 5) [23]. Regarding other hard cardiometabolic outcomes, the Mediterranean diet with nuts resulted in a 49% reduction (HR = 0.51; 95% CI, 0.32, 0.83) in the incidence of peripheral artery disease [117] and a non-significant 18% reduction (HR = 0.82; 95% CI, 0.61, 1.10) in incident T2D [118]. In participants with MetS at baseline (*n* = 3392), the nut-supplemented diet resulted in a 28% (HR = 1.28; CI 1.08, 1.51) higher probability of reversion of MetS compared with the control diet, and this beneficial effect was driven mainly by reduced WC [119]. The risk of heart failure, a secondary outcome of the trial, was unaffected by the Mediterranean diets [120]. A post hoc analysis revealed no effect of the nut-supplemented diet on incident atrial fibrillation (HR = 0.89; 95% CI, 0.65, 1.20) [119]. The trial was not powered to examine mortality risk; however, when considered as an observational cohort, nut consumption was associated with a significantly reduced risk of all-cause mortality: compared to non-consumers, participants consuming nuts >3 servings/week (32% of the cohort) had a 39% lower mortality risk (HR = 0.61; 95% CI, 0.45, 0.83) [121]. A similar protective effect against CVD and cancer mortality was observed.

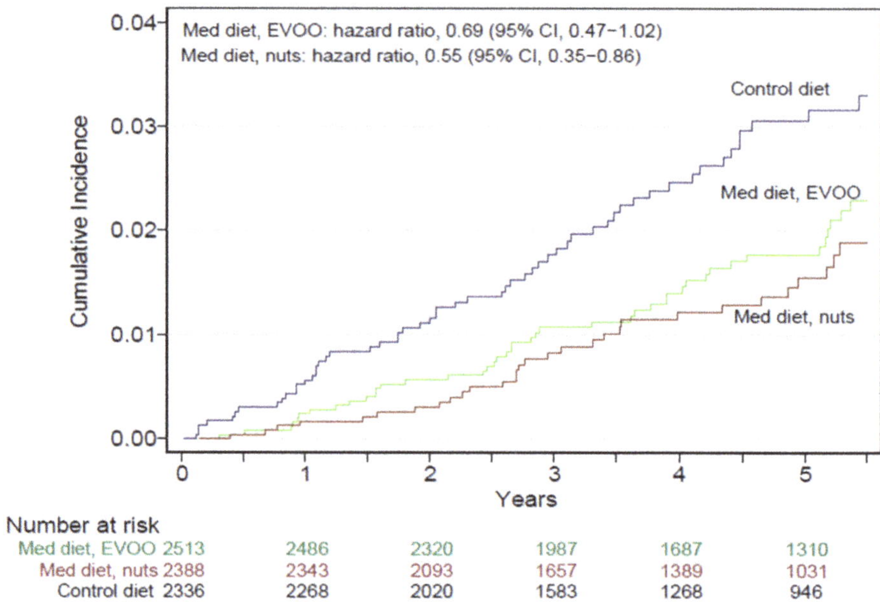

Figure 5. Cumulative incidence of stroke by intervention group in the PREDIMED trial [23]. Copyright © (2018) Massachusetts Medical Society. Reprinted with permission. Med diet, Mediterranean diet; EVOO, extra-virgin olive oil.

Concerning intermediate outcomes, data from the full PREDIMED cohort showed a stable body weight, but WC (which tends to increase with age in older populations) increased less in the Mediterranean diet with nuts group, with and adjusted difference in 5-y changes of –0.92 cm (95% CI, –1.60, –0.24) compared with the control group [122]. These results provide first-level evidence that an ad libitum Mediterranean diet high in fat because of supplementation with nuts does not promote weight gain or visceral adiposity. Data from PREDIMED sub-studies revealed beneficial changes of the Mediterranean diet with nuts on: blood lipids and fasting glucose [122,123]; office BP, insulin resistance and soluble inflammatory markers (except CRP) [122]; BP as assessed by 24 h ambulatory monitoring, with 1 y changes of nearly –4 mm Hg for SBP and –2 mm Hg for DBP compared with the control diet, remarkable given that most participants were hypertensive and received standard anti-hypertensive medications [123]; and carotid plaque regression compared with progression in the control group after intervention for 2.4 y [124]. In another sub-study with 334 participants, a comprehensive neuropsychological test battery was administered at baseline and after a mean follow-up of 4.1 y. The results showed that, compared with the control diet, both Mediterranean diets resulted in delayed age-related cognitive decline, while the nut-supplemented diet performed better in the memory domain [125]. This PREDIMED sub-study is the first RCT demonstrating that a dietary pattern enhances cognitive function.

It must be emphasized that the PREDIMED interventions were meant to improve the overall diet, but the major between-group differences in food consumption were for the provisioned supplemental foods. It follows that nut consumption was probably responsible for most of the observed benefits in the Mediterranean diet with nuts group. The PREDIMED results illustrate the remarkable potential of nuts and other healthy foods such as extra-virgin olive oil to beneficially impact health outcomes. Given the age of PREDIMED participants, an important lesson of the trial is that it is never too late to change dietary habits to improve CV health and brain function.

6. Conclusions

Nuts, by virtue of their beneficial nutrients and phytochemicals, appear to bestow favorable and wide-ranging health dividends. The PREDIMED RCT showed a protective effect of nuts against CVD. Other RCTs have demonstrated that nuts lower LDL-cholesterol concentration, reduce insulin resistance and improve vascular reactivity. Epidemiological studies report largely congruent findings indicating that nut consumption is associated with lower risks for CVD, total mortality, atrial fibrillation, hypertension, and cancer. Habitual nut consumption does not promote obesity and may even result in less weight gain over time, particularly among individuals with overweight/obesity. Table 3 summarizes the main beneficial effects of nuts on health outcomes.

Table 3. Associations of nut consumption with health outcomes and disease risk factors. Summary of scientific evidence.

Disease/Factor	Association	Level of Evidence
Epidemiologic studies		
Cardiovascular disease	Reduction	++
Coronary heart disease	Reduction	++
Stroke	No change/Reduction	+/−
Heart failure	No change/reduction	+/−
Atrial fibrillation	Reduction	+
Hypertension	Reduction	+
Diabetes	No change/reduction	+/−
Cognitive dysfunction	Improvement	+
Depression	No change/reduction	+/−
Cancer	Reduction	++
Obesity	No change/reduction	++
All-cause mortality	Reduction	++
Randomized clinical trials		
Blood lipid profile		
Total cholesterol	Reduction *	++
LDL-cholesterol	Reduction *	++
HDL-cholesterol	No change	++
Triglycerides	Reduction *	++
Insulin sensitivity	No change/increase *	+
Diabetes control	Improvement	+
Blood pressure	No change/reduction *	+/−

Table 3. *Cont.*

Disease/Factor	Association	Level of Evidence
Inflammation	No change/reduction *	+
Vascular reactivity	Improvement	+
Body weight	No change/slight reduction *	++
Waist circumference	No change/slight reduction *	++
Metabolic syndrome	Improvement or reversion **	+
Type-2 diabetes incidence	No change **	+
CVD incidence	Reduction **	+
Stroke incidence	Reduction **	+
PAD incidence	Reduction **	+
Cognitive function	Improvement *	+

Abbreviations: +/−, equivocal evidence; +, limited evidence from few studies; ++, evidence from many studies; HDL, high-density lipoprotein; LDL, low-density lipoprotein; PAD peripheral artery disease. * Evidence collected in the PREDIMED trial, among others. ** Evidence collected only in the PREDIMED trial.

The cumulative scientific evidence indicates that nuts are one of the most wholesome and nutritious foods in the usual diet, but which nuts are best for health? The three nuts that were supplied and consumed by the participants in the Mediterranean diet plus nuts arm of the PREDIMED study were walnuts, almonds and hazelnuts. Thus, these are the only three nuts with first level evidence for improving CV outcomes in the context of a RCT.

However, the nutrient profiles of other nuts make them excellent dietary options as well. For example, Brazil nuts are especially rich in selenium, pecans and peanuts are great sources of polyphenol antioxidants, pistachios are particularly high in carotenoids, tocopherols and phytosterols, and macadamias are replete with monounsaturated fats and flavonoids. Consequently, consuming a mixture of nuts, aiming for a daily dose of at least 30 g/d, is ideal for optimizing health.

As nuts are naturally high in non-sodium minerals and virtually devoid of sodium, lightly salted nuts are still a healthy low-salt snack that many people find more palatable than unsalted nuts. The inner peel between the shell and the nut is rich in polyphenols. Given that the peel and its polyphenols are lost when nuts are roasted, raw, unpeeled nuts are generally the healthiest and those that can be rightly considered as the natural food with most pluripotential nutraceutical properties. It is noteworthy that with the choice of a single wholefood we can positively impact multiple cardiometabolic risk factors, promote healthy aging, and live longer [126]. Regular nut consumption is an indispensable component of any healthy, plant-based dietary pattern.

Author Contributions: E.R. conceptualized the research. E.R. and J.H.O. examined the relevance, quality, and consistency of the scientific literature on nuts and health, and independently extracted and reached a consensus when in doubt about a specific citation. A.S. assisted in retrieving articles and writing. All authors participated in writing of this manuscript. All authors have read and agreed to the published version of the manuscript.

Funding: This research received no external funding.

Institutional Review Board Statement: Not applicable.

Informed Consent Statement: Not applicable.

Acknowledgments: CIBEROBN is an initiative of Instituto de Salud Carlos III, Spain.

Conflicts of Interest: E.R. reports grants, personal fees, non-financial support and other from California Walnut Commission, during the conduct of the study; grants, personal fees, non-financial support and other from Alexion; personal fees and other from Amarin, outside the submitted work. J.H.O.

reports a major ownership interest in CardioTabs, a nutraceutical company. A.S. has not conflicts of interest.

Abbreviations

ALA—α-Linolenic Acid
BP—Blood Pressure
BMI—Body Mass Index
CHD—Coronary Heart Disease
CI—Confidence Interval
CRP—C-Reactive Protein
CV—Cardiovascular
CVD—Cardiovascular Disease
DBP—Diastolic Blood Pressure
FMD—Flow-Mediated Dilation
HOMA-IR—Homeostasis Model Assessment of Insulin Resistance
HPFS—Health Professionals Follow-Up Study
HR—Hazard Ratio
ICAM—Intercellular Adhesion Molecule
IL—Interleukin
MetS—Metabolic syndrome
MUFA—Monounsaturated Fatty Acids
NAFLD—Non-Alcoholic Fatty Liver Disease
NHS—Nurses' Health Study
NO—Nitric Oxide
OR—Odds ratio
PREDIMED—PREvención con DIeta MEDiterránea
PUFA—Polyunsaturated Fatty Acids
PURE—Prospective Urban and Rural Epidemiology
RCT—Randomized Controlled Trial
RR—Relative Risk
SBP—Systolic Blood Pressure
SFA—Saturated Fatty Acids
T2D—Type-2 Diabetes Mellitus
TNF-α—Tumor-necrosis Factor-α
VCAM—Vascular Cell Adhesion Molecule
WAHA—Walnuts and Healthy Aging
WC—Waist Circumference
WMD—Weighted Mean Difference

References

1. Ros, E. Health benefits of nut consumption. *Nutrients* **2010**, *2*, 652–683. [CrossRef] [PubMed]
2. Fraser, G.E.; Sabaté, J.; Beeson, W.L.; Strahan, T.M. A possible protective effect of nut consumption on risk of coronary heart disease. The Adventist Health Study. *Arch. Intern. Med.* **1992**, *152*, 1416–1424. [CrossRef] [PubMed]
3. Sabaté, J.; Fraser, G.E.; Burke, K.; Knutsen, S.F.; Bennett, H.; Lindsted, K.D. Effects of walnuts on serum lipid levels and blood pressure in normal men. *N. Engl. J. Med.* **1993**, *328*, 603–607. [CrossRef] [PubMed]
4. Eaton, S.B.; Konner, M. Paleolithic nutrition. A consideration of its nature and current implications. *N. Engl. J. Med.* **1985**, *312*, 283–289. [CrossRef] [PubMed]
5. Sabaté, J. Nut consumption, vegetarian diets, ischemic heart disease risk, and all-cause mortality: Evidence from epidemiologic studies. *Am. J. Clin. Nutr.* **1999**, *70*, 500S–503S. [CrossRef] [PubMed]
6. Taylor, C.L. *Qualified Health Claims: Letter of Enforcement Discretion—Nuts and Coronary Heart Disease (Docket No 02P-0505)*; Food and Drug Administration, Ed.; Office of Nutritional Products, Labeling and Dietary Supplements: Washington, DC, USA, 2003. Available online: http://wayback.archive-it.org/7993/20171114183724/https://www.fda.gov/Food/IngredientsPackagingLabeling/LabelingNutrition/ucm072926.htm (accessed on 25 July 2021).
7. Lloyd-Jones, D.M.; Hong, Y.; Labarthe, D.; Mozaffarian, D.; Appel, L.J.; Van Horn, L.; Greenlund, K.; Daniels, S.; Nichol, G.; Tomaselli, G.F.; et al. Defining and setting national goals for cardiovascular health promotion and disease reduction: The American Heart Association's strategic Impact Goal through 2020 and beyond. *Circulation* **2010**, *121*, 586–613. [CrossRef]

8.	Aune, D.; Keum, N.; Giovannucci, E.; Fadnes, L.T.; Boffetta, P.; Greenwood, D.C.; Tonstad, S.; Vatten, L.J.; Riboli, E.; Norat, T. Nut consumption and risk of cardiovascular disease, total cancer, all-cause and cause-specific mortality: A systematic review and dose-response meta-analysis of prospective studies. *BMC Med.* **2016**, *14*, 207. [CrossRef]

9.	Del Gobbo, L.C.; Falk, M.C.; Feldman, R.; Lewis, K.; Mozaffarian, D. Effects of tree nuts on blood lipids, apolipoproteins, and blood pressure: Systematic review, meta-analysis, and dose-response of 61 controlled intervention trials. *Am. J. Clin. Nutr.* **2015**, *102*, 1347–1356. [CrossRef]

10.	Jackson, C.L.; Hu, F.B. Long-term associations of nut consumption with body weight and obesity. *Am. J. Clin. Nutr.* **2014**, *100*, 408S–411S. [CrossRef]

11.	Salas-Salvadó, J.; Casas-Agustench, P.; Salas-Huetos, A. Cultural and historical aspects of Mediterranean nuts with emphasis on their attributed healthy and nutritional properties. *Nutr. Metab. Cardiovasc. Dis.* **2011**, *21*, S1–S6. [CrossRef]

12.	Arya, S.S.; Salve, A.R.; Chauhan, S. Peanuts as functional food: A review. *J. Food Sci. Technol.* **2016**, *53*, 31–41. [CrossRef]

13.	Jenab, M.; Sabate, J.; Slimani, N.; Ferrari, P.; Mazuir, M.; Casagrande, C.; Deharveng, G.; Tjonneland, A.; Olsen, A.; Overvad, K.; et al. Consumption and portion sizes of tree nuts, peanuts and seeds in the European Prospective Investigation into Cancer and Nutrition (EPIC) cohorts from 10 European countries. *Br. J. Nutr.* **2006**, *96*, S12–S23. [CrossRef]

14.	Bach-Faig, A.; Berry, E.M.; Lairon, D.; Reguant, J.; Trichopoulou, A.; Dernini, S.; Medina, F.X.; Battino, M.; Belahsen, R.; Miranda, G.; et al. Mediterranean diet pyramid today. Science and cultural updates. *Public Health Nutr.* **2011**, *14*, 2274–2284. [CrossRef]

15.	ARS. *FoodData Central*; US Department of Agriculture, Agricultural Research Service, Ed.; US Government: Washington, DC, USA, 2019. Available online: http://www.fdc.nal.usda.gov (accessed on 25 July 2021).

16.	Kornsteiner-Krenn, M.; Wagner, K.H.; Elmadfa, I. Phytosterol content and fatty acid pattern of ten different nut types. *Int. J. Vitam. Nutr. Res.* **2013**, *83*, 263–270. [CrossRef]

17.	Moncada, S.; Higgs, A. The L-arginine-nitric oxide pathway. *N. Engl. J. Med.* **1993**, *329*, 2002–2012. [CrossRef] [PubMed]

18.	Neveu, V.; Pérez-Jiménez, J.; Vos, F.; Crespy, V.; du Chaffaut, L.; Mennen, L.; Knox, C.; Eisner, R.; Cruz, J.; Wishart, D.; et al. Phenol-Explorer: An online comprehensive database on polyphenol contents in foods. In *Database*; Phenol-Explorer, Ed.; University of Alberta: Alberta, Canada, 2010. [CrossRef]

19.	Ros, E.; Hu, F.B. Consumption of plant seeds and cardiovascular health: Epidemiological and clinical trial evidence. *Circulation* **2013**, *128*, 553–565. [CrossRef]

20.	Moreau, R.A.; Nystrom, L.; Whitaker, B.D.; Winkler-Moser, J.K.; Baer, D.J.; Gebauer, S.K.; Hicks, K.B. Phytosterols and their derivatives: Structural diversity, distribution, metabolism, analysis, and health-promoting uses. *Prog. Lipid. Res.* **2018**, *70*, 35–61. [CrossRef]

21.	Del Gobbo, L.C.; Falk, M.C.; Feldman, R.; Lewis, K.; Mozaffarian, D. Are Phytosterols Responsible for the Low-Density Lipoprotein-Lowering Effects of Tree Nuts?: A Systematic Review and Meta-Analysis. *J. Am. Coll. Cardiol.* **2015**, *65*, 2765–2767. [CrossRef] [PubMed]

22.	Karppanen, H.; Karppanen, P.; Mervaala, E. Why and how to implement sodium, potassium, calcium, and magnesium changes in food items and diets? *J. Hum. Hypertens.* **2005**, *19*, S10–S19. [CrossRef] [PubMed]

23.	Estruch, R.; Ros, E.; Salas-Salvadó, J.; Covas, M.I.; Corella, D.; Arós, F.; Gomez-Gracia, E.; Ruiz-Gutiérrez, V.; Fiol, M.; Lapetra, J.; et al. Primary Prevention of Cardiovascular Disease with a Mediterranean Diet Supplemented with Extra-Virgin Olive Oil or Nuts. *N. Engl. J. Med.* **2018**, *378*, e34. [CrossRef]

24.	WHO. Cardiovascular Diseases (CVDs). Available online: https://www.who.int/news-room/fact-sheets/detail/cardiovascular-diseases-(cvds) (accessed on 29 July 2021).

25.	Becerra-Tomás, N.; Paz-Graniel, I.; Kendall, C.W.C.; Kahleova, H.; Rahelic, D.; Sievenpiper, J.L.; Salas-Salvado, J. Nut consumption and incidence of cardiovascular diseases and cardiovascular disease mortality: A meta-analysis of prospective cohort studies. *Nutr. Rev.* **2019**, *77*, 691–709. [CrossRef]

26.	Liu, G.; Guasch-Ferré, M.; Hu, Y.; Li, Y.; Hu, F.B.; Rimm, E.B.; Manson, J.E.; Rexrode, K.M.; Sun, Q. Nut Consumption in Relation to Cardiovascular Disease Incidence and Mortality Among Patients With Diabetes Mellitus. *Circ. Res.* **2019**, *124*, 920–929. [CrossRef] [PubMed]

27.	Amba, V.; Murphy, G.; Etemadi, A.; Wang, S.; Abnet, C.C.; Hashemian, M. Nut and Peanut Butter Consumption and Mortality in the National Institutes of Health-AARP Diet and Health Study. *Nutrients* **2019**, *11*, 1508. [CrossRef] [PubMed]

28.	De Souza, R.J.; Dehghan, M.; Mente, A.; Bangdiwala, S.I.; Ahmed, S.H.; Alhabib, K.F.; Altuntas, Y.; Basiak-Rasala, A.; Dagenais, G.R.; Diaz, R.; et al. Association of nut intake with risk factors, cardiovascular disease, and mortality in 16 countries from 5 continents: Analysis from the Prospective Urban and Rural Epidemiology (PURE) study. *Am. J. Clin. Nutr.* **2020**, *112*, 208–219. [CrossRef] [PubMed]

29.	Imran, T.F.; Kim, E.; Buring, J.E.; Lee, I.M.; Gaziano, J.M.; Djousse, L. Nut consumption, risk of cardiovascular mortality, and potential mediating mechanisms: The Women's Health Study. *J. Clin. Lipidol.* **2021**, *15*, 266–274. [CrossRef] [PubMed]

30.	Mohammadifard, N.; Ghaderian, N.; Hassannejad, R.; Sajjadi, F.; Sadeghi, M.; Roohafza, H.; Salas-Salvado, J.; Sarrafzadegan, N. Longitudinal Association of Nut Consumption and the Risk of Cardiovascular Events: A Prospective Cohort Study in the Eastern Mediterranean Region. *Front. Nutr.* **2020**, *7*, 610467. [CrossRef]

31.	Yamakawa, M.; Wada, K.; Koda, S.; Uji, T.; Nakashima, Y.; Onuma, S.; Oba, S.; Nagata, C. Associations of total nut and peanut intakes with all-cause and cause-specific mortality in a Japanese community: The Takayama study. *Br. J. Nutr.* **2021**, 1–8. [CrossRef]

32. Liu, X.; Guasch-Ferré, M.; Drouin-Chartier, J.P.; Tobias, D.K.; Bhupathiraju, S.N.; Rexrode, K.M.; Willett, W.C.; Sun, Q.; Li, Y. Changes in Nut Consumption and Subsequent Cardiovascular Disease Risk Among US Men and Women: 3 Large Prospective Cohort Studies. *J. Am. Heart Assoc.* **2020**, *9*, e013877. [CrossRef]

33. Meier, T.; Grafe, K.; Senn, F.; Sur, P.; Stangl, G.I.; Dawczynski, C.; Marz, W.; Kleber, M.E.; Lorkowski, S. Cardiovascular mortality attributable to dietary risk factors in 51 countries in the WHO European Region from 1990 to 2016: A systematic analysis of the Global Burden of Disease Study. *Eur. J. Epidemiol.* **2019**, *34*, 37–55. [CrossRef] [PubMed]

34. Sisa, I.; Abeya-Gilardon, E.; Fisberg, R.M.; Jackson, M.D.; Mangialavori, G.L.; Sichieri, R.; Cudhea, F.; Bannuru, R.R.; Ruthazer, R.; Mozaffarian, D.; et al. Impact of diet on CVD and diabetes mortality in Latin America and the Caribbean: A comparative risk assessment analysis. *Public Health Nutr.* **2021**, *24*, 2577–2591. [CrossRef]

35. Kim, Y.; Keogh, J.; Clifton, P.M. Nuts and Cardio-Metabolic Disease: A Review of Meta-Analyses. *Nutrients* **2018**, *10*, 1935. [CrossRef] [PubMed]

36. Schwingshackl, L.; Schwedhelm, C.; Hoffmann, G.; Knuppel, S.; Iqbal, K.; Andriolo, V.; Bechthold, A.; Schlesinger, S.; Boeing, H. Food Groups and Risk of Hypertension: A Systematic Review and Dose-Response Meta-Analysis of Prospective Studies. *Adv. Nutr.* **2017**, *8*, 793–803. [CrossRef]

37. Schwingshackl, L.; Hoffmann, G.; Lampousi, A.M.; Knuppel, S.; Iqbal, K.; Schwedhelm, C.; Bechthold, A.; Schlesinger, S.; Boeing, H. Food groups and risk of type 2 diabetes mellitus: A systematic review and meta-analysis of prospective studies. *Eur. J. Epidemiol.* **2017**, *32*, 363–375. [CrossRef]

38. Becerra-Tomás, N.; Paz-Graniel, I.; Hernández-Alonso, P.; Jenkins, D.J.A.; Kendall, C.W.C.; Sievenpiper, J.L.; Salas-Salvadó, J. Nut consumption and type 2 diabetes risk: A systematic review and meta-analysis of observational studies. *Am. J. Clin. Nutr.* **2021**, *113*, 960–971. [CrossRef]

39. Pan, A.; Sun, Q.; Manson, J.E.; Willett, W.C.; Hu, F.B. Walnut consumption is associated with lower risk of type 2 diabetes in women. *J. Nutr.* **2013**, *143*, 512–518. [CrossRef]

40. Arab, L.; Dhaliwal, S.K.; Martin, C.J.; Larios, A.D.; Jackson, N.J.; Elashoff, D. Association between walnut consumption and diabetes risk in NHANES. *Diabetes Metab. Res. Rev.* **2018**, *34*, e3031. [CrossRef] [PubMed]

41. Ros, E.; Izquierdo-Pulido, M.; Sala-Vila, A. Beneficial effects of walnut consumption on human health: Role of micronutrients. *Curr. Opin. Clin. Nutr. Metab. Care* **2018**, *21*, 498–504. [CrossRef] [PubMed]

42. WHO. Cancer Tomorrow. Available online: https://gco.iarc.fr/tomorrow/en (accessed on 29 July 2021).

43. Islami, F.; Goding Sauer, A.; Miller, K.D.; Siegel, R.L.; Fedewa, S.A.; Jacobs, E.J.; McCullough, M.L.; Patel, A.V.; Ma, J.; Soerjomataram, I.; et al. Proportion and number of cancer cases and deaths attributable to potentially modifiable risk factors in the United States. *CA Cancer J. Clin.* **2018**, *68*, 31–54. [CrossRef]

44. Schwingshackl, L.; Schwedhelm, C.; Galbete, C.; Hoffmann, G. Adherence to Mediterranean Diet and Risk of Cancer: An Updated Systematic Review and Meta-Analysis. *Nutrients* **2017**, *9*, 1063. [CrossRef] [PubMed]

45. Naghshi, S.; Sadeghian, M.; Nasiri, M.; Mobarak, S.; Asadi, M.; Sadeghi, O. Association of Total Nut, Tree Nut, Peanut, and Peanut Butter Consumption with Cancer Incidence and Mortality: A Comprehensive Systematic Review and Dose-Response Meta-Analysis of Observational Studies. *Adv. Nutr.* **2021**, *12*, 793–808. [CrossRef]

46. NIPHE. Dutch Food Composition Database. Available online: https://www.rivm.nl/en/dutch-food-composition-database (accessed on 29 July 2021).

47. Berkey, C.S.; Tamimi, R.M.; Willett, W.C.; Rosner, B.; Hickey, M.; Toriola, A.T.; Frazier, A.L.; Colditz, G.A. Adolescent alcohol, nuts, and fiber: Combined effects on benign breast disease risk in young women. *NPJ Breast Cancer* **2020**, *6*, 61. [CrossRef]

48. Zhang, D.; Dai, C.; Zhou, L.; Li, Y.; Liu, K.; Deng, Y.J.; Li, N.; Zheng, Y.; Hao, Q.; Yang, S.; et al. Meta-analysis of the association between nut consumption and the risks of cancer incidence and cancer-specific mortality. *Aging* **2020**, *12*, 10772–10794. [CrossRef]

49. Arvanitakis, Z.; Shah, R.C.; Bennett, D.A. Diagnosis and Management of Dementia: Review. *JAMA* **2019**, *322*, 1589–1599. [CrossRef] [PubMed]

50. Kivipelto, M.; Mangialasche, F.; Ngandu, T. Lifestyle interventions to prevent cognitive impairment, dementia and Alzheimer disease. *Nat. Rev. Neurol.* **2018**, *14*, 653–666. [CrossRef]

51. Scarmeas, N.; Anastasiou, C.A.; Yannakoulia, M. Nutrition and prevention of cognitive impairment. *Lancet Neurol.* **2018**, *17*, 1006–1015. [CrossRef]

52. Mecocci, P.; Boccardi, V.; Cecchetti, R.; Bastiani, P.; Scamosci, M.; Ruggiero, C.; Baroni, M. A Long Journey into Aging, Brain Aging, and Alzheimer's Disease Following the Oxidative Stress Tracks. *J. Alzheimers Dis.* **2018**, *62*, 1319–1335. [CrossRef] [PubMed]

53. Santos, C.Y.; Snyder, P.J.; Wu, W.C.; Zhang, M.; Echeverria, A.; Alber, J. Pathophysiologic relationship between Alzheimer's disease, cerebrovascular disease, and cardiovascular risk: A review and synthesis. *Alzheimers Dement.* **2017**, *7*, 69–87. [CrossRef]

54. Theodore, L.E.; Kellow, N.J.; McNeil, E.A.; Close, E.O.; Coad, E.G.; Cardoso, B.R. Nut Consumption for Cognitive Performance: A Systematic Review. *Adv. Nutr.* **2021**, *12*, 777–792. [CrossRef] [PubMed]

55. Chauhan, A.; Chauhan, V. Beneficial Effects of Walnuts on Cognition and Brain Health. *Nutrients* **2020**, *12*, 550. [CrossRef]

56. Jiang, Y.W.; Sheng, L.T.; Feng, L.; Pan, A.; Koh, W.P. Consumption of dietary nuts in midlife and risk of cognitive impairment in late-life: The Singapore Chinese Health Study. *Age Ageing* **2021**, *50*, 1215–1221. [CrossRef] [PubMed]

57. Leonard, B.E. Inflammation and depression: A causal or coincidental link to the pathophysiology? *Acta Neuropsychiatr.* **2018**, *30*, 1–16. [CrossRef] [PubMed]

58. Vaccarino, V.; Badimón, L.; Bremner, J.D.; Cenko, E.; Cubedo, J.; Dorobantu, M.; Duncker, D.J.; Koller, A.; Manfrini, O.; Milicic, D.; et al. Depression and coronary heart disease: 2018 position paper of the ESC working group on coronary pathophysiology and microcirculation. *Eur. Heart J.* **2020**, *41*, 1687–1696. [CrossRef]

59. Su, Q.; Yu, B.; He, H.; Zhang, Q.; Meng, G.; Wu, H.; Du, H.; Liu, L.; Shi, H.; Xia, Y.; et al. Nut Consumption Is Associated with Depressive Symptoms among Chinese Adults. *Depress. Anxiety* **2016**, *33*, 1065–1072. [CrossRef] [PubMed]

60. Arab, L.; Guo, R.; Elashoff, D. Lower Depression Scores among Walnut Consumers in NHANES. *Nutrients* **2019**, *11*, 275. [CrossRef]

61. Elstgeest, L.E.M.; Visser, M.; Penninx, B.; Colpo, M.; Bandinelli, S.; Brouwer, I.A. Bidirectional associations between food groups and depressive symptoms: Longitudinal findings from the Invecchiare in Chianti (InCHIANTI) study. *Br. J. Nutr.* **2019**, *121*, 439–450. [CrossRef]

62. Chen, G.C.; Zhang, R.; Martínez-González, M.A.; Zhang, Z.L.; Bonaccio, M.; van Dam, R.M.; Qin, L.Q. Nut consumption in relation to all-cause and cause-specific mortality: A meta-analysis 18 prospective studies. *Food Funct.* **2017**, *8*, 3893–3905. [CrossRef] [PubMed]

63. Jo, G.; Oh, H.; Singh, G.M.; Park, D.; Shin, M.J. Impact of dietary risk factors on cardiometabolic and cancer mortality burden among Korean adults: Results from nationally representative repeated cross-sectional surveys 1998–2016. *Nutr. Res. Pract.* **2020**, *14*, 384–400. [CrossRef]

64. Fernández-Rodríguez, R.; Mesas, A.E.; Garrido-Miguel, M.; Martínez-Ortega, I.A.; Jiménez-López, E.; Martínez-Vizcaino, V. The Relationship of Tree Nuts and Peanuts with Adiposity Parameters: A Systematic Review and Network Meta-Analysis. *Nutrients* **2021**, *13*, 2251. [CrossRef] [PubMed]

65. Guarneiri, L.L.; Cooper, J.A. Intake of Nuts or Nut Products Does Not Lead to Weight Gain, Independent of Dietary Substitution Instructions: A Systematic Review and Meta-Analysis of Randomized Trials. *Adv. Nutr.* **2021**, *12*, 384–401. [CrossRef]

66. Liu, X.; Li, Y.; Guasch-Ferré, M.; Willett, W.C.; Drouin-Chartier, J.P.; Bhupathiraju, S.N.; Tobias, D.K. Changes in nut consumption influence long-term weight change in US men and women. *BMJ Nutr. Prev. Health* **2019**, *2*, 90–99. [CrossRef]

67. Sabaté, J.; Oda, K.; Ros, E. Nut consumption and blood lipid levels: A pooled analysis of 25 intervention trials. *Arch. Intern. Med.* **2010**, *170*, 821–827. [CrossRef]

68. Cofán, M.; Ros, E. Use of Plant Sterol and Stanol Fortified Foods in Clinical Practice. *Curr. Med. Chem.* **2019**, *26*, 6691–6703. [CrossRef]

69. Schwingshackl, L.; Hoffmann, G.; Iqbal, K.; Schwedhelm, C.; Boeing, H. Food groups and intermediate disease markers: A systematic review and network meta-analysis of randomized trials. *Am. J. Clin. Nutr.* **2018**, *108*, 576–586. [CrossRef]

70. Guasch-Ferré, M.; Li, J.; Hu, F.B.; Salas-Salvadó, J.; Tobias, D.K. Effects of walnut consumption on blood lipids and other cardiovascular risk factors: An updated meta-analysis and systematic review of controlled trials. *Am. J. Clin. Nutr.* **2018**, *108*, 174–187. [CrossRef]

71. Asbaghi, O.; Moodi, V.; Hadi, A.; Eslampour, E.; Shirinbakhshmasoleh, M.; Ghaedi, E.; Miraghajani, M. The effect of almond intake on lipid profile: A systematic review and meta-analysis of randomized controlled trials. *Food Funct.* **2021**, *12*, 1882–1896. [CrossRef] [PubMed]

72. Ghanavati, M.; Rahmani, J.; Clark, C.C.T.; Hosseinabadi, S.M.; Rahimlou, M. Pistachios and cardiometabolic risk factors: A systematic review and meta-analysis of randomized controlled clinical trials. *Complement. Ther. Med.* **2020**, *52*, 102513. [CrossRef] [PubMed]

73. Perna, S.; Giacosa, A.; Bonitta, G.; Bologna, C.; Isu, A.; Guido, D.; Rondanelli, M. Effects of Hazelnut Consumption on Blood Lipids and Body Weight: A Systematic Review and Bayesian Meta-Analysis. *Nutrients* **2016**, *8*, 747. [CrossRef] [PubMed]

74. Jalali, M.; Karamizadeh, M.; Ferns, G.A.; Zare, M.; Moosavian, S.P.; Akbarzadeh, M. The effects of cashew nut intake on lipid profile and blood pressure: A systematic review and meta-analysis of randomized controlled trials. *Complement. Ther. Med.* **2020**, *50*, 102387. [CrossRef] [PubMed]

75. Liu, K.; Hui, S.; Wang, B.; Kaliannan, K.; Guo, X.; Liang, L. Comparative effects of different types of tree nut consumption on blood lipids: A network meta-analysis of clinical trials. *Am. J. Clin. Nutr.* **2020**, *111*, 219–227. [CrossRef] [PubMed]

76. Mohammadifard, N.; Salehi-Abargouei, A.; Salas-Salvado, J.; Guasch-Ferre, M.; Humphries, K.; Sarrafzadegan, N. The effect of tree nut, peanut, and soy nut consumption on blood pressure: A systematic review and meta-analysis of randomized controlled clinical trials. *Am. J. Clin. Nutr.* **2015**, *101*, 966–982. [CrossRef]

77. Eslampour, E.; Asbaghi, O.; Hadi, A.; Abedi, S.; Ghaedi, E.; Lazaridi, A.V.; Miraghajani, M. The effect of almond intake on blood pressure: A systematic review and meta-analysis of randomized controlled trials. *Complement. Ther. Med.* **2020**, *50*, 102399. [CrossRef]

78. Asbaghi, O.; Hadi, A.; Campbell, M.S.; Venkatakrishnan, K.; Ghaedi, E. Effects of pistachios on anthropometric indices, inflammatory markers, endothelial function and blood pressure in adults: A systematic review and meta-analysis of randomised controlled trials. *Br. J. Nutr.* **2021**, *126*, 718–729. [CrossRef]

79. Domènech, M.; Serra-Mir, M.; Roth, I.; Freitas-Simoes, T.; Valls-Pedret, C.; Cofán, M.; López, A.; Sala-Vila, A.; Calvo, C.; Rajaram, S.; et al. Effect of a Walnut Diet on Office and 24-Hour Ambulatory Blood Pressure in Elderly Individuals. *Hypertension* **2019**, *73*, 1049–1057. [CrossRef] [PubMed]

80. Kendall, C.W.; Josse, A.R.; Esfahani, A.; Jenkins, D.J. Nuts, metabolic syndrome and diabetes. *Br. J. Nutr.* **2010**, *104*, 465–473. [CrossRef] [PubMed]

81. Dreher, M.L. A Comprehensive Review of Almond Clinical Trials on Weight Measures, Metabolic Health Biomarkers and Outcomes, and the Gut Microbiota. *Nutrients* **2021**, *13*, 1968. [CrossRef] [PubMed]
82. Tindall, A.M.; Johnston, E.A.; Kris-Etherton, P.M.; Petersen, K.S. The effect of nuts on markers of glycemic control: A systematic review and meta-analysis of randomized controlled trials. *Am. J. Clin. Nutr.* **2019**, *109*, 297–314. [CrossRef]
83. Neale, E.P.; Guan, V.; Tapsell, L.C.; Probst, Y.C. Effect of walnut consumption on markers of blood glucose control: A systematic review and meta-analysis. *Br. J. Nutr.* **2020**, *124*, 641–653. [CrossRef] [PubMed]
84. Matsuzawa, Y.; Kwon, T.G.; Lennon, R.J.; Lerman, L.O.; Lerman, A. Prognostic Value of Flow-Mediated Vasodilation in Brachial Artery and Fingertip Artery for Cardiovascular Events: A Systematic Review and Meta-Analysis. *J. Am. Heart Assoc.* **2015**, *4*, e002270. [CrossRef] [PubMed]
85. Deanfield, J.E.; Halcox, J.P.; Rabelink, T.J. Endothelial function and dysfunction: Testing and clinical relevance. *Circulation* **2007**, *115*, 1285–1295. [CrossRef]
86. Neale, E.P.; Tapsell, L.C.; Guan, V.; Batterham, M.J. The effect of nut consumption on markers of inflammation and endothelial function: A systematic review and meta-analysis of randomised controlled trials. *BMJ Open* **2017**, *7*, e016863. [CrossRef] [PubMed]
87. Xiao, Y.; Huang, W.; Peng, C.; Zhang, J.; Wong, C.; Kim, J.H.; Yeoh, E.K.; Su, X. Effect of nut consumption on vascular endothelial function: A systematic review and meta-analysis of randomized controlled trials. *Clin. Nutr.* **2018**, *37*, 831–839. [CrossRef]
88. Dikariyanto, V.; Smith, L.; Francis, L.; Robertson, M.; Kusaslan, E.; O'Callaghan-Latham, M.; Palanche, C.; D'Annibale, M.; Christodoulou, D.; Basty, N.; et al. Snacking on whole almonds for 6 weeks improves endothelial function and lowers LDL cholesterol but does not affect liver fat and other cardiometabolic risk factors in healthy adults: The ATTIS study, a randomized controlled trial. *Am. J. Clin. Nutr.* **2020**, *111*, 1178–1189. [CrossRef]
89. Morgillo, S.; Hill, A.M.; Coates, A.M. The Effects of Nut Consumption on Vascular Function. *Nutrients* **2019**, *11*, 116. [CrossRef] [PubMed]
90. Libby, P.; Crea, F. Clinical implications of inflammation for cardiovascular primary prevention. *Eur. Heart J.* **2010**, *31*, 777–783. [CrossRef] [PubMed]
91. Libby, P. Inflammation in Atherosclerosis-No Longer a Theory. *Clin. Chem.* **2021**, *67*, 131–142. [CrossRef]
92. Xiao, Y.; Xia, J.; Ke, Y.; Cheng, J.; Yuan, J.; Wu, S.; Lv, Z.; Huang, S.; Kim, J.H.; Wong, S.Y.; et al. Effects of nut consumption on selected inflammatory markers: A systematic review and meta-analysis of randomized controlled trials. *Nutrition* **2018**, *54*, 129–143. [CrossRef] [PubMed]
93. Cofan, M.; Rajaram, S.; Sala-Vila, A.; Valls-Pedret, C.; Serra-Mir, M.; Roth, I.; Freitas-Simoes, T.M.; Bitok, E.; Sabate, J.; Ros, E. Effects of 2-Year Walnut-Supplemented Diet on Inflammatory Biomarkers. *J. Am. Coll. Cardiol.* **2020**, *76*, 2282–2284. [CrossRef]
94. Tsai, C.J.; Leitzmann, M.F.; Hu, F.B.; Willett, W.C.; Giovannucci, E.L. Frequent nut consumption and decreased risk of cholecystectomy in women. *Am. J. Clin. Nutr.* **2004**, *80*, 76–81. [CrossRef] [PubMed]
95. Tsai, C.J.; Leitzmann, M.F.; Hu, F.B.; Willett, W.C.; Giovannucci, E.L. A prospective cohort study of nut consumption and the risk of gallstone disease in men. *Am. J. Epidemiol.* **2004**, *160*, 961–968. [CrossRef]
96. Alberti, K.G.; Eckel, R.H.; Grundy, S.M.; Zimmet, P.Z.; Cleeman, J.I.; Donato, K.A.; Fruchart, J.C.; James, W.P.; Loria, C.M.; Smith, S.C., Jr.; et al. Harmonizing the metabolic syndrome: A joint interim statement of the International Diabetes Federation Task Force on Epidemiology and Prevention; National Heart, Lung, and Blood Institute; American Heart Association; World Heart Federation; International Atherosclerosis Society; and International Association for the Study of Obesity. *Circulation* **2009**, *120*, 1640–1645. [CrossRef]
97. Perez-Martinez, P.; Mikhailidis, D.P.; Athyros, V.G.; Bullo, M.; Couture, P.; Covas, M.I.; de Koning, L.; Delgado-Lista, J.; Diaz-Lopez, A.; Drevon, C.A.; et al. Lifestyle recommendations for the prevention and management of metabolic syndrome: An international panel recommendation. *Nutr. Rev.* **2017**, *75*, 307–326. [CrossRef]
98. Blanco Mejia, S.; Kendall, C.W.; Viguiliouk, E.; Augustin, L.S.; Ha, V.; Cozma, A.I.; Mirrahimi, A.; Maroleanu, A.; Chiavaroli, L.; Leiter, L.A.; et al. Effect of tree nuts on metabolic syndrome criteria: A systematic review and meta-analysis of randomised controlled trials. *BMJ Open* **2014**, *4*, e004660. [CrossRef]
99. Al Abdrabalnabi, A.; Rajaram, S.; Bitok, E.; Oda, K.; Beeson, W.L.; Kaur, A.; Cofán, M.; Serra-Mir, M.; Roth, I.; Ros, E.; et al. Effects of Supplementing the Usual Diet with a Daily Dose of Walnuts for Two Years on Metabolic Syndrome and Its Components in an Elderly Cohort. *Nutrients* **2020**, *12*, 451. [CrossRef] [PubMed]
100. Younossi, Z.; Anstee, Q.M.; Marietti, M.; Hardy, T.; Henry, L.; Eslam, M.; George, J.; Bugianesi, E. Global burden of NAFLD and NASH: Trends, predictions, risk factors and prevention. *Nat. Rev. Gastroenterol. Hepatol.* **2018**, *15*, 11–20. [CrossRef]
101. Plaz Torres, M.C.; Bodini, G.; Furnari, M.; Marabotto, E.; Zentilin, P.; Giannini, E.G. Nuts and Non-Alcoholic Fatty Liver Disease: Are Nuts Safe for Patients with Fatty Liver Disease? *Nutrients* **2020**, *12*, 3363. [CrossRef] [PubMed]
102. Arias-Fernández, L.; Machado-Fragua, M.D.; Graciani, A.; Guallar-Castillón, P.; Banegas, J.R.; Rodríguez-Artalejo, F.; Lana, A.; López-García, E. Prospective Association Between Nut Consumption and Physical Function in Older Men and Women. *J. Gerontol. A Biol. Sci. Med. Sci.* **2019**, *74*, 1091–1097. [CrossRef]
103. Freitas-Simoes, T.M.; Wagner, M.; Samieri, C.; Sala-Vila, A.; Grodstein, F. Consumption of Nuts at Midlife and Healthy Aging in Women. *J. Aging Res.* **2020**, *2020*, 5651737. [CrossRef] [PubMed]
104. Munoz-Garach, A.; Garcia-Fontana, B.; Munoz-Torres, M. Nutrients and Dietary Patterns Related to Osteoporosis. *Nutrients* **2020**, *12*, 1986. [CrossRef] [PubMed]

105. Howitz, K.T.; Bitterman, K.J.; Cohen, H.Y.; Lamming, D.W.; Lavu, S.; Wood, J.G.; Zipkin, R.E.; Chung, P.; Kisielewski, A.; Zhang, L.L.; et al. Small molecule activators of sirtuins extend Saccharomyces cerevisiae lifespan. *Nature* **2003**, *425*, 191–196. [CrossRef]

106. Griel, A.E.; Kris-Etherton, P.M.; Hilpert, K.F.; Zhao, G.; West, S.G.; Corwin, R.L. An increase in dietary n-3 fatty acids decreases a marker of bone resorption in humans. *Nutr. J.* **2007**, *6*, 2. [CrossRef] [PubMed]

107. Agarwal, A.; Baskaran, S.; Parekh, N.; Cho, C.L.; Henkel, R.; Vij, S.; Arafa, M.; Panner Selvam, M.K.; Shah, R. Male infertility. *Lancet* **2021**, *397*, 319–333. [CrossRef]

108. Salas-Huetos, A.; James, E.R.; Aston, K.I.; Jenkins, T.G.; Carrell, D.T. Diet and sperm quality: Nutrients, foods and dietary patterns. *Reprod. Biol.* **2019**, *19*, 219–224. [CrossRef]

109. Robbins, W.A.; Xun, L.; FitzGerald, L.Z.; Esguerra, S.; Henning, S.M.; Carpenter, C.L. Walnuts improve semen quality in men consuming a Western-style diet: Randomized control dietary intervention trial. *Biol. Reprod.* **2012**, *87*, 101. [CrossRef] [PubMed]

110. Salas-Huetos, A.; Moraleda, R.; Giardina, S.; Antón, E.; Blanco, J.; Salas-Salvadó, J.; Bulló, M. Effect of nut consumption on semen quality and functionality in healthy men consuming a Western-style diet: A randomized controlled trial. *Am. J. Clin. Nutr.* **2018**, *108*, 953–962. [CrossRef] [PubMed]

111. Allen, M.S.; Walter, E.E. Erectile Dysfunction: An Umbrella Review of Meta-Analyses of Risk-Factors, Treatment, and Prevalence Outcomes. *J. Sex. Med.* **2019**, *16*, 531–541. [CrossRef] [PubMed]

112. Ostfeld, R.J.; Allen, K.E.; Aspry, K.; Brandt, E.J.; Spitz, A.; Liberman, J.; Belardo, D.; O'Keefe, J.H.; Aggarwal, M.; Miller, M.; et al. Vasculogenic Erectile Dysfunction: The Impact of Diet and Lifestyle. *Am. J. Med.* **2021**, *134*, 310–316. [CrossRef]

113. Salas-Huetos, A.; Muralidharan, J.; Galie, S.; Salas-Salvado, J.; Bullo, M. Effect of Nut Consumption on Erectile and Sexual Function in Healthy Males: A Secondary Outcome Analysis of the FERTINUTS Randomized Controlled Trial. *Nutrients* **2019**, *11*, 1372. [CrossRef]

114. Lynch, S.V.; Pedersen, O. The Human Intestinal Microbiome in Health and Disease. *N. Engl. J. Med.* **2016**, *375*, 2369–2379. [CrossRef]

115. Lamuel-Raventos, R.M.; Onge, M.S. Prebiotic nut compounds and human microbiota. *Crit. Rev. Food Sci. Nutr.* **2017**, *57*, 3154–3163. [CrossRef] [PubMed]

116. Creedon, A.C.; Hung, E.S.; Berry, S.E.; Whelan, K. Nuts and their Effect on Gut Microbiota, Gut Function and Symptoms in Adults: A Systematic Review and Meta-Analysis of Randomised Controlled Trials. *Nutrients* **2020**, *12*, 2347. [CrossRef] [PubMed]

117. Ruiz-Canela, M.; Estruch, R.; Corella, D.; Salas-Salvado, J.; Martinez-Gonzalez, M.A. Association of Mediterranean diet with peripheral artery disease: The PREDIMED randomized trial. *JAMA* **2014**, *311*, 415–417. [CrossRef] [PubMed]

118. Salas-Salvadó, J.; Bulló, M.; Estruch, R.; Ros, E.; Covas, M.I.; Ibarrola-Jurado, N.; Corella, D.; Arós, F.; Gómez-Gracia, E.; Ruiz-Gutiérrez, V.; et al. Prevention of diabetes with Mediterranean diets: A subgroup analysis of a randomized trial. *Ann. Intern. Med.* **2014**, *160*, 1–10. [CrossRef]

119. Martínez-González, M.A.; Toledo, E.; Arós, F.; Fiol, M.; Corella, D.; Salas-Salvadó, J.; Ros, E.; Covas, M.I.; Fernández-Crehuet, J.; Lapetra, J.; et al. Extravirgin olive oil consumption reduces risk of atrial fibrillation: The PREDIMED (Prevencion con Dieta Mediterranea) trial. *Circulation* **2014**, *130*, 18–26. [CrossRef] [PubMed]

120. Papadaki, A.; Martínez-González, M.A.; Alonso-Gómez, A.; Rekondo, J.; Salas-Salvadó, J.; Corella, D.; Ros, E.; Fitó, M.; Estruch, R.; Lapetra, J.; et al. Mediterranean diet and risk of heart failure: Results from the PREDIMED randomized controlled trial. *Eur. J. Heart Fail.* **2017**, *19*, 1179–1185. [CrossRef]

121. Guasch-Ferré, M.; Bulló, M.; Martínez-González, M.A.; Ros, E.; Corella, D.; Estruch, R.; Fitó, M.; Arós, F.; Warnberg, J.; Fiol, M.; et al. Frequency of nut consumption and mortality risk in the PREDIMED nutrition intervention trial. *BMC Med.* **2013**, *11*, 164. [CrossRef]

122. Estruch, R.; Martínez-González, M.A.; Corella, D.; Salas-Salvadó, J.; Fitó, M.; Chiva-Blanch, G.; Fiol, M.; Gómez-Gracia, E.; Arós, F.; Lapetra, J.; et al. Effect of a high-fat Mediterranean diet on bodyweight and waist circumference: A prespecified secondary outcomes analysis of the PREDIMED randomised controlled trial. *Lancet Diabetes Endocrinol.* **2019**, *7*, e6–e17. [CrossRef]

123. Domènech, M.; Roman, P.; Lapetra, J.; García de la Corte, F.J.; Sala-Vila, A.; de la Torre, R.; Corella, D.; Salas-Salvadó, J.; Ruiz-Gutiérrez, V.; Lamuela-Raventós, R.M.; et al. Mediterranean diet reduces 24-hour ambulatory blood pressure, blood glucose, and lipids: One-year randomized, clinical trial. *Hypertension* **2014**, *64*, 69–76. [CrossRef] [PubMed]

124. Sala-Vila, A.; Romero-Mamani, E.S.; Gilabert, R.; Núñez, I.; de la Torre, R.; Corella, D.; Ruiz-Gutiérrez, V.; López-Sabater, M.C.; Pintó, X.; Rekondo, J.; et al. Changes in ultrasound-assessed carotid intima-media thickness and plaque with a Mediterranean diet: A substudy of the PREDIMED trial. *Arterioscler. Thromb. Vasc. Biol.* **2014**, *34*, 439–445. [CrossRef] [PubMed]

125. Valls-Pedret, C.; Sala-Vila, A.; Serra-Mir, M.; Corella, D.; de la Torre, R.; Martínez-González, M.A.; Martínez-Lapiscina, E.H.; Fitó, M.; Pérez-Heras, A.; Salas-Salvadó, J.; et al. Mediterranean Diet and Age-Related Cognitive Decline: A Randomized Clinical Trial. *JAMA Intern. Med.* **2015**, *175*, 1094–1103. [CrossRef]

126. Ros, E. Eat Nuts, Live Longer. *J. Am. Coll. Cardiol.* **2017**, *70*, 2533–2535. [CrossRef] [PubMed]

Review

A Systematic Review of Carotenoids in the Management of Diabetic Retinopathy

Drake W. Lem [1,†], Dennis L. Gierhart [2] and Pinakin Gunvant Davey [1,*,†]

[1] College of Optometry, Western University of Health Sciences, 309 E Second St, Pomona, CA 91766, USA; drake.lem@westernu.edu
[2] ZeaVision, LLC, Chesterfield, MO 63005, USA; dgierhart@zeavision.com
[*] Correspondence: contact@pinakin-gunvant.com; Tel.: +1-909-469-8473
[†] These authors contributed equally to this work.

Citation: Lem, D.W.; Gierhart, D.L.; Davey, P.G. A Systematic Review of Carotenoids in the Management of Diabetic Retinopathy. *Nutrients* **2021**, *13*, 2441. https://doi.org/10.3390/nu13072441

Academic Editors: Paolo Magni, Andrea Baragetti and Andrea Poli

Received: 30 May 2021
Accepted: 12 July 2021
Published: 16 July 2021

Publisher's Note: MDPI stays neutral with regard to jurisdictional claims in published maps and institutional affiliations.

Abstract: Diabetic retinopathy, which was primarily regarded as a microvascular disease, is the leading cause of irreversible blindness worldwide. With obesity at epidemic proportions, diabetes-related ocular problems are exponentially increasing in the developed world. Oxidative stress due to hyperglycemic states and its associated inflammation is one of the pathological mechanisms which leads to depletion of endogenous antioxidants in retina in a diabetic patient. This contributes to a cascade of events that finally leads to retinal neurodegeneration and irreversible vision loss. The xanthophylls lutein and zeaxanthin are known to promote retinal health, improve visual function in retinal diseases such as age-related macular degeneration that has oxidative damage central in its etiopathogenesis. Thus, it can be hypothesized that dietary supplements with xanthophylls that are potent antioxidants may regenerate the compromised antioxidant capacity as a consequence of the diabetic state, therefore ultimately promoting retinal health and visual improvement. We performed a comprehensive literature review of the National Library of Medicine and Web of Science databases, resulting in 341 publications meeting search criteria, of which, 18 were found eligible for inclusion in this review. Lutein and zeaxanthin demonstrated significant protection against capillary cell degeneration and hyperglycemia-induced changes in retinal vasculature. Observational studies indicate that depletion of xanthophyll carotenoids in the macula may represent a novel feature of DR, specifically in patients with type 2 or poorly managed type 1 diabetes. Meanwhile, early interventional trials with dietary carotenoid supplementation show promise in improving their levels in serum and macular pigments concomitant with benefits in visual performance. These findings provide a strong molecular basis and a line of evidence that suggests carotenoid vitamin therapy may offer enhanced neuroprotective effects with therapeutic potential to function as an adjunct nutraceutical strategy for management of diabetic retinopathy.

Keywords: diabetic retinopathy; macular xanthophylls; carotenoids; macular pigment; macular pigment optical density; MPOD; lutein; zeaxanthin; *meso*-zeaxanthin; diabetes; diabetic retinopathy; retinal neurodegeneration; neuroprotection

1. Introduction

Although half a billion individuals are estimated to be living with this condition globally, diabetes remains severely underdiagnosed, with one in every two individuals living with the disease unaware [1–3]. It is further projected that the prevalence of diabetes is likely to increase to 700 million by the year 2045 [2–4]. The systemic disease of endocrine origin leads to progressive damage throughout the body with all end-organs suffering damage [5–10]. Chronic hyperglycemia causes irreversible damage to all parts of the eye. Both the anterior segment structures, cornea, conjunctiva, and lens as well as the posterior segment become damaged [6,11,12]. In the posterior segment, particularly the retina in an individual shows pathognomonic damage, leading to diabetic retinopathy (DR) [6,7,11,12].

The prevalence of diabetes mellitus (DM) has reached epidemic proportions [4,12]. Increased life expectancy and the chronic nature of diabetes with no "true" cure has led to and will continue being a massive health care and socio-economic burden [2,3,5,13,14]. Consequently, it is expected that annual global expenditures will exceed USD 825 billion by the year 2030 [15].

The natural history of DR features retinal capillary degeneration and subsequent significant visual impairment [16], when poorly managed, causes vasoproliferative disease in retina and/or edema in the central macular region; these complications may arise consecutively or simultaneously [11,12]. Approximately one in three individuals with diabetes is affected by retinopathy [4–7]. The severity of DR is associated with both with the duration of diabetes and glycemic control [17,18]. An estimated 4.1 million individuals in the US are afflicted with DR, of which approximately 899,000 have vision-threatening retinopathy [1]. It is estimated globally that 146 million adults have DR with a projected increase to 191 million by 2030 [2,3,14]. The vision loss due to hyperglycemia-induced retinopathy is irreversible as the retinal tissue does not regenerate. However, the damage due to diabetes and DR is preventable, and thus allows for a potential of improvement in the quality of life, decrease in susceptibility to further complications, and reducing health care expenditures [4,7].

Hyperglycemia-induced damage to other parts of the body has been shown to correlate with the severity of DR, including peripheral neuropathy, nephropathy and cardiovascular complications [5–10]. It is well known that chronic hyperglycemic states promote oxidative damage particularly in highly susceptible regions with corresponding high metabolic demands. The extremely metabolically active retinal tissue is particularly susceptible to oxidative damage due to constant exposure to light [19,20]. Recent work strongly implicate that neurodegeneration in retina is proliferated by pro-oxidative and pro-inflammatory mechanisms prior to indications of clinical retinopathy [5,7,10,18,20–24]. Inherent defense mechanisms against oxidative damage in the retina involve constant neutralization of reactive oxygen species (ROS). Congruously, both endogenous and exogenous antioxidants are essential in maintaining cellular redox homeostasis [20,25,26]. Quite appropriately, it is postulated that the interdependence between prolonged hyperglycemia, oxidative stress, and changes in redox homeostasis is a key factor contributing to the pathogenesis of diabetic retinopathy [19,25].

More than 750 naturally occurring phytochemical carotenoids have been identified and characterized, of which, approximately 20 types are present in serum and tissue [27–30]. Among them, the only dietary carotenoids which accumulate in the human eye are lutein and zeaxanthin [27,30]. They belong to the xanthophyll class of carotenoids which contain oxygen in their polyene chain structure and are more lipophilic in comparison with the other subgroup of carotenoids known as carotenes, which do not contain oxygen and are purely hydrocarbons [27,31]. Three isomeric xanthophyll carotenoids—lutein, zeaxanthin, and *meso*-zeaxanthin (Figure 1)—are believed to possess significant antioxidant and anti-inflammatory properties in the retina and have been shown to benefit in prevention of age-related macular degeneration (AMD) [25,27,32–35]. Oxidative insult contributing to retinal neurodegeneration is common to the pathogenesis of both DR and AMD. Hence, it is hypothesized that xanthophyll carotenoids may be clinically beneficial in management of DR.

To the best of our knowledge, the neuroprotective potential afforded by these xanthophylls in clinical management of DR has not been thoroughly reviewed. The primary objective of this systematic review focuses on summarizing the evidence from animal models, clinical observational studies, and randomized controlled trials that have reported on the putative relationship between DR and carotenoids lutein, zeaxanthin, and/or *meso*-zeaxanthin. Thus, the goal of this systematic review is to determine the degree of clinical benefits of carotenoids as an adjunct therapy for the management of DR.

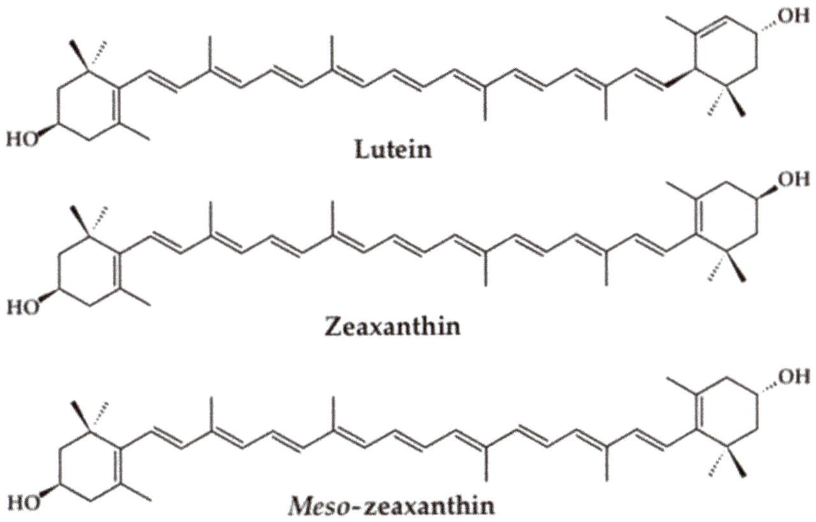

Figure 1. Chemical structures of isomeric xanthophyll carotenoids lutein, zeaxanthin, and *meso*-zeaxanthin.

Retinal Changes in Diabetics

Retinal changes in diabetes are graded by fundoscopic lesions as outlined by the International Clinical Disease Severity Scale [12,13,16,36–38]. Large-scale clinical trials established the severity classification system (Table 1) that is currently used: The Early Treatment Diabetic Retinopathy Study (ETDRS) and the Wisconsin Epidemiological Study of DR (WESDR) [16,37–39]. Non-proliferative diabetic retinopathy (NPDR) is seen as microvascular abnormalities limited to the retinal surface. Additionally, some other features visible are intraretinal hemorrhages ("dot and blot" shaped), microaneurysms, hard exudates, and intraretinal microvascular abnormalities (i.e., tortuous sinus shunt vessels) [37–39]. The degeneration of capillaries and apoptosis in the endothelium are an outcome of progressive oxidative damage in this stage that leads to capillary nonperfusion and vascular occlusion leading to retinal ischemia/hypoxia. This compromises oxygenation and further aggravates oxidative and pro-inflammatory processes in the extremely metabolically-active retina [17,18,36]. These events promote angiogenesis due to the release of vascular endothelial growth factor (VEGF) [17,18,36,40]. The manifestation of cotton wool spots represents hypoxic retina that leads to neurodegeneration [18]. Subsequent retinal neovascularization with aberrant angiogenesis marks disease progression to proliferative diabetic retinopathy (PDR). The new blood vessel formation is an ineffectual attempt to re-establish vascular perfusion and restore homeostasis. However, the response mechanism itself paradoxically further threatens function and viability of the retina ensuing leakage or hemorrhaging into the vitreous cavity, which can lead to retinal detachment and irreversible vision loss [12,13,16,36].

Structural and cellular changes to the retinal architecture enhance permeability, contributing to the break in the blood–retinal barrier that leads to diabetic macular edema (DME); the primary cause of significant vision loss in DR [17,36]. Signs of overt edema are seen during fundoscopic exam. However, subtle edema, evidenced by thickening of basement membrane and presence of exudates, is best visible using optical coherence tomography (OCT) [41,42]. It is extremely important to note, the onset of DME can occur at any stage of DR [5,36].

Table 1. International Clinical Diabetic Retinopathy Disease Severity Scale [37].

Disease Severity Scale	Clinical Features
No apparent retinopathy	No fundus abnormalities present
Mild NPDR	Microaneurysms only
Moderate NPDR	More than just MAs, but less than severe NPDR
Severe NPDR	Any of the following: (with no signs of PDR) extensive DBH in each of 4 quadrants ($\geq$20/quadrants), venous beading in at least 2 quadrants, and/or IRMA in at least 1 quadrant
PDR	One or more of the following: neovascularization, tractional retinal detachment, or vitreous/preretinal hemorrhage

Abbreviations: NPDR, non-proliferative diabetic retinopathy; MA, microaneurysms; PDR, proliferative diabetic retinopathy; DBH, dot blot hemorrhages; IRMA, intraretinal microvascular abnormalities.

2. Diabetic Retinopathy and Macular Pigment

2.1. Basics of Macular Pigment

The yellow spot that is visible during ophthalmoscopy is due to macular pigment, which contains three carotenoids—(1) lutein, (2) zeaxanthin, and (3) a stereo isomer of zeaxanthin called *meso*-zeaxanthin [43,44]—which are known as macular xanthophylls. They are uniquely concentrated in the fovea centralis. A recent study that used confocal resonance Raman microscopy showed that although both lutein and zeaxanthin are concentrated in the fovea, zeaxanthin mainly accumulates in the inner plexiform, outer plexiform and outer nuclear layers of the retina [43–47]. Lutein is more diffusely distributed throughout the macula and is present at lower concentrations in comparison to zeaxanthin at the fovea [47]. Humans have lost the ability to synthesize lutein and zeaxanthin in vivo and thus lutein and zeaxanthin can only be acquired through dietary intake [27]. Common food sources that can provide these xanthophylls are green leafy cruciferous vegetables and egg yolks [44,48–50]. Unless artificially supplemented, *meso*-zeaxanthin found in the retina is an outcome of biochemical conversion of lutein via RPE65 isomerase in the retinal pigment epithelium (RPE) [44,47,48,51–54]. The biological processes involving the uptake, metabolism, and transport of xanthophyll carotenoids to the retina have been explored in greater depth in these review articles [27,28,44,48,51,53–56]. Supplementation of macular xanthophylls improves their levels in the serum [44,48,52,57] and is well known to accumulate in the human retina [27,43,57–68].

Clinical measurement of the macular pigment optical density (MPOD) is as close as we can get to quantification of macular carotenoids. The level of MPOD is indeed a biomarker and is strongly associated with maintenance of retinal health and optimal visual function in both health and disease [44,46,50,59]. Prior reports have demonstrated that carotenoids afford enhanced protection in the retina, specifically in the central region, via two proposed mechanisms: (1) acting as a naturally occurring blue light filter or blocker, and (2) a potent antioxidant and anti-inflammatory substance in the retina [44,50,59,69–72]. The short-wavelength (blue) light triggers production of ROS due to photo-oxidation that leads to damage of the lipid bilayer in cell membranes, proteins, and DNA, in addition to mitochondrial dysfunction which leads to cellular necrosis [44,70–74]. Absorption of the blue light by macular pigment prevents formation of ROS and the consequent oxidative injury triggered by photo-oxidation [43,72,73]. These properties of carotenoids in macular pigment may in part explain how MPOD levels provide neuroprotective capabilities in the retina.

2.2. Measuring MPOD

There are several techniques available to effectively quantify MPOD in vivo [27,46,50,75–80]. The techniques can be broadly divided into two types: (1) subjective—that is, requiring patient response or participation and (2) objective—that is, requiring minimal to no participant involvement to collect measurements [46,50,75,77,81–84].

Heterochromatic flicker photometry (HFP) is the most widely used technique to measure MPOD [46,50,75,76,78,79]. The precise mechanism used to measure macular pigment levels by HFP devices may vary based on the manufacturer, which has been described in prior literature [27,45,46,50,69,77,78,85–87]. Briefly, current HFP devices adjust the intensity of the blue to green ratio in the target stimuli, which is perceived as a flicker. Steady light is observed when the blue component is fully absorbed by the macula, and only green is visible. This is the lowest point in the absorption curve that is measured and converted to MPOD density units [46,50,77,79,80,88,89].

Fundus reflectometry [61,83,90–93], fundus autofluorescence [81,82,94] and resonance Raman spectroscopy [47,95,96] are all non-invasive, objective imaging modalities that can measure MPOD [50,75]. Details regarding both subjective and objective techniques to measure MPOD can be found in these review articles [27,45,46,50,75,82,94,97,98].

3. Materials and Methods

This systematic review was conducted in accordance with the Preferred Reporting Items for Systematic Reviews and Meta-Analysis (PRISMA) reporting guidelines [99].

3.1. Literature Search and Selection Strategy

Two authors (PGD and DWL) performed a wide-ranging search of the scientific databases National Library of Medicine and Web of Science to identify all relevant publications reporting on the association between macular carotenoids and DR until 21 December 2020. Under the guidance of the university librarian, the two authors conducted the full search strategy and data collection together using the following keywords and the combination of their variants during the search query: carotenoids, lutein, zeaxanthin, *meso*-zeaxanthin, macular pigment, macular pigment optical density, MPOD, diabetes, diabetic eye disease, and diabetic retinopathy. The database selection strategy was limited to records pertaining to macular carotenoids (i.e., lutein and/or zeaxanthin and/or *meso*-zeaxanthin) and diabetic retinopathy only. Primary search results were identified for initial screening according to titles and abstracts available in English by PGD and DWL. Among the eligible records, full-text publications were retrieved and evaluated for study inclusion or exclusion criteria. To ensure all relevant studies were included in this review, we individually conducted backward and forward searches of the eligible publications by reviewing reference lists and cited references, respectively. All records retrieved in full text were individually screened and evaluated by two authors (PGD and DWL) for inclusion/exclusion and any discrepancies were resolved through discussion involving the third author (DLG). Selected publications were quantitative research articles evaluating the association between MPOD/carotenoids (including lutein and/or zeaxanthin and/or *meso*-zeaxanthin) and diabetic retinopathy. Additional records involving other forms of diabetes-associated ocular disease were not considered in this review (such as diabetic cataract, diabetic anterior segment or corneal changes associated with hyperglycemia). The full inclusion criteria for eligible publications from experimental and clinical studies are outlined below.

3.2. Study Selection

Experimental animal studies included in this review met the following criteria: (1) evaluating the effects of treatment with carotenoids (including lutein, L and/or zeaxanthin, Z) on outcomes of retinal neurodegeneration, such as markers of oxidative stress, cell viability and visual performance in murine models of DR; (2) carotenoid interventions include powder diet supplemented with L and/or Z only, nutraceutical diet containing L/Z, and powder diet supplemented with micronutrient formula containing L/Z; (3) presentation of DR pathology induced using standard induction methods (i.e., administration of the drug alloxan/streptozotocin, high-sugar diet, and surgical or chemically-induced damage) or genetic models (namely the *Lepr*^{db} model) in rodents only; and (4) experimental models of type 1 or type 2 diabetes in rodents were included.

Inclusion criteria for this systematic review were: (1) observational studies evaluating the association among macular xanthophylls and DR; (2) prospective randomized clinical trials assessing the benefits of carotenoid vitamin therapy in diabetic patients; (3) interventions include dietary carotenoid supplementation (containing L and/or Z) or in a multivitamin formula containing micronutrients and antioxidants; (4) assessment of macular carotenoid levels reported by serum/plasma concentrations of L/Z, or by validated MPOD measurement techniques; (5) cohorts of diabetic patients (type 1 diabetes mellitus, T1DM; and/or type 2 diabetes mellitus, T2DM); and (6) study cohorts of both T1DM and T2DM with either no retinopathy present or mild/moderate NPDR.

Exclusion criteria were based on the following: (1) carotenoid treatment did not include either lutein and/or zeaxanthin in formulation/design; (2) carotenoid treatment included other types of carotenoids; (3) experimental diabetes pathology (as listed previously) were not standard methods of induction; (4) inclusion of adults with other forms of diabetes associated eye disease; and (5) publications were not available in English.

3.3. Data Extraction, Reliability and Risk of Bias Assessment

The PRISMA reporting guidelines were carefully followed as closely as possible, as discussed previously [99]. The risk of bias was assessed using standard metrics established to evaluate the intervention studies and randomized controlled trials. The SYRCLE's RoB tool which is an adaptation of Cochrane RoB tool was used to evaluate the risk of bias for the animal studies [100]. The Cochrane Collaboration's tool for assessing risk of bias for the randomized controlled trials [101].

4. Results

4.1. Search and Selection of Studies

In total, 397 studies were identified during the primary search from scientific databases (Figure 2). After removing duplicate records and including additional records retrieved from reference list searches, 281 studies remained for titles and abstract screening. Consequently, 103 records were excluded based on article type, with an additional 148 records excluded due to the aforementioned inclusion criteria for clinical and preclinical studies. Finally, 30 records were identified to be eligible for full-text assessment, of which, 18 studies were included in the final review: seven preclinical studies [102–108], nine observational clinical studies [19,25,109–115] and two interventional clinical trials [34,116].

4.2. Carotenoids in the Management of Diabetic Retinopathy—Animal Studies

Figure 3 provides a summary of the assessment of risk of bias using the SYRCLE's RoB tool [100]. The studies were unclear on performance bias blinding and outcome assessment blinding was not performed (see Figure 3). However, given that studies have utilized laboratory analysis and histology and not psychophysical response measured in animals or subjective interpretations we can overall safely conclude that the overall risk of bias in these studies were low.

There is an increasing amount of research and animal trials that substantiate the neuroprotective effects of carotenoids lutein and zeaxanthin in rodent models of DR using either chemical induction or genetic modes to engender diabetic state (Table 2) [102–108]. Pharmacological injection of alloxan or streptozotocin (STZ) are often used to recapitulate T1DM pathology in both mice and rats through death of pancreatic beta cells and subsequent insulin deficiency [102–106,117–120]. Genetic modes offer unique models to examine pathophysiological mechanisms of metabolic perturbations that may contribute to incident retinopathy; in particular, leptin receptor deficient (db/db) mice develop morbid obesity and hypoinsulinemia, making them a desirable model for replicating conditions found in T2DM [107,108,118–120]. Importantly, these murine models mimic the characteristic pathological changes induced by hyperglycemia, including oxidative stress driven by free radicals, chronic low-grade inflammation, morphological abnormalities from capillary cell death, and visual dysfunction. Results from these studies are congruous, indicating that lutein and zeaxanthin supplementation has significant potential to protect the retina from the onset of DR.

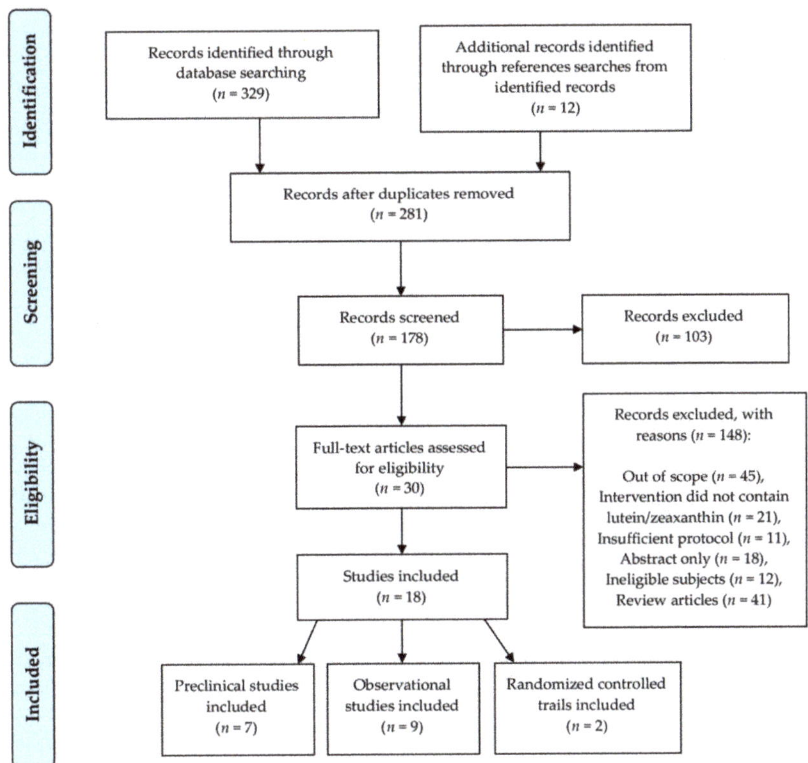

Figure 2. Flow diagram of literature search and selection criteria.

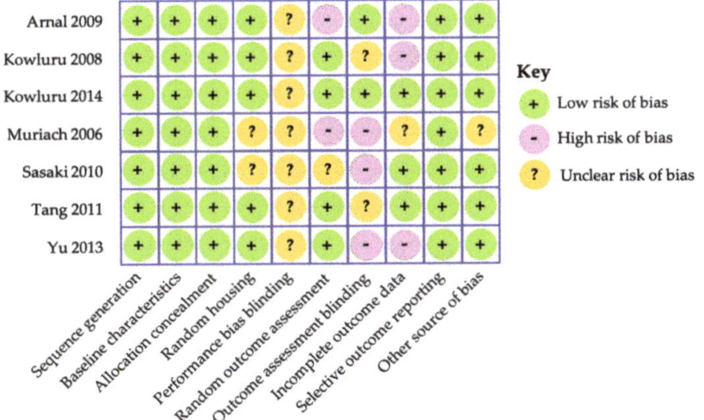

Figure 3. SYRCLE's risk of bias assessment for animal studies [100].

Table 2. Animal studies of carotenoid treatment in diabetic retinopathy.

Author (Year)	DM Study Design	Duration	Treatment	Results
Arnal (2009) [102]	T1DM, via STZ-injection in Wistar rats	12 wks	L (0.5 mg/kg)	Significantly improved GSH and GPx activity
Kowluru (2008) [103]	T1DM, via STZ-injection in Lewis rats	2 months	Z (8.4 ± 1.6 mg/d); Z (44 ± 8 mg/d)	Enhanced MnSOD and complex III expression
Kowluru (2014) [104]	T1DM, via STZ-injection in Wistar rats	11 months	L (1 mg/d) and Z (2 mg/d) *	Augmented retinal cell viability and survival
Muriach (2006) [105]	T1DM, via A-injection in Albino mice	2 wks	L (0.2 mg/kg)	Re-established levels of MDA, GSH and GPx
Sasaki (2010) [106]	T1DM, via STZ-injection in C57BL/6 mice	4 months	L (0.1% diet)	Protected visual function of inner retina
Tang (2011) [107]	T2DM, via genetic db/db mice ($Lepr^{db}$)	8 wks	L (0.05 mg/g fruits) and Z (1.76 mg/g fruits) [†]	Attenuated ER stress and ganglion cell loss
Yu (2013) [108]	T2DM, via genetic db/db mice ($Lepr^{db}$)	8 wks	L and Z (*values not available*) [†]	Ameliorated hypoxia and mitochondrial stress

Abbreviations: DM, diabetes mellitus; T1DM, type 1 diabetes mellitus; T2DM, type 2 diabetes mellitus; L, lutein; Z, zeaxanthin; STZ, streptozotocin; A, alloxan; db/db, leptin receptor deficient ($Lepr^{db}$); GSH, glutathione; GPx, glutathione peroxidase; MnSOD, manganese superoxide dismutase; MDA, malondialdehyde; ER, endoplasmic reticulum * Multivitamin supplement formula; [†] Wolfberry nutraceutical.

The importance of macular carotenoid's antioxidant properties is evident by their enhanced capacity to ameliorate the extent of oxidative injury caused by hyperglycemia in diabetic retina. Supplementation with lutein and/or zeaxanthin was shown to protect against measures of oxidative and nitrosative stress, marked by significant reductions in malondialdehyde, 8-OHdG (oxidatively-modified DNA), and nitrotyrosine, respectively [102,103,105,117,121–123]. Additionally, one study found that micronutrients containing carotenoids prevented a significant rise in retinal ROS levels in T1DM rats following treatment with the EyePromise Diabetes and Visual Function Study (DVS) formula (ZeaVision LLC, Chesterfield, MO, USA) [104,124]. These findings suggest that the mechanism of protection against oxidative damage to the retina may involve improving mitochondrial dysfunction, the primary source of aberrant free radical production as a consequence of hyperglycemia [26,125–129]. In fact, lutein and zeaxanthin were shown to protect against mitochondrial stress induced by T1DM pathology, and improved retinal expression of mtDNA-encoded proteins involved in oxidative phosphorylation and mitochondrial biogenesis [26,102,103,108,117]. Thus, dietary treatment using lutein and zeaxanthin supplementation may prevent early lesions of retinopathy by alleviating pro-oxidant stressors and redox imbalance propagated by hyperglycemic state.

Dietary augmentation of the compromised endogenous antioxidant defenses has been considered the key modulator in the pathogenesis of DR. Multiple studies found that lutein and zeaxanthin recovered enzymatic activity and expression levels of glutathione, glutathione peroxidase and manganese superoxide dismutase [102,103,105,107,117]; indicating a reversal of hyperglycemic-induced impairment in free radical detoxification and clearance mechanisms [26,121,130,131]. Similarly, one animal model demonstrated that an AREDS-based micronutrient formulation improved total antioxidant capacity in the retina, as well as metabolic abnormalities associated with early stages of retinopathy progression [104]. By regenerating endogenous antioxidant capacity, dietary supplementation with lutein and zeaxanthin may serve to reduce the proliferation of consequent damage brought on by oxidative stress and inflammation in diabetic retina [104,121,130–136].

Macular carotenoids may further protect against retinal neurodegeneration by limiting activation of low-grade inflammatory pathways triggered by metabolic and oxidative insults concomitant with hyperglycemic conditions [17,18,21,22,132,133,137,138]. Consistent with this, carotenoid supplementation was shown to mitigate T1DM-induced increase in retinal pro-inflammatory mediators, such as nuclear transcriptional factor-B (NF-kB), interleukin-1β and intercellular adhesion molecule-1 [103–105,137,139–145]. In addition, several studies found that carotenoids demonstrated significant potential to offset pathogenic factor associated with pivotal changes observed in early and advanced stages of retinopathy [17,21,22,104,108,117]; namely, increased cell permeability and neovascularization, respectively [133,135,136,142,146,147]. This neuroprotection following lutein and zeaxanthin administration was evidenced by attenuating the upregulation of pro-angiogenic factor VEGF in diabetic retina of mice and rats [104,108,117]. Preliminary reports suggest carotenoids may protect the local retinal tissue by reducing pro-inflammatory signaling, thereby limiting exacerbation of the inflammatory response to surrounding tissues [138,143,144,148].

The neuroprotective potential of lutein and zeaxanthin positively influencing the pathogenesis of DR was most substantial preventing changes in retinal morphology as a consequence of accelerated capillary cell loss induced by hyperglycemia; regarded as hallmark features of early-stage retinopathy [17,18,148–153]. Lutein and zeaxanthin improved cell viability and markedly enhanced cell survival of the retinal vasculature, which was marked by significant reduction in apoptotic nuclei and formation of degenerative (acellular) capillaries [102–104,106,154,155]. Similarly, carotenoid treatment completely reversed significant loss of ganglion cells caused by hyperglycemic state in murine model [102,106,107]. Studies found lutein and zeaxanthin effectively protected against DM-induced alterations in retinal histology, such as accelerated thinning of the ganglion cell layer (GCL), inner plexiform layer (IPL), inner nuclear layer (INL), outer nuclear layer (ONL), and the photoreceptor layer (inner and outer segment) [102,106,107]. It is important to note, improvement in the photoreceptor layer indicate that the extent of augmentation in cell survival following lutein and zeaxanthin supplementation can be seen maintaining both vascular and non-vascular cells throughout the retina.

Experimental studies strongly suggest that carotenoids may sufficiently protect against the cumulative effect of hyperglycemic-induced retinopathy, or rather progressive neurodegeneration in retinal function made evident by abnormal or delayed response on electroretinogram (ERG). Studies found that lutein and zeaxanthin preserved measures of inner retinal function at the post-receptor level, attenuating DM-induced reduction in oscillatory potentials and the amplitudes of both a- and b-waves on ERG [102,104–106,156–159]. Increased retinal expression of synaptophysin and brain-derived nuclear factor (BDNF) seem to corroborate these findings, wherein greater synaptic activity and cell survival in the inner retina were observed following supplementation with lutein and zeaxanthin [106,160–163]. Thus, preliminary findings offer substantial evidence demonstrating neuroprotective effects of macular carotenoids preventing vision loss in models of both type 1 and type 2 diabetic retina.

Although results from these animal models are promising, interpretation of the immediate translative potential for clinical application must be performed with prudence. Briefly, accumulation of carotenoids in the macula is unique to primate retinas, and therefore macular pigments cannot be fully studied using only rodent models of DR [27,164,165]. It is important to note the potential limitations depending on the method of DM-induction utilized in rodents; namely, pathophysiological differences in T1DM (via pharmacological injection with STZ/Alloxan) compared to T2DM (using genetic modes) [118]. For instance, while models of T1DM using STZ are more common since it results in the fastest rate of disease progression, evidence from these reports is not directly comparable between animal models of DR, and therefore each induction method contains its own set of advantages and limitations [118]. In light of this, when accounting for average body weight and daily food consumption in these rodent models, the concentrations of carotenoids and antioxidants used in some reports [104,117] are largely equivalent to the dosage of lutein and zeaxanthin used in clinical intervention trials [34,116,166]. Thus, findings from these preclinical studies are encouraging since the observed protective effects are not due primarily as a consequence of inflated carotenoid concentrations that are beyond clinical relevance for humans. Nonetheless, we can conclude there is a significant and growing body of evidence in agreement with the neuroprotective benefits of lutein and zeaxanthin in ameliorating the onset and progression of hyperglycemia-induced retinopathy.

4.3. Clinical Studies Using Carotenoids in the Management of Diabetic Retinopathy

Clinical studies implicate MPOD depletion, as well as low serum levels of lutein and zeaxanthin, may represent a novel clinical feature of DR; one that is likely contingent upon several metabolic perturbations associated with chronic hyperglycemia in type 1 and type 2 diabetes. Reports from observational studies are consistent in demonstrating carotenoid levels (measured both in serum and the macular pigment) are further reduced among diabetic patients with clinically evident retinopathy (Table 3) [19,25,34,109–116]. In fact, one study found that lower plasma concentrations of lutein and zeaxanthin were significantly associated with greater risk of incident maculopathy as well as disease progression in patients with T2DM [109]. Macular pigment data seem to mirror these findings, providing a strong line of evidence that MPOD levels are substantially lower in diabetic retina [34,110–115] and in particular, individuals with T2DM with retinopathy [19,25,112]. Several studies have also shown the severity of diabetic maculopathy was significantly associated with lower MPOD levels [110–113]. Moreover, preliminary findings are largely comparable and suggest that the relationship between compromised macular pigment and incident retinopathy may vary between diabetes types [19,25,112].

Table 3. A summary of the observational trials.

Author (Year)	Participants	DR Present	Results
Brazionis (2009) [109]	111 patients with T2DM, aged 44–77 years in USA	78 No DR, 33 DR	Lower risk of DR with greater serum levels of non-pro-vitamin A (including L/Z) carotenoids ($p = 0.039$)
Cennamo (2019) [110]	59 patients with T1DM, aged (38.2 ± 13.4) years; 40 healthy controls, aged (31.6 ± 7.4) years in Italy	59 DR	Significantly reduced MPOD ($p < 0.001$) measured by fundus reflectometry
Davies (2002) [111]	34 patients with DM (24 T2DM, 10 T1DM), aged (48.1 ± 11.6) years; 34 healthy controls, aged (36.7 ± 15.1) in United Kingdom	Not specified	Significant lower MPOD among patients with grade 2 maculopathy ($p = 0.016$)
Lima (2010) [112]	29 patients with T2DM, aged (60.7 ± 10.7) years; 14 healthy controls, aged (56.2 ± 11.7) years in USA	17 No DR, 12 NPDR	T2DM patients with or without retinopathy showed reduced MPOD ($p < 0.001$) measured by autofluorescence
Mares (2006) [113]	1698 women from CAREDS, aged 53–86 years (108 patients with diabetes) in USA	Not specified	MPOD measured by HFP ($p < 0.01$) significantly inversely related to diabetes and waist circumference
Scanlon (2015) [19]	102 patients with DM (34 T1DM, 68 T2DM), aged (53.2 ± 12.2) years; 48 healthy controls, aged (52.5 ± 16) years in Ireland	55 No DR, 47 NPDR	MPOD measured by cHFP significantly lower among T2DM ($p = 0.04$) compared to T1DM and controls
Scanlon (2019) [25]	188 patients with T2DM, aged (64.7 ± 8.3) years; 2594 healthy controls, aged (61.4 ± 7.6) years in Ireland	152 No DR, 10 NPDR	T2DM patients saw lower MPOD ($p = 0.047$) measured by cHFP compared to non-diabetic controls
She (2016) [114]	182 patients with DM, aged (62.5 ± 7.2) years; 219 healthy controls, aged (63.6 ± 7.4) years in China	134 No DR, 48 NPDR	MPOD level measured by HFP was significantly associated with central foveal thickness ($p = 0.001$)
Zagers (2005) [115]	14 patients with DM, aged (46 ± 11) years; 14 healthy controls, aged (47 ± 11) years in Netherlands	Not specified	Diabetic eyes showed significant reduction in fundus reflectance MPOD measurement ($p < 0.001$) compared to controls

Abbreviations: NPDR, non-proliferative diabetic retinopathy; DM, diabetes mellitus; T1DM, type 1 diabetes mellitus; T2DM, type 2 diabetes mellitus; L, lutein; Z, zeaxanthin; HFP, heterochromatic flicker photometry; cHFP, customized heterochromatic flicker photometry.

There is limited evidence of RCTs evaluating the benefits of carotenoids in management of diabetic retinopathy. We used the Cochrane Collaboration's tool for assessing risk of bias, which covers the following domains—selection bias, performance bias, detection bias, attrition bias, reporting bias, and other bias [101]. Figure 4 provides a summary of the risk assessed using the Cochrane Collaboration's tool.

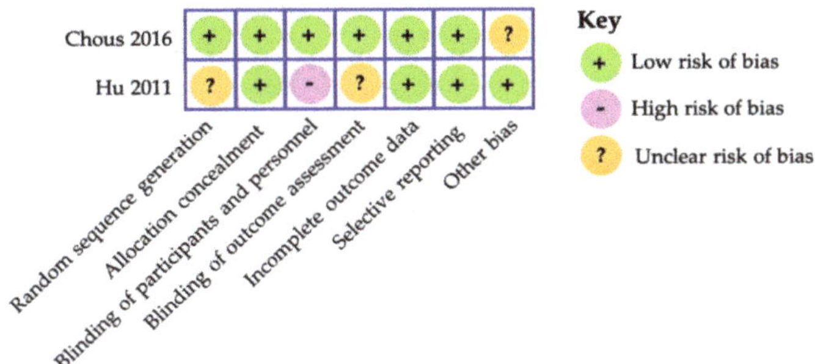

Figure 4. The Cochrane Collaboration's tool for assessing risk of bias in randomized controlled trials [101].

It is well known that both type 1 and type 2 diabetic patients with mild NPDR or no retinopathy exhibit a notable range of visual function impairment, even in the absence of clinically relevant lesions of neurodegeneration in the retina [167–173]. Following active oral supplementation containing lutein and/or zeaxanthin micronutrients, two interventional studies (Table 4) observed marked improvements in serum carotenoids and MPOD levels. Additionally, clinically meaningful improvements in visual performance were also observed in these short-term supplementation trials between three and six months, respectively [34,116]. Most notably, the randomized, placebo-controlled Diabetes Visual Function Supplement Study (DiVFuSS) demonstrated an average increase of 27% in MPOD levels (measured by HFP) after six months of active supplementation [34]. This study revealed that daily supplementation of 4 mg lutein and 8 mg zeaxanthin plus antioxidants offered significant improvement in contrast sensitivity, color discrimination error score and mean visual field sensitivity in diabetic patients presenting with or without mild-to-moderate NPDR [34]. Thus, these results suggest that carotenoid vitamin therapy formulation may offer protection against diabetes-induced retinal neurodegenerative pathology with concomitant effects on visual performance measures in both type 1 and type 2 diabetes. In fact, the enhanced neuroprotective capacity of a similar carotenoid formula has been shown in experimental model of DR using chemical induction to recapitulate pathology observed in T1DM, discussed previously [34,104,166]. The risk of bias was low for this trial as assessed by the Cochrane Collaboration's tool.

Table 4. Characteristics of the eligible randomized clinical trials.

Author (Year)	Participants	DM Subtype	Duration	Interventions	Results
Chous (2016) [34]	67 patients with no retinopathy or mild/moderate NPDR, aged (56.1 ± 13.2) years in USA	27 T1DM, 40 T2DM	6 months	Daily: 4 mg L and 8 mg Z (n = 39, multivitamin oral supplementation) [†]; placebo (n = 28)	Significant increase in MPOD ($p < 0.001$), contrast sensitivity ($p < 0.01$, for all) and color error score ($p < 0.001$)
Hu (2011) [116]	60 patients with NPDR, aged (59.5 ± 14.5) years; 30 healthy controls aged (55 ± 9.0) years in China	10 T1DM, 50 T2DM	3 months	Daily: 6 mg L and 0.5 mg Z (n = 30 NPDR); placebo oral supplementation (n = 30 NPDR, 30 controls)	Significant increase in serum L/Z ($p < 0.001$), visual acuity ($p < 0.001$) and contrast sensitivity ($p < 0.05$, for all)

Abbreviations: NPDR, non-proliferative diabetic retinopathy; T1DM, type 1 diabetes mellitus; T2DM, type 2 diabetes mellitus; L, lutein; Z, zeaxanthin; [†] EyePromise DVS multivitamin supplement.

Various reports seem to suggest these improvements in visual performance following increases in serum carotenoid levels and MPOD concentrations may be attributed, at least in part, to the enhanced functional capacity of the macular pigments to preferentially absorb short-wavelength blue light [27,174–183]. Greater MPOD levels may provide neuroprotective, pre-receptoral filtration against harmful blue light thereby attenuating the deleterious effects of chromatic aberration [27,178,180–183]. One school of thought argues that MPOD status may represent a sine qua non for improvements in visual function; namely, that significant benefit in visual performance will occur only after MPOD density has been maintained at greater concentrations for a period of time [62,178,184]. Alternatively, carotenoid vitamin therapy is also believed to augment total antioxidant capacity which may ameliorate intracellular redox homeostasis in the surrounding tissue including the photoreceptor cells of the neurosensory retina [26,127,185,186]. Further implications of greater carotenoid levels in the macula are also thought to improve metabolic efficiency of the visual cycle thereby promoting enhancement of the post-receptoral circuitry [187]. Indeed, the neuroprotective benefits in ganglion cells and photoreceptors observed in experimental models [102,106,107] are also implicated in humans marked by restoring clinical measures of both inner and outer retinal function, respectively [34,116,169,188]. By augmenting their levels in the diet through oral supplementation, the potent antioxidant and anti-inflammatory properties of xanthophyll carotenoids likely counteract the compounding insult from oxidative stress and chronic inflammation in the diabetic retina, as discussed previously [10,55,189–204]. However, future studies are required in order to elucidate the precise mechanisms responsible for the visual improvements in diabetic retina using carotenoid vitamin therapy.

In view of these findings, available reports among diabetic patients with and without non-proliferative retinopathy are encouraging in demonstrating the potential for carotenoid supplementation as an adjunct nutraceutical approach to offer enhanced protection against further hyperglycemia-induced injury to the retina. Figure 5 illustrates major causative mechanisms which have been postulated in diabetic retinopathy onset, of which, several interconnected processes are believed to represent key drivers among those with type 2 diabetes or poorly-managed type 1 diabetes [34,205]. One mechanism of action involves systemic, atherogenic metabolic imbalance which is believed to play a significant role in macular pigment depletion [49,189,190,192,195,202,205–208]. Prior to exerting their nutraceutical effects, lutein and zeaxanthin acquired from the diet must first be released and then absorbed from food matrices before being transported into circulation [56,206,209]. The bioavailability of these dietary xanthophylls in the blood has been shown to fluctuate greatly as a consequence of high-glycemic-index foods [205,206,210–213]. It is known that dietary behaviors such as those in the Western diet contribute significantly to the onset of metabolic syndrome and may also contribute to MPOD depletion in DR. Thus, metabolic perturbations typically present in patients with T2DM or poorly controlled T1DM, such as obesity, dyslipidemia, insulin deficiency and hyperglycemia are believed to substantially compromise the bioavailability and assimilation of dietary lutein and zeaxanthin to the retina [55,189–195,206]. The bioavailability of dietary carotenoids is also strongly influenced by age, gender, and racial/ethnic origin, in addition to these anthropometric measures [55,112,175,176,189,190,192–195,206].

While there are no established recommendations currently regarding daily intake levels of lutein and zeaxanthin consumption, oral supplementation with these carotenoids has a relatively high safety profile, with low risk for adverse effects and are appropriately considered by the US Food and Drug Administration to be Generally Regarded as Safe (GRAS) [214]. Large-scale epidemiological studies are needed to elucidate the putative role of dietary carotenoid intake and risk of DR along with disease progression among cohorts of both type 1 and type 2 diabetic patients. To this point, population data in healthy individuals on dietary intake levels of lutein and zeaxanthin is fairly limited and likely varies significantly among populations based on their dietary behaviors, as mentioned previously [28]. However, one may speculate that individuals whose diet primarily consist of foods rich in refined carbohydrates and artificially sweetened beverages containing high-fructose corn syrup, such as those with T2DM or poorly controlled T1DM for instance, are likely to have significantly lower levels of daily carotenoid intake when compared to those following a Mediterranean-style diet [28,205,215–217]. This may be explained, at least in part, by the disparities in regular consumption of various functional food groups (i.e., fresh fruit, nuts, leafy vegetables, and unrefined cereals), of which, several possess relatively high concentrations of lutein and zeaxanthin content per serving (Table 5) [27,28,44,215,218,219]. Based on the available evidence, it remains unclear whether relying solely upon dietary consumption of these carotenoid-rich food is sufficient to achieve the neuroprotective benefits with greater MPOD levels observed in patients with type 1 and type 2 DM following the use of carotenoid vitamin therapy.

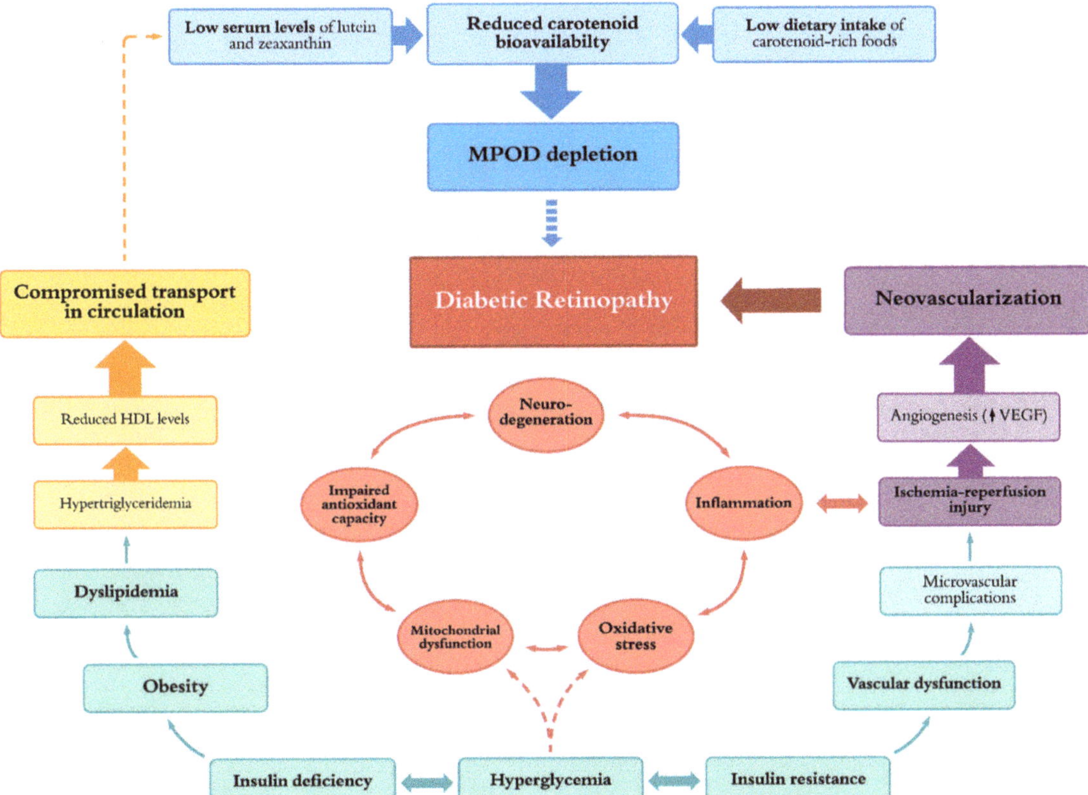

Figure 5. Schematic overview of proposed causative mechanisms and metabolic perturbations implicated in diabetic retinopathy. MPOD, macular pigment optical density; HDL, high-density lipoprotein; VEGF, vascular endothelial growth factor.

It is important to note that these clinically meaningful benefits in diabetic patients with or without DR were independent of any changes in hyperglycemic status or in relation to blood glucose control. Moreover, based on these results, there is a considerable body of preliminary evidence to substantiate the neuroprotective capacity of macular carotenoids to inhibit or reverse disease progression by ameliorating the metabolic correlates and comorbidities often seen in patients with type 2 or poorly controlled type 1 diabetes. Encouraging results from early interventional studies offer scientific justification for renewed clinical trials thereby corroborating the potential use of carotenoid vitamin therapy as an adjunctive therapeutic approach in the management of diabetic retinopathy for patients with either type 1 or type 2 diabetes.

Table 5. Common dietary sources of xanthophylls lutein and zeaxanthin [218,219].

Foods	Serving Size	Lutein + Zeaxanthin Content (mg)
Spinach, frozen (cooked)	1 cup	29.8
Kale, frozen (cooked)	1 cup	25.6
Swiss chard (cooked)	1 cup	11.0
Collard greens, frozen (cooked)	1 cup	8.9
Summer squash (cooked)	1 cup	4.0
Peas, frozen (cooked)	1 cup	3.8
Brussel sprouts, frozen (cooked)	1 cup	2.4
Broccoli, frozen (cooked)	1 cup	2.0
Edamame, frozen	1 cup	1.6
Sweet yellow corn (boiled)	1 cup	1.5
Asparagus (boiled)	0.5 cup	0.7
Avocado, raw	1 medium-size	0.4
Egg yolk, raw	1 large	0.2

However, there are several limitations currently that must be addressed in future clinical studies should carotenoid supplementation be used for this purpose. First, there is a growing need for further studies to investigate the potential implications associated with long-term use of adjunctive carotenoid vitamin therapy in larger cohorts of individuals with T1DM and T2DM. Second, additional randomized placebo-controlled trials are needed to determine the optimal dosage of lutein and zeaxanthin necessary to achieve clinically meaningful benefits, in addition to whether all three xanthophyll carotenoids found in the retina should be included in formulation. A recent systematic review in healthy adult eyes, found that lutein and zeaxanthin intake of less than 5 mg per day (by oral supplement or food sources) was insufficient dosage to significantly raise MPOD levels during trials up to six months [220]. Additionally, there have been no clinical trials investigating the effects of oral supplementation with *meso*-zeaxanthin in diabetic patients with or without DR to date. Further investigations are required to better understand if the addition of *meso*-zeaxanthin in combination with lutein and zeaxanthin may offer greater benefit or ascertain whether formulations with the two dietary xanthophylls are sufficient to elicit protective effect in diabetic retina. One of the limitations of this systematic review is that the number of databases searched was limited to National Library of Medicine and Web of Science. Additionally, the articles evaluated were limited to those published in the English language.

Lastly, given the systemic etiopathogenesis of diabetes which can manifest in the eye as vascular endotheliopathy, future strategies may focus on ameliorating early microvasculature complications such as retinal vascular occlusion as a consequence of capillary nonperfusion. While experimental models have shown that lutein and zeaxanthin offer protection against retinal capillary degeneration triggered by ischemic-reperfusion injury, it is unclear whether these xanthophylls can prevent microvasculature alterations which ultimately lead to vascular dysregulation. However, oral supplementation with a similar xanthophyll carotenoid known as astaxanthin has been shown in healthy adults has shown to exert benefits on retinal hemodynamic measures including capillary blood flow and velocity of choroidal circulation [221–224]. Given that astaxanthins retinal uptake has not been clearly demonstrated, its similar neuroprotective properties comparable to those of lutein and zeaxanthin provide scientific rationale for including astaxanthin into carotenoid vitamin therapy formulations in future nutraceutical trials of diabetic retinopathy [221–224].

5. Conclusions

Substantial efforts are necessary in developing early prophylactic measures that offer synergistic protection against several pathogenic mechanisms contributing to retinal neurodegeneration and subsequently preventing irreversible vision loss. To this accord, there is robust preclinical evidence and at least early clinical trials supporting the potential use of carotenoid vitamin supplementation in diabetics with and without retinopathy. Chronic hyperglycemia significantly compromises the endogenous defense systems in a diabetic individual. The metabolic changes due to diabetes possibly lead to depletion of macular carotenoids lutein, zeaxanthin, and *meso*-zeaxanthin, in addition to other potent antioxidants that are pertinent for maintaining retinal health as seen in various observational studies. MPOD measurements may also have a role to play in screening high-risk individuals prior to overt changes in retina due to DR pathology. Further randomized placebo-controlled trials are needed to support and solidify its use more universally as a first line of defense in combination with routine systemic management of diabetes and in susceptible individuals that are at risk of diabetes or pre-diabetics.

Author Contributions: Conceptualization, D.W.L. and P.G.D.; methodology, D.W.L. and P.G.D.; data curation, D.W.L.; writing—original draft preparation, D.W.L. and P.G.D.; writing—review and editing, D.W.L., D.L.G. and P.G.D. All authors have read and agreed to the published version of the manuscript.

Funding: This research received no external funding.

Acknowledgments: The authors would like to thank the Assessment and Public Relations Librarian.

Conflicts of Interest: Drake W. Lem, none. Pinakin Davey is a consultant and has received research grants from ZeaVision and Guardion Health Sciences for his prior research not related to current manuscript. Dennis L. Gierhart is an Employee, Chief Scientific Officer for ZeaVision manufacturer of various nutritional supplements including the DVS.

References

1. Centers for Disease Control and Prevention. *National Diabetes Statistics Report 2020*; Centers for Disease Control and Prevention, US Department of Health and Human Services: Atlanta, GA, USA, 2020; pp. 12–15.
2. Saeedi, P.; Petersohn, I.; Salpea, P.; Malanda, B.; Karuranga, S.; Unwin, N.; Colagiuri, S.; Guariguata, L.; Motala, A.A.; Ogurtsova, K.; et al. Global and regional diabetes prevalence estimates for 2019 and projections for 2030 and 2045: Results from the International Diabetes Federation Diabetes Atlas, 9th edition. *Diabetes Res. Clin. Pract.* **2019**, *157*, 107843. [CrossRef]
3. Lin, X.; Xu, Y.; Pan, X.; Xu, J.; Ding, Y.; Sun, X.; Song, X.; Ren, Y.; Shan, P.F. Global, regional, and national burden and trend of diabetes in 195 countries and territories: An analysis from 1990 to 2025. *Sci. Rep.* **2020**, *10*, 14790. [CrossRef] [PubMed]
4. Wong, T.Y.; Sabanayagam, C. Strategies to Tackle the Global Burden of Diabetic Retinopathy: From Epidemiology to Artificial Intelligence. *Ophthalmologica* **2020**, *243*, 9–20. [CrossRef]
5. Lee, R.; Wong, T.Y.; Sabanayagam, C. Epidemiology of diabetic retinopathy, diabetic macular edema and related vision loss. *Eye Vis.* **2015**, *2*, 17. [CrossRef] [PubMed]
6. Pearce, I.; Simo, R.; Lovestam-Adrian, M.; Wong, D.T.; Evans, M. Association between diabetic eye disease and other complications of diabetes: Implications for care. A systematic review. *Diabetes Obes. Metab.* **2019**, *21*, 467–478. [CrossRef]
7. Simo-Servat, O.; Hernandez, C.; Simo, R. Diabetic Retinopathy in the Context of Patients with Diabetes. *Ophthalmic Res.* **2019**, *62*, 211–217. [CrossRef] [PubMed]
8. He, F.; Xia, X.; Wu, X.F.; Yu, X.Q.; Huang, F.X. Diabetic retinopathy in predicting diabetic nephropathy in patients with type 2 diabetes and renal disease: A meta-analysis. *Diabetologia* **2013**, *56*, 457–466. [CrossRef] [PubMed]
9. Kawasaki, R.; Tanaka, S.; Tanaka, S.; Abe, S.; Sone, H.; Yokote, K.; Ishibashi, S.; Katayama, S.; Ohashi, Y.; Akanuma, Y.; et al. Risk of cardiovascular diseases is increased even with mild diabetic retinopathy: The Japan Diabetes Complications Study. *Ophthalmology* **2013**, *120*, 574–582. [CrossRef]
10. Tong, L.; Vernon, S.A.; Kiel, W.; Sung, V.; Orr, G.M. Association of macular involvement with proliferative retinopathy in Type 2 diabetes. *Diabet. Med.* **2001**, *18*, 388–394. [CrossRef]
11. Antonetti, D.A.; Klein, R.; Gardner, T.W. Diabetic retinopathy. *N. Engl. J. Med.* **2012**, *366*, 1227–1239. [CrossRef] [PubMed]
12. Cheloni, R.; Gandolfi, S.A.; Signorelli, C.; Odone, A. Global prevalence of diabetic retinopathy: Protocol for a systematic review and meta-analysis. *BMJ Open* **2019**, *9*, e022188. [CrossRef] [PubMed]
13. Solomon, S.D.; Chew, E.; Duh, E.J.; Sobrin, L.; Sun, J.K.; VanderBeek, B.L.; Wykoff, C.C.; Gardner, T.W. Diabetic Retinopathy: A Position Statement by the American Diabetes Association. *Diabetes Care* **2017**, *40*, 412–418. [CrossRef] [PubMed]

14. World Health Organization. World Report on Vision 2019. 2020. Available online: https://www.who.int/publications/i/item/world-report-on-vision (accessed on 15 December 2020).

15. International Diabetes Federation. IDF Diabetes Atlas 9th Edition 2019. Available online: https://www.diabetesatlas.org/en/ (accessed on 12 December 2020).

16. Wu, L.; Fernandez-Loaiza, P.; Sauma, J.; Hernandez-Bogantes, E.; Masis, M. Classification of diabetic retinopathy and diabetic macular edema. *World J. Diabetes* **2013**, *4*, 290–294. [CrossRef]

17. Cheung, N.; Mitchell, P.; Wong, T.Y. Diabetic retinopathy. *Lancet* **2010**, *376*, 124–136. [CrossRef]

18. Duh, E.J.; Sun, J.K.; Stitt, A.W. Diabetic retinopathy: Current understanding, mechanisms, and treatment strategies. *JCI Insight* **2017**, *2*. [CrossRef]

19. Scanlon, G.; Connell, P.; Ratzlaff, M.; Foerg, B.; McCartney, D.; Murphy, A.; O'Connor, K.; Loughman, J. Macular Pigment Optical Density Is Lower in Type 2 Diabetes, Compared with Type 1 Diabetes and Normal Controls. *Retina* **2015**, *35*, 1808–1816. [CrossRef] [PubMed]

20. Kowluru, R.A.; Chan, P.S. Oxidative stress and diabetic retinopathy. *Exp. Diabetes Res.* **2007**, *2007*, 43603. [CrossRef]

21. Simo, R.; Stitt, A.W.; Gardner, T.W. Neurodegeneration in diabetic retinopathy: Does it really matter? *Diabetologia* **2018**, *61*, 1902–1912. [CrossRef]

22. Simo, R.; Hernandez, C.; European Consortium for the Early Treatment of Diabetic Retinopathy. Neurodegeneration in the diabetic eye: New insights and therapeutic perspectives. *Trends Endocrinol. Metab.* **2014**, *25*, 23–33. [CrossRef]

23. Bek, T. Diameter Changes of Retinal Vessels in Diabetic Retinopathy. *Curr. Diab. Rep.* **2017**, *17*, 82. [CrossRef]

24. Gardner, T.W.; Davila, J.R. The neurovascular unit and the pathophysiologic basis of diabetic retinopathy. *Graefes Arch. Clin. Exp. Ophthalmol.* **2017**, *255*, 1–6. [CrossRef]

25. Scanlon, G.; McCartney, D.; Butler, J.S.; Loskutova, E.; Loughman, J. Identification of Surrogate Biomarkers for the Prediction of Patients at Risk of Low Macular Pigment in Type 2 Diabetes. *Curr. Eye Res.* **2019**, *44*, 1369–1380. [CrossRef] [PubMed]

26. Georgieva, E.; Ivanova, D.; Zhelev, Z.; Bakalova, R.; Gulubova, M.; Aoki, I. Mitochondrial Dysfunction and Redox Imbalance as a Diagnostic Marker of "Free Radical Diseases". *Anticancer Res.* **2017**, *37*, 5373–5381. [CrossRef]

27. Bernstein, P.S.; Li, B.; Vachali, P.P.; Gorusupudi, A.; Shyam, R.; Henriksen, B.S.; Nolan, J.M. Lutein, zeaxanthin, and meso-zeaxanthin: The basic and clinical science underlying carotenoid-based nutritional interventions against ocular disease. *Prog. Retin. Eye Res.* **2016**, *50*, 34–66. [CrossRef]

28. Eisenhauer, B.; Natoli, S.; Liew, G.; Flood, V.M. Lutein and Zeaxanthin-Food Sources, Bioavailability and Dietary Variety in Age-Related Macular Degeneration Protection. *Nutrients* **2017**, *9*, 120. [CrossRef]

29. Maoka, T. Carotenoids as natural functional pigments. *J. Nat. Med.* **2020**, *74*, 1–16. [CrossRef]

30. Rao, A.V.; Rao, L.G. Carotenoids and human health. *Pharmacol. Res.* **2007**, *55*, 207–216. [CrossRef] [PubMed]

31. Arunkumar, R.; Gorusupudi, A.; Bernstein, P.S. The macular carotenoids: A biochemical overview. *Biochim. Biophys. Acta Mol. Cell Biol. Lipids* **2020**, *1865*, 158617. [CrossRef]

32. Age-Related Eye Disease Study 2 Research Group. Lutein + zeaxanthin and omega-3 fatty acids for age-related macular degeneration: The Age-Related Eye Disease Study 2 (AREDS2) randomized clinical trial. *JAMA* **2013**, *309*, 2005–2015. [CrossRef]

33. SanGiovanni, J.P.; Chew, E.Y.; Clemons, T.E.; Ferris, F.L., 3rd; Gensler, G.; Lindblad, A.S.; Milton, R.C.; Seddon, J.M.; Sperduto, R.D. The relationship of dietary carotenoid and vitamin A, E, and C intake with age-related macular degeneration in a case-control study: AREDS Report No. 22. *Arch. Ophthalmol.* **2007**, *125*, 1225–1232. [CrossRef]

34. Chous, A.P.; Richer, S.P.; Gerson, J.D.; Kowluru, R.A. The Diabetes Visual Function Supplement Study (DiVFuSS). *Br. J. Ophthalmol.* **2016**, *100*, 227–234. [CrossRef]

35. Davey, P.G.; Henderson, T.; Lem, D.W.; Weis, R.; Amonoo-Monney, S.; Evans, D.W. Visual Function and Macular Carotenoid Changes in Eyes with Retinal Drusen-An Open Label Randomized Controlled Trial to Compare a Micronized Lipid-Based Carotenoid Liquid Supplementation and AREDS-2 Formula. *Nutrients* **2020**, *12*, 3271. [CrossRef]

36. Berco, E.; Rappoport, D.; Pollack, A.; Kleinmann, G.; Greenwald, Y. Management of Diabetic Retinopathy and Other Ocular Complications in Type 1 Diabetes. In *Major Topics in Type 1 Diabetes*; IntechOpen: London, UK, 2015.

37. Wilkinson, C.P.; Ferris, F.L., 3rd; Klein, R.E.; Lee, P.P.; Agardh, C.D.; Davis, M.; Dills, D.; Kampik, A.; Pararajasegaram, R.; Verdaguer, J.T.; et al. Proposed international clinical diabetic retinopathy and diabetic macular edema disease severity scales. *Ophthalmology* **2003**, *110*, 1677–1682. [CrossRef]

38. Yonekawa, Y.; Modi, Y.S.; Kim, L.A.; Skondra, D.; Kim, J.E.; Wykoff, C.C. American Society of Retina Specialists Clinical Practice Guidelines: Management of Nonproliferative and Proliferative Diabetic Retinopathy Without Diabetic Macular Edema. *J. Vitr. Dis.* **2020**, *4*, 125–135. [CrossRef]

39. Early Treatment Diabetic Retinopathy Study Research Group. Grading diabetic retinopathy from stereoscopic color fundus photographs–an extension of the modified Airlie House classification. ETDRS report number 10. Early Treatment Diabetic Retinopathy Study Research Group. *Ophthalmology* **1991**, *98*, 786–806.

40. Ishibazawa, A.; De Pretto, L.R.; Alibhai, A.Y.; Moult, E.M.; Arya, M.; Sorour, O.; Mehta, N.; Baumal, C.R.; Witkin, A.J.; Yoshida, A.; et al. Retinal Nonperfusion Relationship to Arteries or Veins Observed on Widefield Optical Coherence Tomography Angiography in Diabetic Retinopathy. *Investig. Ophthalmol. Vis. Sci.* **2019**, *60*, 4310–4318. [CrossRef]

41. Acon, D.; Wu, L. Multimodal Imaging in Diabetic Macular Edema. *Asia Pac. J. Ophthalmol.* **2018**, *7*, 22–27. [CrossRef]

42. You, Q.S.; Bartsch, D.U.; Espina, M.; Alam, M.; Camacho, N.; Mendoza, N.; Freeman, W.R. Reproducibility of Macular Pigment Optical Density Measurement by Two-Wavelength Autofluorescence in a Clinical Setting. *Retina* **2016**, *36*, 1381–1387. [CrossRef]

43. Bernstein, P.S.; Delori, F.C.; Richer, S.; van Kuijk, F.J.; Wenzel, A.J. The value of measurement of macular carotenoid pigment optical densities and distributions in age-related macular degeneration and other retinal disorders. *Vis. Res.* **2010**, *50*, 716–728. [CrossRef]

44. Gruszecki, W.I.; Sielewiesiuk, J. Orientation of xanthophylls in phosphatidylcholine multibilayers. *Biochim. Biophys. Acta* **1990**, *1023*, 405–412. [CrossRef]

45. de Kinkelder, R.; van der Veen, R.L.; Verbaak, F.D.; Faber, D.J.; van Leeuwen, T.G.; Berendschot, T.T. Macular pigment optical density measurements: Evaluation of a device using heterochromatic flicker photometry. *Eye* **2011**, *25*, 105–112. [CrossRef] [PubMed]

46. Leung, I.Y. Macular pigment: New clinical methods of detection and the role of carotenoids in age-related macular degeneration. *Optometry* **2008**, *79*, 266–272. [CrossRef] [PubMed]

47. Li, B.; George, E.W.; Rognon, G.T.; Gorusupudi, A.; Ranganathan, A.; Chang, F.Y.; Shi, L.; Frederick, J.M.; Bernstein, P.S. Imaging lutein and zeaxanthin in the human retina with confocal resonance Raman microscopy. *Proc. Natl. Acad. Sci. USA* **2020**, *117*, 12352–12358. [CrossRef] [PubMed]

48. Bone, R.A.; Landrum, J.T.; Hime, G.W.; Cains, A.; Zamor, J. Stereochemistry of the human macular carotenoids. *Investig. Ophthalmol. Vis. Sci.* **1993**, *34*, 2033–2040.

49. Scripsema, N.K.; Hu, D.N.; Rosen, R.B. Lutein, Zeaxanthin, and meso-Zeaxanthin in the Clinical Management of Eye Disease. *J. Ophthalmol.* **2015**, *2015*, 865179. [CrossRef]

50. Howells, O.; Eperjesi, F.; Bartlett, H. Measuring macular pigment optical density in vivo: A review of techniques. *Graefes Arch. Clin. Exp. Ophthalmol.* **2011**, *249*, 315–347. [CrossRef]

51. Shyam, R.; Gorusupudi, A.; Nelson, K.; Horvath, M.P.; Bernstein, P.S. RPE65 has an additional function as the lutein to meso-zeaxanthin isomerase in the vertebrate eye. *Proc. Natl. Acad. Sci. USA* **2017**, *114*, 10882–10887. [CrossRef]

52. Bone, R.A.; Landrum, J.T.; Mayne, S.T.; Gomez, C.M.; Tibor, S.E.; Twaroska, E.E. Macular pigment in donor eyes with and without AMD: A case-control study. *Investig. Ophthalmol. Vis. Sci.* **2001**, *42*, 235–240.

53. Gorusupudi, A.; Shyam, R.; Li, B.; Vachali, P.; Subhani, Y.K.; Nelson, K.; Bernstein, P.S. Developmentally Regulated Production of meso-Zeaxanthin in Chicken Retinal Pigment Epithelium/Choroid and Retina. *Investig. Ophthalmol. Vis. Sci.* **2016**, *57*, 1853–1861. [CrossRef]

54. Khachik, F.; de Moura, F.F.; Zhao, D.Y.; Aebischer, C.P.; Bernstein, P.S. Transformations of selected carotenoids in plasma, liver, and ocular tissues of humans and in nonprimate animal models. *Investig. Ophthalmol. Vis. Sci.* **2002**, *43*, 3383–3392.

55. Connor, W.E.; Duell, P.B.; Kean, R.; Wang, Y. The prime role of HDL to transport lutein into the retina: Evidence from HDL-deficient WHAM chicks having a mutant ABCA1 transporter. *Investig. Ophthalmol. Vis. Sci.* **2007**, *48*, 4226–4231. [CrossRef] [PubMed]

56. Nagao, A. Absorption and metabolism of dietary carotenoids. *BioFactors* **2011**, *37*, 83–87. [CrossRef] [PubMed]

57. Bone, R.A.; Davey, P.G.; Roman, B.O.; Evans, D.W. Efficacy of Commercially Available Nutritional Supplements: Analysis of Serum Uptake, Macular Pigment Optical Density and Visual Functional Response. *Nutrients* **2020**, *12*, 1321. [CrossRef] [PubMed]

58. Landrum, J.T.; Bone, R.A.; Joa, H.; Kilburn, M.D.; Moore, L.L.; Sprague, K.E. A one year study of the macular pigment: The effect of 140 days of a lutein supplement. *Exp. Eye Res.* **1997**, *65*, 57–62. [CrossRef] [PubMed]

59. Bone, R.A.; Landrum, J.T.; Cao, Y.; Howard, A.N.; Alvarez-Calderon, F. Macular pigment response to a supplement containing meso-zeaxanthin, lutein and zeaxanthin. *Nutr. Metab.* **2007**, *4*, 12. [CrossRef]

60. Akuffo, K.O.; Beatty, S.; Stack, J.; Dennison, J.; O'Regan, S.; Meagher, K.A.; Peto, T.; Nolan, J. Central Retinal Enrichment Supplementation Trials (CREST): Design and methodology of the CREST randomized controlled trials. *Ophthalmic Epidemiol.* **2014**, *21*, 111–123. [CrossRef]

61. Berendschot, T.T.; Goldbohm, R.A.; Klopping, W.A.; van de Kraats, J.; van Norel, J.; van Norren, D. Influence of lutein supplementation on macular pigment, assessed with two objective techniques. *Investig. Ophthalmol. Vis. Sci.* **2000**, *41*, 3322–3326.

62. Huang, Y.M.; Dou, H.L.; Huang, F.F.; Xu, X.R.; Zou, Z.Y.; Lin, X.M. Effect of supplemental lutein and zeaxanthin on serum, macular pigmentation, and visual performance in patients with early age-related macular degeneration. *Biomed. Res. Int.* **2015**, *2015*, 564738. [CrossRef]

63. Khachik, F.; de Moura, F.F.; Chew, E.Y.; Douglass, L.W.; Ferris, F.L., 3rd; Kim, J.; Thompson, D.J. The effect of lutein and zeaxanthin supplementation on metabolites of these carotenoids in the serum of persons aged 60 or older. *Investig. Ophthalmol. Vis. Sci.* **2006**, *47*, 5234–5242. [CrossRef]

64. Koh, H.H.; Murray, I.J.; Nolan, D.; Carden, D.; Feather, J.; Beatty, S. Plasma and macular responses to lutein supplement in subjects with and without age-related maculopathy: A pilot study. *Exp. Eye Res.* **2004**, *79*, 21–27. [CrossRef]

65. Ma, L.; Yan, S.F.; Huang, Y.M.; Lu, X.R.; Qian, F.; Pang, H.L.; Xu, X.R.; Zou, Z.Y.; Dong, P.C.; Xiao, X.; et al. Effect of lutein and zeaxanthin on macular pigment and visual function in patients with early age-related macular degeneration. *Ophthalmology* **2012**, *119*, 2290–2297. [CrossRef]

66. Richer, S.; Devenport, J.; Lang, J.C. LAST II: Differential temporal responses of macular pigment optical density in patients with atrophic age-related macular degeneration to dietary supplementation with xanthophylls. *Optometry* **2007**, *78*, 213–219. [CrossRef]

67. Trieschmann, M.; Beatty, S.; Nolan, J.M.; Hense, H.W.; Heimes, B.; Austermann, U.; Fobker, M.; Pauleikhoff, D. Changes in macular pigment optical density and serum concentrations of its constituent carotenoids following supplemental lutein and zeaxanthin: The LUNA study. *Exp. Eye Res.* **2007**, *84*, 718–728. [CrossRef] [PubMed]
68. Weigert, G.; Kaya, S.; Pemp, B.; Sacu, S.; Lasta, M.; Werkmeister, R.M.; Dragostinoff, N.; Simader, C.; Garhofer, G.; Schmidt-Erfurth, U.; et al. Effects of lutein supplementation on macular pigment optical density and visual acuity in patients with age-related macular degeneration. *Investig. Ophthalmol. Vis. Sci.* **2011**, *52*, 8174–8178. [CrossRef]
69. Howells, O.; Eperjesi, F.; Bartlett, H. Improving the repeatability of heterochromatic flicker photometry for measurement of macular pigment optical density. *Graefes Arch. Clin. Exp. Ophthalmol.* **2013**, *251*, 871–880. [CrossRef]
70. Krinsky, N.I.; Johnson, E.J. Carotenoid actions and their relation to health and disease. *Mol. Asp. Med.* **2005**, *26*, 459–516. [CrossRef] [PubMed]
71. Junghans, A.; Sies, H.; Stahl, W. Macular pigments lutein and zeaxanthin as blue light filters studied in liposomes. *Arch. Biochem. Biophys.* **2001**, *391*, 160–164. [CrossRef]
72. Kijlstra, A.; Tian, Y.; Kelly, E.R.; Berendschot, T.T. Lutein: More than just a filter for blue light. *Prog. Retin. Eye Res.* **2012**, *31*, 303–315. [CrossRef] [PubMed]
73. Landrum, J.T.; Bone, R.A. *Mechanistic Evidence for Eye Disease and Carotenoids*; Krinsky, N.I., Mayne, S.T., Sies, H., Eds.; CRC Press: New York, NY, USA, 2004.
74. Li, S.Y.; Fu, Z.J.; Lo, A.C. Hypoxia-induced oxidative stress in ischemic retinopathy. *Oxid. Med. Cell Longev.* **2012**, *2012*, 426769. [CrossRef]
75. Suarez-Berumen, K.; Davey, P.G. Macular Pigments Optical Density: A Review of Techniques of Measurements and Factors Influencing their Levels. *JSM Ophthalmol.* **2014**, *3*, 4.
76. Bartlett, H.; Eperjesi, F. Apparent motion photometry: Evaluation and reliability of a novel method for the measurement of macular pigment. *Br. J. Ophthalmol.* **2011**, *95*, 662–665. [CrossRef] [PubMed]
77. Bartlett, H.; Howells, O.; Eperjesi, F. The role of macular pigment assessment in clinical practice: A review. *Clin. Exp. Optom.* **2010**, *93*, 300–308. [CrossRef]
78. Davey, P.G.; Alvarez, S.D.; Lee, J.Y. Macular pigment optical density: Repeatability, intereye correlation, and effect of ocular dominance. *Clin. Ophthalmol.* **2016**, *10*, 1671–1678. [CrossRef] [PubMed]
79. Moreland, J.D. Macular pigment assessment by motion photometry. *Arch. Biochem. Biophys.* **2004**, *430*, 143–148. [CrossRef]
80. Lee, B.B.; Martin, P.R.; Valberg, A. The physiological basis of heterochromatic flicker photometry demonstrated in the ganglion cells of the macaque retina. *J. Physiol.* **1988**, *404*, 323–347. [CrossRef] [PubMed]
81. Delori, F.C. Autofluorescence method to measure macular pigment optical densities fluorometry and autofluorescence imaging. *Arch. Biochem. Biophys.* **2004**, *430*, 156–162. [CrossRef]
82. Delori, F.C.; Goger, D.G.; Hammond, B.R.; Snodderly, D.M.; Burns, S.A. Macular pigment density measured by autofluorescence spectrometry: Comparison with reflectometry and heterochromatic flicker photometry. *J. Opt. Soc. Am. A Opt. Image Sci. Vis.* **2001**, *18*, 1212–1230. [CrossRef] [PubMed]
83. Berendschot, T.T.; van Norren, D. Objective determination of the macular pigment optical density using fundus reflectance spectroscopy. *Arch. Biochem. Biophys.* **2004**, *430*, 149–155. [CrossRef]
84. Bernstein, P.S.; Yoshida, M.D.; Katz, N.B.; McClane, R.W.; Gellermann, W. Raman detection of macular carotenoid pigments in intact human retina. *Investig. Ophthalmol. Vis. Sci.* **1998**, *39*, 2003–2011.
85. Bartlett, H.; Stainer, L.; Singh, S.; Eperjesi, F.; Howells, O. Clinical evaluation of the MPS 9000 Macular Pigment Screener. *Br. J. Ophthalmol.* **2010**, *94*, 753–756. [CrossRef]
86. Loughman, J.; Scanlon, G.; Nolan, J.M.; O'Dwyer, V.; Beatty, S. An evaluation of a novel instrument for measuring macular pigment optical density: The MPS 9000. *Acta Ophthalmol.* **2012**, *90*, e90–e97. [CrossRef] [PubMed]
87. van der Veen, R.L.; Berendschot, T.T.; Hendrikse, F.; Carden, D.; Makridaki, M.; Murray, I.J. A new desktop instrument for measuring macular pigment optical density based on a novel technique for setting flicker thresholds. *Ophthalmic Physiol. Opt.* **2009**, *29*, 127–137. [CrossRef] [PubMed]
88. Bone, R.A.; Mukherjee, A. Innovative Troxler-free Measurement of Macular Pigment and Lens Density with Correction of the Former for the Aging Lens. *J. Biomed. Opt.* **2013**, *18*, 9. [CrossRef]
89. Obana, A.; Gohto, Y.; Moriyama, T.; Seto, T.; Sasano, H.; Okazaki, S. Reliability of a commercially available heterochromatic flicker photometer, the MPS2, for measuring the macular pigment optical density of a Japanese population. *Jpn. J. Ophthalmol.* **2018**, *62*, 473–480. [CrossRef]
90. Davey, P.G.; Ngo, A.; Cross, J.; Gierhart, D.L. Macular Pigment Reflectometry: Development and evaluation of a novel clinical device for rapid objective assessment of the macular carotenoids. In *Ophthalmic Technologies Xxix*; Manns, F., Soderberg, P.G., Ho, A., Eds.; Spie-Int Soc Optical Engineering: Bellingham, DC, USA, 2019; Volume 10858.
91. Huang, H.; Guan, C.; Ng, D.S.; Liu, X.; Chen, H. Macular Pigment Optical Density Measured by a Single Wavelength Reflection Photometry with and without Mydriasis. *Curr. Eye Res.* **2019**, *44*, 324–328. [CrossRef]
92. Kilbride, P.E.; Alexander, K.R.; Fishman, M.; Fishman, G.A. Human macular pigment assessed by imaging fundus reflectometry. *Vis. Res.* **1989**, *29*, 663–674. [CrossRef]
93. Sanabria, J.C.; Bass, J.; Spors, F.; Gierhart, D.L.; Davey, P.G. Measurement of Carotenoids in Perifovea using the Macular Pigment Reflectometer. *J. Vis. Exp.* **2020**, 9. [CrossRef]

94. Dennison, J.L.; Stack, J.; Beatty, S.; Nolan, J.M. Concordance of macular pigment measurements obtained using customized heterochromatic flicker photometry, dual-wavelength autofluorescence, and single-wavelength reflectance. *Exp. Eye Res.* **2013**, *116*, 190–198. [CrossRef]

95. Bernstein, P.S.; Zhao, D.Y.; Sharifzadeh, M.; Ermakov, I.V.; Gellermann, W. Resonance Raman measurement of macular carotenoids in the living human eye. *Arch. Biochem. Biophys.* **2004**, *430*, 163–169. [CrossRef] [PubMed]

96. Gellermann, W.; Ermakov, I.V.; Ermakova, M.R.; McClane, R.W.; Zhao, D.Y.; Bernstein, P.S. In vivo resonant Raman measurement of macular carotenoid pigments in the young and the aging human retina. *J. Opt. Soc. Am. A Opt. Image Sci. Vis.* **2002**, *19*, 1172–1186. [CrossRef]

97. Lapierre-Landry, M.; Carroll, J.; Skala, M.C. Imaging retinal melanin: A review of current technologies. *J. Biol. Eng.* **2018**, *12*, 29. [CrossRef]

98. Yung, M.; Klufas, M.A.; Sarraf, D. Clinical applications of fundus autofluorescence in retinal disease. *Int. J. Retin. Vitr.* **2016**, *2*, 12. [CrossRef] [PubMed]

99. Moher, D.; Liberati, A.; Tetzlaff, J.; Altman, D.G.; Group, P. Preferred reporting items for systematic reviews and meta-analyses: The PRISMA statement. *J. Clin. Epidemiol.* **2009**, *62*, 1006–1012. [CrossRef] [PubMed]

100. Hooijmans, C.R.; Rovers, M.M.; de Vries, R.B.; Leenaars, M.; Ritskes-Hoitinga, M.; Langendam, M.W. SYRCLE's risk of bias tool for animal studies. *BMC Med. Res. Methodol.* **2014**, *14*, 43. [CrossRef]

101. Higgins, J.P.; Altman, D.G.; Gotzsche, P.C.; Juni, P.; Moher, D.; Oxman, A.D.; Savovic, J.; Schulz, K.F.; Weeks, L.; Sterne, J.A.; et al. The Cochrane Collaboration's tool for assessing risk of bias in randomised trials. *BMJ* **2011**, *343*, d5928. [CrossRef] [PubMed]

102. Arnal, E.; Miranda, M.; Johnsen-Soriano, S.; Alvarez-Nolting, R.; Diaz-Llopis, M.; Araiz, J.; Cervera, E.; Bosch-Morell, F.; Romero, F.J. Beneficial effect of docosahexanoic acid and lutein on retinal structural, metabolic, and functional abnormalities in diabetic rats. *Curr. Eye Res.* **2009**, *34*, 928–938. [CrossRef]

103. Kowluru, R.A.; Menon, B.; Gierhart, D.L. Beneficial effect of zeaxanthin on retinal metabolic abnormalities in diabetic rats. *Investig. Ophthalmol. Vis. Sci.* **2008**, *49*, 1645–1651. [CrossRef]

104. Kowluru, R.A.; Zhong, Q.; Santos, J.M.; Thandampallayam, M.; Putt, D.; Gierhart, D.L. Beneficial effects of the nutritional supplements on the development of diabetic retinopathy. *Nutr. Metab.* **2014**, *11*, 8. [CrossRef]

105. Muriach, M.; Bosch-Morell, F.; Alexander, G.; Blomhoff, R.; Barcia, J.; Arnal, E.; Almansa, I.; Romero, F.J.; Miranda, M. Lutein effect on retina and hippocampus of diabetic mice. *Free Radic. Biol. Med.* **2006**, *41*, 979–984. [CrossRef]

106. Sasaki, M.; Ozawa, Y.; Kurihara, T.; Kubota, S.; Yuki, K.; Noda, K.; Kobayashi, S.; Ishida, S.; Tsubota, K. Neurodegenerative influence of oxidative stress in the retina of a murine model of diabetes. *Diabetologia* **2010**, *53*, 971–979. [CrossRef]

107. Tang, L.; Zhang, Y.; Jiang, Y.; Willard, L.; Ortiz, E.; Wark, L.; Medeiros, D.; Lin, D. Dietary wolfberry ameliorates retinal structure abnormalities in db/db mice at the early stage of diabetes. *Exp. Biol. Med.* **2011**, *236*, 1051–1063. [CrossRef]

108. Yu, H.; Wark, L.; Ji, H.; Willard, L.; Jaing, Y.; Han, J.; He, H.; Ortiz, E.; Zhang, Y.; Medeiros, D.M.; et al. Dietary wolfberry upregulates carotenoid metabolic genes and enhances mitochondrial biogenesis in the retina of db/db diabetic mice. *Mol. Nutr. Food Res.* **2013**, *57*, 1158–1169. [CrossRef]

109. Brazionis, L.; Rowley, K.; Itsiopoulos, C.; O'Dea, K. Plasma carotenoids and diabetic retinopathy. *Br. J. Nutr.* **2009**, *101*, 270–277. [CrossRef] [PubMed]

110. Cennamo, G.; Lanni, V.; Abbate, R.; Velotti, N.; Fossataro, F.; Sparnelli, F.; Romano, M.R.; de Crecchio, G.; Cennamo, G. The Relationship between Macular Pigment and Vessel Density in Patients with Type 1 Diabetes Mellitus. *Ophthalmic Res.* **2019**, *61*, 19–25. [CrossRef]

111. Davies, N.P.; Morland, A.B. Color matching in diabetes: Optical density of the crystalline lens and macular pigments. *Investig. Ophthalmol. Vis. Sci.* **2002**, *43*, 281–289.

112. Lima, V.C.; Rosen, R.B.; Maia, M.; Prata, T.S.; Dorairaj, S.; Farah, M.E.; Sallum, J. Macular pigment optical density measured by dual-wavelength autofluorescence imaging in diabetic and nondiabetic patients: A comparative study. *Investig. Ophthalmol. Vis. Sci.* **2010**, *51*, 5840–5845. [CrossRef]

113. Mares, J.A.; LaRowe, T.L.; Snodderly, D.M.; Moeller, S.M.; Gruber, M.J.; Klein, M.L.; Wooten, B.R.; Johnson, E.J.; Chappell, R.J.; Group, C.M.P.S.; et al. Predictors of optical density of lutein and zeaxanthin in retinas of older women in the Carotenoids in Age-Related Eye Disease Study, an ancillary study of the Women's Health Initiative. *Am. J. Clin. Nutr.* **2006**, *84*, 1107–1122. [CrossRef]

114. She, C.Y.; Gu, H.; Xu, J.; Yang, X.F.; Ren, X.T.; Liu, N.P. Association of macular pigment optical density with early stage of non-proliferative diabetic retinopathy in Chinese patients with type 2 diabetes mellitus. *Int. J. Ophthalmol.* **2016**, *9*, 1433–1438. [CrossRef] [PubMed]

115. Zagers, N.P.; Pot, M.C.; van Norren, D. Spectral and directional reflectance of the fovea in diabetes mellitus: Photoreceptor integrity, macular pigment and lens. *Vis. Res.* **2005**, *45*, 1745–1753. [CrossRef]

116. Hu, B.J.; Hu, Y.N.; Lin, S.; Ma, W.J.; Li, X.R. Application of Lutein and Zeaxanthin in nonproliferative diabetic retinopathy. *Int. J. Ophthalmol.* **2011**, *4*, 303–306. [CrossRef] [PubMed]

117. Kowluru, R.A.; Kanwar, M.; Chan, P.S.; Zhang, J.P. Inhibition of retinopathy and retinal metabolic abnormalities in diabetic rats with AREDS-based micronutrients. *Arch. Ophthalmol.* **2008**, *126*, 1266–1272. [CrossRef] [PubMed]

118. Olivares, A.M.; Althoff, K.; Chen, G.F.; Wu, S.; Morrisson, M.A.; DeAngelis, M.M.; Haider, N. Animal Models of Diabetic Retinopathy. *Curr. Diab. Rep.* **2017**, *17*, 93. [CrossRef]

119. Rakieten, N.; Rakieten, M.L.; Nadkarni, M.R. Studies on the diabetogenic action of streptozotocin (NSC-37917). *Cancer Chemother. Rep.* **1963**, *29*, 91–98. [PubMed]

120. Robinson, R.; Barathi, V.A.; Chaurasia, S.S.; Wong, T.Y.; Kern, T.S. Update on animal models of diabetic retinopathy: From molecular approaches to mice and higher mammals. *Dis. Model. Mech.* **2012**, *5*, 444–456. [CrossRef] [PubMed]

121. Cobb, C.A.; Cole, M.P. Oxidative and nitrative stress in neurodegeneration. *Neurobiol. Dis.* **2015**, *84*, 4–21. [CrossRef]

122. Aslan, M.; Yucel, I.; Akar, Y.; Yucel, G.; Ciftcioglu, M.A.; Sanlioglu, S. Nitrotyrosine formation and apoptosis in rat models of ocular injury. *Free Radic. Res.* **2006**, *40*, 147–153. [CrossRef]

123. Toda, N.; Nakanishi-Toda, M. Nitric oxide: Ocular blood flow, glaucoma, and diabetic retinopathy. *Prog. Retin. Eye Res.* **2007**, *26*, 205–238. [CrossRef] [PubMed]

124. ZeaVision LLC. EyePromise DVS. Available online: http://www.eyepromise.com/doctors/products/eyepromise-dvs/?utm_source=Website&utm_medium=Blog-Post&utm_campaign=Doctor-Blog&utm_content=Diabetes-Awareness-Series-3-Sept2018 (accessed on 10 December 2020).

125. Kanwar, M.; Chan, P.S.; Kern, T.S.; Kowluru, R.A. Oxidative damage in the retinal mitochondria of diabetic mice: Possible protection by superoxide dismutase. *Investig. Ophthalmol. Vis. Sci.* **2007**, *48*, 3805–3811. [CrossRef]

126. Kowluru, R.A.; Kowluru, A.; Mishra, M.; Kumar, B. Oxidative stress and epigenetic modifications in the pathogenesis of diabetic retinopathy. *Prog. Retin. Eye Res.* **2015**, *48*, 40–61. [CrossRef]

127. Droge, W. Free radicals in the physiological control of cell function. *Physiol. Rev.* **2002**, *82*, 47–95. [CrossRef] [PubMed]

128. Halliwell, B. Biochemistry of oxidative stress. *Biochem. Soc. Trans.* **2007**, *35*, 1147–1150. [CrossRef]

129. Sena, L.A.; Chandel, N.S. Physiological roles of mitochondrial reactive oxygen species. *Mol. Cell* **2012**, *48*, 158–167. [CrossRef]

130. Kowluru, R.; Kern, T.S.; Engerman, R.L. Abnormalities of retinal metabolism in diabetes or galactosemia. II. Comparison of gamma-glutamyl transpeptidase in retina and cerebral cortex, and effects of antioxidant therapy. *Curr. Eye Res.* **1994**, *13*, 891–896. [CrossRef] [PubMed]

131. Kowluru, R.A.; Kern, T.S.; Engerman, R.L. Abnormalities of retinal metabolism in diabetes or experimental galactosemia. IV. Antioxidant defense system. *Free Radic. Biol. Med.* **1997**, *22*, 587–592. [CrossRef]

132. Neelam, K.; Goenadi, C.J.; Lun, K.; Yip, C.C.; Au Eong, K.G. Putative protective role of lutein and zeaxanthin in diabetic retinopathy. *Br. J. Ophthalmol.* **2017**, *101*, 551–558. [CrossRef] [PubMed]

133. Izumi-Nagai, K.; Nagai, N.; Ohgami, K.; Satofuka, S.; Ozawa, Y.; Tsubota, K.; Umezawa, K.; Ohno, S.; Oike, Y.; Ishida, S. Macular pigment lutein is antiinflammatory in preventing choroidal neovascularization. *Arterioscler. Thromb. Vasc. Biol.* **2007**, *27*, 2555–2562. [CrossRef]

134. Sarıkaya, E.; Doğan, S. Glutathione Peroxidase in Health and Diseases. In *Glutathione System and Oxidative Stress in Health and Disease*; IntechOpen: London, UK, 2020.

135. Bian, Q.; Gao, S.; Zhou, J.; Qin, J.; Taylor, A.; Johnson, E.J.; Tang, G.; Sparrow, J.R.; Gierhart, D.; Shang, F. Lutein and zeaxanthin supplementation reduces photooxidative damage and modulates the expression of inflammation-related genes in retinal pigment epithelial cells. *Free Radic. Biol. Med.* **2012**, *53*, 1298–1307. [CrossRef]

136. Li, S.Y.; Fung, F.K.; Fu, Z.J.; Wong, D.; Chan, H.H.; Lo, A.C. Anti-inflammatory effects of lutein in retinal ischemic/hypoxic injury: In vivo and in vitro studies. *Investig. Ophthalmol. Vis. Sci.* **2012**, *53*, 5976–5984. [CrossRef]

137. Tarr, J.M.; Kaul, K.; Chopra, M.; Kohner, E.M.; Chibber, R. Pathophysiology of diabetic retinopathy. *ISRN Ophthalmol.* **2013**, *2013*, 343560. [CrossRef] [PubMed]

138. Joussen, A.M.; Poulaki, V.; Le, M.L.; Koizumi, K.; Esser, C.; Janicki, H.; Schraermeyer, U.; Kociok, N.; Fauser, S.; Kirchhof, B.; et al. A central role for inflammation in the pathogenesis of diabetic retinopathy. *FASEB J.* **2004**, *18*, 1450–1452. [CrossRef]

139. Chan, P.S.; Kanwar, M.; Kowluru, R.A. Resistance of retinal inflammatory mediators to suppress after reinstitution of good glycemic control: Novel mechanism for metabolic memory. *J. Diabetes Complicat.* **2010**, *24*, 55–63. [CrossRef]

140. Kowluru, R.A.; Odenbach, S. Role of interleukin-1beta in the development of retinopathy in rats: Effect of antioxidants. *Investig. Ophthalmol. Vis. Sci.* **2004**, *45*, 4161–4166. [CrossRef]

141. Kowluru, R.A.; Koppolu, P.; Chakrabarti, S.; Chen, S. Diabetes-induced activation of nuclear transcriptional factor in the retina, and its inhibition by antioxidants. *Free Radic. Res.* **2003**, *37*, 1169–1180. [CrossRef]

142. Abcouwer, S.F. Angiogenic Factors and Cytokines in Diabetic Retinopathy. *J. Clin. Cell Immunol.* **2013**, *7* (Suppl. 1). [CrossRef]

143. Klein, B.E.; Knudtson, M.D.; Tsai, M.Y.; Klein, R. The relation of markers of inflammation and endothelial dysfunction to the prevalence and progression of diabetic retinopathy: Wisconsin epidemiologic study of diabetic retinopathy. *Arch. Ophthalmol.* **2009**, *127*, 1175–1182. [CrossRef] [PubMed]

144. Schroder, S.; Palinski, W.; Schmid-Schonbein, G.W. Activated monocytes and granulocytes, capillary nonperfusion, and neovascularization in diabetic retinopathy. *Am. J. Pathol.* **1991**, *139*, 81–100. [PubMed]

145. Kowluru, R.A.; Mohammad, G.; Santos, J.M.; Tewari, S.; Zhong, Q. Interleukin-1beta and mitochondria damage, and the development of diabetic retinopathy. *J. Ocul. Biol. Dis. Inform.* **2011**, *4*, 3–9. [CrossRef]

146. Feit-Leichman, R.A.; Kinouchi, R.; Takeda, M.; Fan, Z.; Mohr, S.; Kern, T.S.; Chen, D.F. Vascular damage in a mouse model of diabetic retinopathy: Relation to neuronal and glial changes. *Investig. Ophthalmol. Vis. Sci.* **2005**, *46*, 4281–4287. [CrossRef] [PubMed]

147. Kern, T.S. Interrelationships between the Retinal Neuroglia and Vasculature in Diabetes. *Diabetes Metab. J.* **2014**, *38*, 163–170. [CrossRef]

148. Aiello, L.P.; Wong, J.S. Role of vascular endothelial growth factor in diabetic vascular complications. *Kidney Int. Suppl.* **2000**, *77*, S113–S119. [CrossRef]
149. Kern, T.S.; Engerman, R.L. Pharmacological inhibition of diabetic retinopathy: Aminoguanidine and aspirin. *Diabetes* **2001**, *50*, 1636–1642. [CrossRef]
150. Kern, T.S.; Tang, J.; Mizutani, M.; Kowluru, R.A.; Nagaraj, R.H.; Romeo, G.; Podesta, F.; Lorenzi, M. Response of capillary cell death to aminoguanidine predicts the development of retinopathy: Comparison of diabetes and galactosemia. *Investig. Ophthalmol. Vis. Sci.* **2000**, *41*, 3972–3978.
151. Mizutani, M.; Kern, T.S.; Lorenzi, M. Accelerated death of retinal microvascular cells in human and experimental diabetic retinopathy. *J. Clin. Investig.* **1996**, *97*, 2883–2890. [CrossRef]
152. Curtis, T.M.; Gardiner, T.A.; Stitt, A.W. Microvascular lesions of diabetic retinopathy: Clues towards understanding pathogenesis? *Eye* **2009**, *23*, 1496–1508. [CrossRef]
153. Kohzaki, K.; Vingrys, A.J.; Bui, B.V. Early inner retinal dysfunction in streptozotocin-induced diabetic rats. *Investig. Ophthalmol. Vis. Sci.* **2008**, *49*, 3595–3604. [CrossRef]
154. Fung, F.K.; Law, B.Y.; Lo, A.C. Lutein Attenuates Both Apoptosis and Autophagy upon Cobalt (II) Chloride-Induced Hypoxia in Rat Muller Cells. *PLoS ONE* **2016**, *11*, e0167828. [CrossRef] [PubMed]
155. Li, S.Y.; Fu, Z.J.; Ma, H.; Jang, W.C.; So, K.F.; Wong, D.; Lo, A.C. Effect of lutein on retinal neurons and oxidative stress in a model of acute retinal ischemia/reperfusion. *Investig. Ophthalmol. Vis. Sci.* **2009**, *50*, 836–843. [CrossRef]
156. Dong, C.J.; Agey, P.; Hare, W.A. Origins of the electroretinogram oscillatory potentials in the rabbit retina. *Vis. Neurosci.* **2004**, *21*, 533–543. [CrossRef]
157. Hancock, H.A.; Kraft, T.W. Oscillatory potential analysis and ERGs of normal and diabetic rats. *Investig. Ophthalmol. Vis. Sci.* **2004**, *45*, 1002–1008. [CrossRef]
158. Kizawa, J.; Machida, S.; Kobayashi, T.; Gotoh, Y.; Kurosaka, D. Changes of oscillatory potentials and photopic negative response in patients with early diabetic retinopathy. *Jpn. J. Ophthalmol.* **2006**, *50*, 367–373. [CrossRef] [PubMed]
159. Yonemura, D.; Tsuzuki, K.; Aoki, T. Clinical importance of the oscillatory potential in the human ERG. *Acta Ophthalmol. Suppl.* **1962**, *40* (Suppl. 70), 115–123. [CrossRef]
160. Binder, D.K.; Scharfman, H.E. Brain-derived neurotrophic factor. *Growth Factors* **2004**, *22*, 123–131. [CrossRef]
161. Martin, P.M.; Roon, P.; Van Ells, T.K.; Ganapathy, V.; Smith, S.B. Death of retinal neurons in streptozotocin-induced diabetic mice. *Investig. Ophthalmol. Vis. Sci.* **2004**, *45*, 3330–3336. [CrossRef]
162. Kern, T.S.; Barber, A.J. Retinal ganglion cells in diabetes. *J. Physiol.* **2008**, *586*, 4401–4408. [CrossRef]
163. Seki, M.; Tanaka, T.; Nawa, H.; Usui, T.; Fukuchi, T.; Ikeda, K.; Abe, H.; Takei, N. Involvement of brain-derived neurotrophic factor in early retinal neuropathy of streptozotocin-induced diabetes in rats: Therapeutic potential of brain-derived neurotrophic factor for dopaminergic amacrine cells. *Diabetes* **2004**, *53*, 2412–2419. [CrossRef] [PubMed]
164. Xue, C.; Rosen, R.; Jordan, A.; Hu, D.N. Management of Ocular Diseases Using Lutein and Zeaxanthin: What Have We Learned from Experimental Animal Studies? *J. Ophthalmol.* **2015**, *2015*, 523027. [CrossRef]
165. Li, B.; Vachali, P.P.; Gorusupudi, A.; Shen, Z.; Sharifzadeh, H.; Besch, B.M.; Nelson, K.; Horvath, M.M.; Frederick, J.M.; Baehr, W.; et al. Inactivity of human beta,beta-carotene-9′,10′-dioxygenase (BCO2) underlies retinal accumulation of the human macular carotenoid pigment. *Proc. Natl. Acad. Sci. USA* **2014**, *111*, 10173–10178. [CrossRef] [PubMed]
166. Kowluru, R.A.; Zhong, Q. Beyond AREDS: Is there a place for antioxidant therapy in the prevention/treatment of eye disease? *Investig. Ophthalmol. Vis. Sci.* **2011**, *52*, 8665–8671. [CrossRef]
167. Daley, M.L.; Watzke, R.C.; Riddle, M.C. Early loss of blue-sensitive color vision in patients with type I diabetes. *Diabetes Care* **1987**, *10*, 777–781. [CrossRef] [PubMed]
168. Han, Y.; Adams, A.J.; Bearse, M.A., Jr.; Schneck, M.E. Multifocal electroretinogram and short-wavelength automated perimetry measures in diabetic eyes with little or no retinopathy. *Arch. Ophthalmol.* **2004**, *122*, 1809–1815. [CrossRef] [PubMed]
169. Jackson, G.R.; Barber, A.J. Visual dysfunction associated with diabetic retinopathy. *Curr. Diab. Rep.* **2010**, *10*, 380–384. [CrossRef]
170. Takahashi, H.; Chihara, E. Impact of diabetic retinopathy on quantitative retinal nerve fiber layer measurement and glaucoma screening. *Investig. Ophthalmol. Vis. Sci.* **2008**, *49*, 687–692. [CrossRef] [PubMed]
171. van Dijk, H.W.; Kok, P.H.; Garvin, M.; Sonka, M.; Devries, J.H.; Michels, R.P.; van Velthoven, M.E.; Schlingemann, R.O.; Verbraak, F.D.; Abramoff, M.D. Selective loss of inner retinal layer thickness in type 1 diabetic patients with minimal diabetic retinopathy. *Investig. Ophthalmol. Vis. Sci.* **2009**, *50*, 3404–3409. [CrossRef]
172. van Dijk, H.W.; Verbraak, F.D.; Kok, P.H.; Garvin, M.K.; Sonka, M.; Lee, K.; Devries, J.H.; Michels, R.P.; van Velthoven, M.E.; Schlingemann, R.O.; et al. Decreased retinal ganglion cell layer thickness in patients with type 1 diabetes. *Investig. Ophthalmol. Vis. Sci.* **2010**, *51*, 3660–3665. [CrossRef] [PubMed]
173. Greenstein, V.C.; Hood, D.C.; Ritch, R.; Steinberger, D.; Carr, R.E. S (blue) cone pathway vulnerability in retinitis pigmentosa, diabetes and glaucoma. *Investig. Ophthalmol. Vis. Sci.* **1989**, *30*, 1732–1737.
174. Johnson, E.J.; Avendano, E.E.; Mohn, E.S.; Raman, G. The association between macular pigment optical density and visual function outcomes: A systematic review and meta-analysis. *Eye* **2020**, *35*, 1620–1628. [CrossRef]
175. Lem, D.W.; Gierhart, D.L.; Davey, P.G. Carotenoids in the Management of Glaucoma: A Systematic Review of the Evidence. *Nutrients* **2021**, *13*, 1949. [CrossRef] [PubMed]

176. Lem, D.W.; Gierhart, D.L.; Gunvant Davey, P. Management of Diabetic Eye Disease using Carotenoids and Nutrients. In *Antioxidants-Benefits, Sources, and Mechanisms of Action*; Waisundara, V.Y., Ed.; IntechOpen: London, UK, 2021.

177. Liu, R.; Wang, T.; Zhang, B.; Qin, L.; Wu, C.; Li, Q.; Ma, L. Lutein and zeaxanthin supplementation and association with visual function in age-related macular degeneration. *Investig. Ophthalmol. Vis. Sci.* **2014**, *56*, 252–258. [CrossRef]

178. Loughman, J.; Akkali, M.C.; Beatty, S.; Scanlon, G.; Davison, P.A.; O'Dwyer, V.; Cantwell, T.; Major, P.; Stack, J.; Nolan, J.M. The relationship between macular pigment and visual performance. *Vis. Res.* **2010**, *50*, 1249–1256. [CrossRef]

179. Loughman, J.; Nolan, J.M.; Beatty, S. Impact of dietary carotenoid deprivation on macular pigment and serum concentrations of lutein and zeaxanthin. *Br. J. Nutr.* **2012**, *108*, 2102–2103. [CrossRef] [PubMed]

180. Loughman, J.; Nolan, J.M.; Howard, A.N.; Connolly, E.; Meagher, K.; Beatty, S. The impact of macular pigment augmentation on visual performance using different carotenoid formulations. *Investig. Ophthalmol. Vis. Sci.* **2012**, *53*, 7871–7880. [CrossRef] [PubMed]

181. Hammond, B.R., Jr.; Fletcher, L.M.; Elliott, J.G. Glare disability, photostress recovery, and chromatic contrast: Relation to macular pigment and serum lutein and zeaxanthin. *Investig. Ophthalmol. Vis. Sci.* **2013**, *54*, 476–481. [CrossRef] [PubMed]

182. Hammond, B.R.; Fletcher, L.M.; Roos, F.; Wittwer, J.; Schalch, W. A double-blind, placebo-controlled study on the effects of lutein and zeaxanthin on photostress recovery, glare disability, and chromatic contrast. *Investig. Ophthalmol. Vis. Sci.* **2014**, *55*, 8583–8589. [CrossRef] [PubMed]

183. Kleiner, R.C.; Enger, C.; Alexander, M.F.; Fine, S.L. Contrast sensitivity in age-related macular degeneration. *Arch. Ophthalmol.* **1988**, *106*, 55–57. [CrossRef]

184. Nolan, J.M.; Loughman, J.; Akkali, M.C.; Stack, J.; Scanlon, G.; Davison, P.; Beatty, S. The impact of macular pigment augmentation on visual performance in normal subjects: COMPASS. *Vis. Res.* **2011**, *51*, 459–469. [CrossRef] [PubMed]

185. Almasieh, M.; Wilson, A.M.; Morquette, B.; Cueva Vargas, J.L.; Di Polo, A. The molecular basis of retinal ganglion cell death in glaucoma. *Prog. Retin. Eye Res.* **2012**, *31*, 152–181. [CrossRef]

186. Pinazo-Durán, M.D.; Zanón-Moreno, V.; Gallego-Pinazo, R.; García-Medina, J.J. Chapter 6-Oxidative stress and mitochondrial failure in the pathogenesis of glaucoma neurodegeneration. In *Progress in Brain Research*; Bagetta, G., Nucci, C., Eds.; Elsevier: Amsterdam, The Netherlands, 2015; Volume 220, pp. 127–153.

187. Stringham, J.M.; O'Brien, K.J.; Stringham, N.T. Contrast Sensitivity and Lateral Inhibition Are Enhanced With Macular Carotenoid Supplementation. *Investig. Ophthalmol. Vis. Sci.* **2017**, *58*, 2291–2295. [CrossRef]

188. Jackson, G.R.; Scott, I.U.; Quillen, D.A.; Walter, L.E.; Gardner, T.W. Inner retinal visual dysfunction is a sensitive marker of non-proliferative diabetic retinopathy. *Br. J. Ophthalmol.* **2012**, *96*, 699–703. [CrossRef]

189. Johnson, E.J.; Hammond, B.R.; Yeum, K.J.; Qin, J.; Wang, X.D.; Castaneda, C.; Snodderly, D.M.; Russell, R.M. Relation among serum and tissue concentrations of lutein and zeaxanthin and macular pigment density. *Am. J. Clin. Nutr.* **2000**, *71*, 1555–1562. [CrossRef] [PubMed]

190. Hammond, B.R., Jr.; Ciulla, T.A.; Snodderly, D.M. Macular pigment density is reduced in obese subjects. *Investig. Ophthalmol. Vis. Sci.* **2002**, *43*, 47–50.

191. Thomson, L.R.; Toyoda, Y.; Langner, A.; Delori, F.C.; Garnett, K.M.; Craft, N.; Nichols, C.R.; Cheng, K.M.; Dorey, C.K. Elevated retinal zeaxanthin and prevention of light-induced photoreceptor cell death in quail. *Investig. Ophthalmol. Vis. Sci.* **2002**, *43*, 3538–3549. [CrossRef]

192. Nolan, J.; O'Donovan, O.; Kavanagh, H.; Stack, J.; Harrison, M.; Muldoon, A.; Mellerio, J.; Beatty, S. Macular pigment and percentage of body fat. *Investig. Ophthalmol. Vis. Sci.* **2004**, *45*, 3940–3950. [CrossRef]

193. Wang, W.; Connor, S.L.; Johnson, E.J.; Klein, M.L.; Hughes, S.; Connor, W.E. Effect of dietary lutein and zeaxanthin on plasma carotenoids and their transport in lipoproteins in age-related macular degeneration. *Am. J. Clin. Nutr.* **2007**, *85*, 762–769. [CrossRef]

194. Lim, C.; Kim, D.-w.; Sim, T.; Hoang, N.H.; Lee, J.W.; Lee, E.S.; Youn, Y.S.; Oh, K.T. Preparation and characterization of a lutein loading nanoemulsion system for ophthalmic eye drops. *J. Drug Deliv. Sci. Technol.* **2016**, *36*, 168–174. [CrossRef]

195. Broekmans, W.M.; Berendschot, T.T.; Klopping-Ketelaars, I.A.; de Vries, A.J.; Goldbohm, R.A.; Tijburg, L.B.; Kardinaal, A.F.; van Poppel, G. Macular pigment density in relation to serum and adipose tissue concentrations of lutein and serum concentrations of zeaxanthin. *Am. J. Clin. Nutr.* **2002**, *76*, 595–603. [CrossRef]

196. Loane, E.; Nolan, J.M.; Beatty, S. The respective relationships between lipoprotein profile, macular pigment optical density, and serum concentrations of lutein and zeaxanthin. *Investig. Ophthalmol. Vis. Sci.* **2010**, *51*, 5897–5905. [CrossRef]

197. Goldberg, I.J. Clinical review 124: Diabetic dyslipidemia: Causes and consequences. *J. Clin. Endocrinol. Metab.* **2001**, *86*, 965–971. [CrossRef] [PubMed]

198. Al-Aubaidy, H.A.; Jelinek, H.F. Oxidative DNA damage and obesity in type 2 diabetes mellitus. *Eur. J. Endocrinol.* **2011**, *164*, 899–904. [CrossRef]

199. Castro, A.M.; Macedo-de la Concha, L.E.; Pantoja-Meléndez, C.A. Low-grade inflammation and its relation to obesity and chronic degenerative diseases. *Rev. Médica Hosp. Gen. México* **2017**, *80*, 101–105. [CrossRef]

200. Fernandez-Sanchez, A.; Madrigal-Santillan, E.; Bautista, M.; Esquivel-Soto, J.; Morales-Gonzalez, A.; Esquivel-Chirino, C.; Durante-Montiel, I.; Sanchez-Rivera, G.; Valadez-Vega, C.; Morales-Gonzalez, J.A. Inflammation, oxidative stress, and obesity. *Int. J. Mol. Sci.* **2011**, *12*, 3117–3132. [CrossRef] [PubMed]

201. Kwon, H.; Pessin, J.E. Adipokines mediate inflammation and insulin resistance. *Front. Endocrinol.* **2013**, *4*, 71. [CrossRef]

202. Nieves, D.J.; Cnop, M.; Retzlaff, B.; Walden, C.E.; Brunzell, J.D.; Knopp, R.H.; Kahn, S.E. The atherogenic lipoprotein profile associated with obesity and insulin resistance is largely attributable to intra-abdominal fat. *Diabetes* **2003**, *52*, 172–179. [CrossRef]
203. Savini, I.; Catani, M.V.; Evangelista, D.; Gasperi, V.; Avigliano, L. Obesity-associated oxidative stress: Strategies finalized to improve redox state. *Int. J. Mol. Sci.* **2013**, *14*, 10497–10538. [CrossRef] [PubMed]
204. Tilg, H.; Moschen, A.R. Adipocytokines: Mediators linking adipose tissue, inflammation and immunity. *Nat. Rev. Immunol.* **2006**, *6*, 772–783. [CrossRef] [PubMed]
205. Scanlon, G.; Loughman, J.; Farrell, D.; McCartney, D. A review of the putative causal mechanisms associated with lower macular pigment in diabetes mellitus. *Nutr. Res. Rev.* **2019**, *32*, 247–264. [CrossRef] [PubMed]
206. van Het Hof, K.H.; West, C.E.; Weststrate, J.A.; Hautvast, J.G. Dietary factors that affect the bioavailability of carotenoids. *J. Nutr.* **2000**, *130*, 503–506. [CrossRef]
207. Nolan, J.M.; Feeney, J.; Kenny, R.A.; Cronin, H.; O'Regan, C.; Savva, G.M.; Loughman, J.; Finucane, C.; Connolly, E.; Meagher, K.; et al. Education is positively associated with macular pigment: The Irish Longitudinal Study on Ageing (TILDA). *Investig. Ophthalmol. Vis. Sci.* **2012**, *53*, 7855–7861. [CrossRef]
208. Bovier, E.R.; Lewis, R.D.; Hammond, B.R., Jr. The relationship between lutein and zeaxanthin status and body fat. *Nutrients* **2013**, *5*, 750–757. [CrossRef]
209. Bohn, T. Bioavailability of non-provitamin A carotenoids. *Curr. Nutr. Food Sci.* **2008**, 240–258. [CrossRef]
210. Bray, G.A.; Nielsen, S.J.; Popkin, B.M. Consumption of high-fructose corn syrup in beverages may play a role in the epidemic of obesity. *Am. J. Clin. Nutr.* **2004**, *79*, 537–543. [CrossRef]
211. Gross, L.S.; Li, L.; Ford, E.S.; Liu, S. Increased consumption of refined carbohydrates and the epidemic of type 2 diabetes in the United States: An ecologic assessment. *Am. J. Clin. Nutr.* **2004**, *79*, 774–779. [CrossRef]
212. Ludwig, D.S.; Majzoub, J.A.; Al-Zahrani, A.; Dallal, G.E.; Blanco, I.; Roberts, S.B. High glycemic index foods, overeating, and obesity. *Pediatrics* **1999**, *103*, E26. [CrossRef] [PubMed]
213. Marshall, J.A.; Hamman, R.F.; Baxter, J. High-fat, low-carbohydrate diet and the etiology of non-insulin-dependent diabetes mellitus: The San Luis Valley Diabetes Study. *Am. J. Epidemiol.* **1991**, *134*, 590–603. [CrossRef]
214. Ranard, K.M.; Jeon, S.; Mohn, E.S.; Griffiths, J.C.; Johnson, E.J.; Erdman, J.W., Jr. Dietary guidance for lutein: Consideration for intake recommendations is scientifically supported. *Eur. J. Nutr.* **2017**, *56*, 37–42. [CrossRef]
215. Abdel-Aal, E.S.; Akhtar, H.; Zaheer, K.; Ali, R. Dietary sources of lutein and zeaxanthin carotenoids and their role in eye health. *Nutrients* **2013**, *5*, 1169–1185. [CrossRef] [PubMed]
216. Djuric, Z.; Ren, J.; Blythe, J.; VanLoon, G.; Sen, A. A Mediterranean dietary intervention in healthy American women changes plasma carotenoids and fatty acids in distinct clusters. *Nutr. Res.* **2009**, *29*, 156–163. [CrossRef]
217. Marhuenda-Munoz, M.; Hurtado-Barroso, S.; Tresserra-Rimbau, A.; Lamuela-Raventos, R.M. A review of factors that affect carotenoid concentrations in human plasma: Differences between Mediterranean and Northern diets. *Eur. J. Clin. Nutr.* **2019**, *72*, 18–25. [CrossRef]
218. Stringham, J.M.; Johnson, E.J.; Hammond, B.R. Lutein across the Lifespan: From Childhood Cognitive Performance to the Aging Eye and Brain. *Curr. Dev. Nutr.* **2019**, *3*, nzz066. [CrossRef]
219. US Department of Agriculture; Agricultural Research Service; Library, N.D. USDA National Nutrient Database for Standard Reference. Available online: https://www.ars.usda.gov/northeast-area/beltsville-md-bhnrc/beltsville-human-nutrition-research-center/methods-and-application-of-food-composition-laboratory (accessed on 10 April 2021).
220. Wilson, L.M.; Tharmarajah, S.; Jia, Y.; Semba, R.D.; Schaumberg, D.A.; Robinson, K.A. The Effect of Lutein/Zeaxanthin Intake on Human Macular Pigment Optical Density: A Systematic Review and Meta-Analysis. *Adv. Nutr.* **2021**. [CrossRef]
221. Donoso, A.; Gonzalez-Duran, J.; Munoz, A.A.; Gonzalez, P.A.; Agurto-Munoz, C. Therapeutic uses of natural astaxanthin: An evidence-based review focused on human clinical trials. *Pharmacol. Res.* **2021**, *166*, 105479. [CrossRef] [PubMed]
222. Fakhri, S.; Abbaszadeh, F.; Dargahi, L.; Jorjani, M. Astaxanthin: A mechanistic review on its biological activities and health benefits. *Pharmacol. Res.* **2018**, *136*, 1–20. [CrossRef]
223. Giannaccare, G.; Pellegrini, M.; Senni, C.; Bernabei, F.; Scorcia, V.; Cicero, A.F.G. Clinical Applications of Astaxanthin in the Treatment of Ocular Diseases: Emerging Insights. *Mar. Drugs* **2020**, *18*, 239. [CrossRef] [PubMed]
224. Saito, M.; Yoshida, K.; Saito, W.; Fujiya, A.; Ohgami, K.; Kitaichi, N.; Tsukahara, H.; Ishida, S.; Ohno, S. Astaxanthin increases choroidal blood flow velocity. *Graefes Arch. Clin. Exp. Ophthalmol.* **2012**, *250*, 239–245. [CrossRef]

Article

Reduction of Cardio-Metabolic Risk and Body Weight through a Multiphasic Very-Low Calorie Ketogenic Diet Program in Women with Overweight/Obesity: A Study in a Real-World Setting

Elena Tragni [1], Luisella Vigna [2], Massimiliano Ruscica [1], Chiara Macchi [1], Manuela Casula [1,3], Alfonso Santelia [1], Alberico L. Catapano [1,3] and Paolo Magni [1,3,*]

[1] Dipartimento di Scienze Farmacologiche e Biomolecolari, Università degli Studi di Milano, 20133 Milan, Italy; elena.tragni@unimi.it (E.T.); massimiliano.ruscica@unimi.it (M.R.); chiara.macchi@unimi.it (C.M.); manuela.casula@unimi.it (M.C.); the.alfsan@gmail.com (A.S.); alberico.catapano@unimi.it (A.L.C.)
[2] Center of Obesity and Work EASO Collaborating Centers for Obesity Management, Occupational Health Unit, Fondazione Ca' Granda Ospedale Maggiore Policlinico, 20122 Milan, Italy; luisella.vigna@policlinico.mi.it
[3] IRCCS MultiMedica, Sesto S. Giovanni, 20099 Milan, Italy
* Correspondence: paolo.magni@unimi.it; Tel.: +39-02-50318229

Citation: Tragni, E.; Vigna, L.; Ruscica, M.; Macchi, C.; Casula, M.; Santelia, A.; Catapano, A.L.; Magni, P. Reduction of Cardio-Metabolic Risk and Body Weight through a Multiphasic Very-Low Calorie Ketogenic Diet Program in Women with Overweight/Obesity: A Study in a Real-World Setting. *Nutrients* **2021**, *13*, 1804. https://doi.org/10.3390/nu13061804

Academic Editor: Rosa Casas

Received: 9 April 2021
Accepted: 22 May 2021
Published: 26 May 2021

Publisher's Note: MDPI stays neutral with regard to jurisdictional claims in published maps and institutional affiliations.

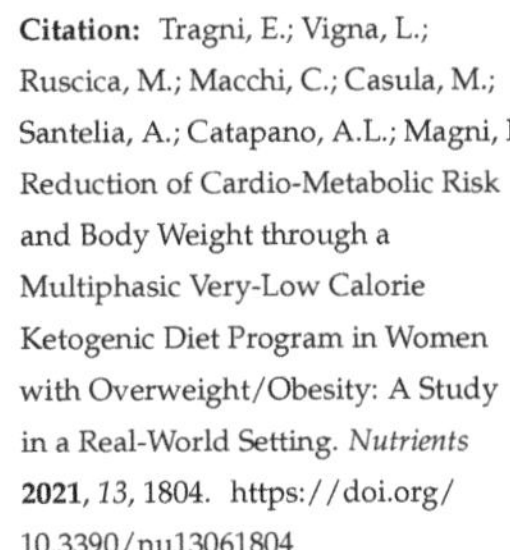

Abstract: Background: The prevention and treatment of obesity and its cardio-metabolic complications are relevant issues worldwide. Among lifestyle approaches, very low-calorie ketogenic diets (VLCKD) have been shown to lead to rapid initial weight loss, resulting in better long-term weight loss maintenance. As no information on VLCKD studies carried on in a real-world setting are available, we conducted this multi-centre study in a real-world setting, aiming at assessing the efficacy and the safety of a specific multiphasic VLCKD program in women with overweight or obesity. Methods: A multi-center, prospective, uncontrolled trial was conducted in 33 outpatient women (age range 27–60 y) with overweight or obesity (BMI: 30.9 ± 2.7 kg/m^2; waist circumference: 96.0 ± 9.4 cm) who started a VLCKD dietary program (duration: 24 weeks), divided into four phases. The efficacy of VLCKD was assessed by evaluating anthropometric measures and cardiometabolic markers; liver and kidney function biomarkers were assessed as safety parameters. Results: The VLCKD program resulted in a significant decrease of body weight and BMI (−14.6%) and waist circumference (−12.4%). At the end of the protocol, 33.3% of the participants reached a normal weight and the subjects in the obesity range were reduced from 70% to 16.7%. HOMA-IR was markedly reduced from 3.17 ± 2.67 to 1.73 ± 1.23 already after phase 2 and was unchanged thereafter. Systolic blood pressure decreased after phase 1 (−3.5 mmHg) and remained unchanged until the end of the program. Total and LDL cholesterol and triglycerides were significantly reduced by VLCKD along with a significant HDL cholesterol increase. Liver, kidney and thyroid function markers did not change and remained within the reference range. Conclusions: The findings of a multi-center VLCKD program conducted in a real-world setting in a cohort of overweight/obese women indicate that it is safe and effective, as it results in a major improvement of cardiometabolic parameters, thus leading to benefits that span well beyond the mere body weight/adiposity reduction.

Keywords: very-low calorie ketogenic diet; obesity; cardiovascular risk; insulin resistance; nutraceutical

1. Introduction

Prevention and treatment of obesity and its cardio-metabolic complications are growing public health problems worldwide since this condition affects a relevant part of the world population across both genders and all ages and ethnic groups, and its prevalence is now maintained or even accelerated in most industrialized countries [1–4]. In recent years, the prevalence of obesity has reached epidemic proportions, and, therefore, the identification of effective lifestyle tools, including nutritional ones [5], able to produce significant weight loss and to maintain it over time is mandatory, in order to limit its progression from the uncomplicated stage to that characterized by cardiovascular and metabolic complications [6–8], as well as oncologic diseases [9]. In this context, cardiovascular disease (CVD) risk and unhealthy lifestyle habits [10] are often underdiagnosed and undertreated, therefore highly contributing to atherosclerotic CVD (ASCVD) prevalence [11]. The current treatment options for obesity include balanced hypocaloric diets, exercise, lifestyle modifications, drugs, use of endoscopic devices (e.g., intragastric balloon) and bariatric surgery [12–15]. The therapeutic benefit of all currently available anti-obesity interventions is often limited by their subjective efficacy, variable tolerability, safety profiles and poor compliance, with the latter being a strongly limiting variable, especially when long-term treatments are needed [4,16]. Many dietary regimens that operate through various mechanisms have been proposed to reduce appetite or for weight control [17,18] and the leading non-pharmacological approach is the use of diets, particularly low-calorie and very low-calorie ketogenic diets (VLCKD) [19–21]. VLCKD has been endorsed by the European Food Safety Agency (EFSA) for reduction of body weight in subjects with obesity, according to a specific Scientific Opinion (https://www.efsa.europa.eu/it/efsajournal/pub/2271, accessed 2 April 2021), and a specific consensus statement discussing the appropriate use of VLCKD has been recently published by a scientific panel of the Italian Society of Endocrinology [22]. Indeed, in studies conducted in hospital settings, the VLCKD approach has been shown to lead to a rapid initial weight loss, which results in better long-term weight loss maintenance [23] although in some cases an adequate weight reduction at the beginning of the diet program is followed by a shotdown of weight decrease. This problem may depend upon different factors, including individual metabolic rate and patient compliance. The VLCKD approach generally includes an initial phase with a complete replacement of regular meals with food or formulations that provide 400–800 kCal/day. This type of diet may be better defined as a "therapeutical approach" since it is commonly followed under medical supervision in patients with BMI > 30 kg/m^2 or in subjects needing a rapid weight loss in preparation to other medical procedures [21,24] and is usually associated with the use of specific food supplements. Since, to our best knowledge, no information on VLCKD studies conducted in a real-world setting are available, the present multi-centre study, conducted in a real-world setting, was aimed at assessing the efficacy, according to anthropometric and cardiometabolic changes, and safety of a specific multiphasic VLCKD program in women with overweight or obesity.

2. Materials and Methods

2.1. Study Design and Population

The study was designed as a multi-center, prospective, uncontrolled trial in a real-life setting and included Caucasian outpatient women with overweight or obesity and some features of the metabolic syndrome, including increased waist circumference (WC) and pharmacologically controlled arterial hypertension [25]; 11/33 subjects were on pharmacological therapy for arterial hypertension (Table S1). All patients were consecutively admitted to one of the 5 participating clinical centers in the Milan area (Italy) in the period 2016–2018. Each clinical center is specialized in the medical management of obesity, with a specific expertise in VLCKD program, and includes expert physicians; 2 centres also included a trained dietician. The inclusion criteria were: female sex upper-range overweight or grade 1 or 2 obesity (body mass index (BMI) range: 27–37 kg/m^2), age between 25 and 65 years, negative for pregnancy test, and having signed an informed consent. The main exclusion criteria were: current or previous smoking, pregnancy and nursing, history of diabetes mellitus, renal disease or severe renal impairment (plasma creatinine >1.5 mg/dL), severe liver disease, HIV infection, nervous system and cardiovascular diseases (including uncontrolled arterial hypertension), blood diseases, cancer or any progressive severe disease, osteoporosis, eating disorders or any psychiatric disease, uncontrolled thyroid diseases, menopause hormonal replacement therapy, pharmacological treatments known to interfere with the study treatment, history of bariatric surgery, and patients who were enrolled in another research study in the last 12 months. At the screening visit, all patients underwent fasting blood sampling and a full clinical examination, to evaluate height (in standing position and without shoes and corrected to the closer 0.5 cm), body weight, WC and hip (HC) circumferences (in standing position, measured with a flexible tape), heart rate (HR) and arterial blood pressure. These parameters were also recorded at all subsequent visits. A total of 44 eligible patients (age 49.5 ± 7.2 yrs, and BMI 30.9 ± 2.7 kg/m^2 (mean $\pm$ SD)) were enrolled in the study and started a VLCKD dietary program (Pentadiet program, Figure 1) with a total intervention duration of 24 weeks. Eleven patients were on chronic therapy known not to interfere with VLCKD treatment (Table S1). The concomitant medications of the study subjects at baseline are reported in Table S1. The indicated treatments were carried on until the end of the VLCKD program, under appropriate monitoring for possible adverse effects. The study was conducted in accordance with the guidelines of the declaration of Helsinki (http://www.wma.net/en/30publications/10policies/b3/, accessed 2 April 2021), and the study protocol was approved by the Institutional Ethics Committee (approval N°441/2011). Patients were informed about all aspects related to the study, possible benefits and risks were explained at the beginning of the study and subjects were informed about the possibility to leave the study at any time without penalty. Written informed consent was obtained from each subject before starting the VLCKD program.

Phase 1	Phase 2	Phase 3	Phase 4
700 Kcal/day 50 g/day of carbohydrate 4 or 5 meal replacement (Protiligne), PentaCal	820 Kcal/day 3 meal replacement (Protiligne), 1 protein dish, PentaCal	1100 Kcal/day 2 meal replacement (Protiligne) 2 protein dish, small amount of carbohydrates	1250 Kcal/day 1 meal replacement (Protiligne) 2 protein dish, reintroduction of carbohydrates
V0-BS1 **start**	**V1** **4 week**	**V2-BS2** **8 week**	**V3-BS3** **16 week**	**V4** **24 week**

Figure 1. Outline of the VLCKD program. The program included 4 separate phases, with a total duration of 24 weeks.

2.2. Clinical Procedures

The overall duration of the study was 24 weeks, divided into 4 sequential phases: two "active phases" (phases 1 and 2) and two "stabilization phases" (phases 3 and 4) (Figure 1). Each phase had a standard duration, and the daily plan included 3 main meals (breakfast, lunch, dinner) and 1 snack in the afternoon in all phases. All low-carbohydrate foods (Protiligne) and a food supplement (PentaCal) used in the VLCKD program were provided by New Penta srl (Milan, Italy). The average daily food intake, including pre-prepared meals (Protiligne, Table S2B), varied according to each phase. As reported in Table S2, the energy and macronutrient content of meal replacement portions were within the indicated range, and varied according to each specific type (i.e., soups, cakes, meat plates, etc.). Thus, during each phase of the program, the daily target of energy and macronutrients was reached combining different meal replacement portions and the allowed foods. The daily intake of protein, carbohydrate, linoleic acid, γ-linoleic acid and micronutrients during all the phases of the VLCKD program was above the minimum content recommended by EFSA, according to a specific Scientific Opinion (https://www.efsa.europa.eu/it/efsajournal/pub/2271, accessed 2 April 2021). Patients were instructed to drink not less than 1.5–2 L of water daily and to avoid ingestion of any sweets, sugarfree chewing gums and soft drinks, herbal tea with fruit, and preserved vegetables. The program included the use of a vitamin and mineral supplement (PentaCal, Table S2A) during phases 1 and 2. At the end of the study, a compliance survey was submitted to all patients.

The efficacy of VLCKD was assessed by evaluating anthropometric measures (height, weight, BMI, WC and HC), SBP/DBP, HR and glucose metabolism markers, whereas liver and kidney function biomarkers were assessed as safety parameters.

2.3. Blood and Urinary Biochemistry

Before starting the VLCKD program and at the end of phases 2 and 4, urine and fasting blood samples from an antecubital vein were collected at 8:00–10:00 a.m. after an overnight fast. The following haematological and biochemical parameters, used as efficacy and safety end-points, were evaluated using standard automated clinical procedures (Cobas system, Roche, Italy): complete blood count, electrolytes (chloride, potassium, calcium, magnesium, sodium), fasting plasma glucose (FPG) and insulin, HbA1c, plasma protein concentration, lipids (total cholesterol (TC), HDL cholesterol (HDL-C) and triglycerides (TG)), uric acid, blood urea nitrogen, creatinine, alanine transferase (AST), aspartate transaminase (AST), γ-glutamyl transpeptidase (γ-GT), high-sensitivity C-reactive protein (hs-CRP) and TSH reflex. Urinary ketones were evaluated using Ketostix strips (Bayer, Germany). All biochemical analyses were conducted in 3 certified clinical laboratories in the Milan area. All samples from each participating subject were collected and analyzed in the same laboratory. LDL cholesterol (LDL-C) was calculated according to the Friedewald formula [26]. The homeostasis model assessment of insulin resistance (HOMA-IR) index was calculated as follows: HOMA-IR = (fasting glucose (mmol/L) × insulin (mU(mL))/22.5) [27]. The triglyceride-glucose (TyG) index was calculated as follows: ln (TG × FPG/2). Creatinine clearance was calculated according to the Cockroft-Gault formula [28].

2.4. Statistical Analysis

Sample size calculation. A sample size of at least 26 subjects in the study group achieves 90% power to detect a reduction of 10% in body weight vs. population of obese women in the same range of BMI (from 27 to 37 kg/m^2; mean body weight = 85 kg; standard deviation = 13 kg), with a type I error rate of 5%. The cardiovascular risk score was calculated according to the Framingham Risk Score using lipid values (FRS lipids) and using BMI (FRS BMI) [29] and the EAS/ESC SCORE for low-risk countries (like Italy) [30]. A per protocol analysis was performed. Quantitative variables are presented as mean values ± standard deviation, SD), while qualitative variables are presented as frequencies. Comparisons between continuous variables across visits were performed by using the non-parametric Friedman test for k mutually related samples. All reported *p*-values are based on two-sided tests and compared to a significance level of 5%. All statistical analyses were performed using IBM SPSS Statistics software package for Windows, Version 25.0. Armonk, NY, USA: IBM Corp.

3. Results

3.1. Study Population

The study included women with upper-range overweight or grade 1–2 obesity and was conducted in a real-life setting. Among the 44 eligible patients, 11 were excluded before the start of the VLCKD program, due to personal reasons or duties, such as lack of motivation in undergoing the dietary plan or family problems (Figure 2). Therefore, 33 subjects were allocated to the VLCKD program, and, since 3 participants dropped-out during phase 1 (n = 2) or phase 2 (n = 1) by directly declaring to exit from the VLCKD program, due to lack of interest/motivation, 30 subjects completed the study (Figure 2) and their baseline data are reported in Table 1. The study subjects had a BMI of 30.9 ± 2.7 kg/m^2, with a relevant abdominal adiposity (WC: 96.0 ± 9.4 cm), mild dyslipidemia (LDL-C: 144.0 ± 33.6 mg/dL; non HDL-cholesterol (non-HDL-C): 164.9 ± 35.7 mg/dL) and some degree of insulin resistance, as shown by a moderately elevated HOMA-IR (3.17 ± 2.67).

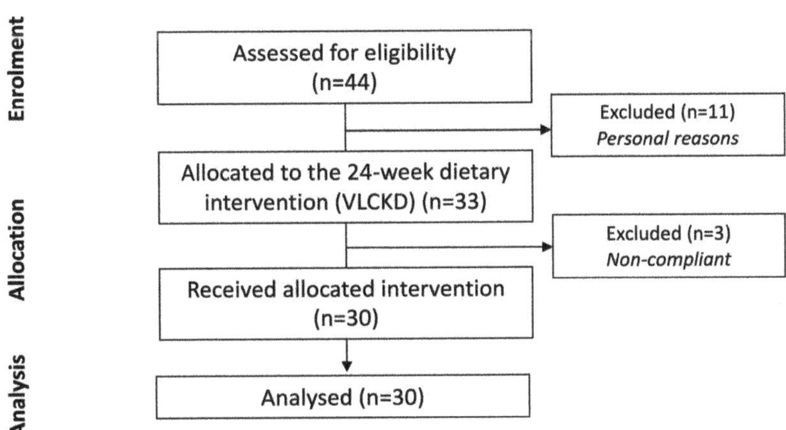

Figure 2. CONSORT statement flow diagram.

Table 1. Baseline data (*n* = 30).

	MEAN ± SD	MINIMUM	MAXIMUM
Age (years)	49.5 ± 7.2	27	60
Weight (kg)	81.8 ± 10.9	63.0	104.6
Height (m)	1.62 ± 0.07	1.48	1.78
BMI (kg/m^2)	30.9 ± 2.7	26.96	36.06
Waist circumference (cm)	96.0 ± 9.4	80.0	114.0
Hip circumference (cm)	113.1 ± 7.7	100.0	130.0
Waist-to-hip ratio	0.85 ± 0.08	0.72	1.04
SBP (mmHg)	127.2 ± 10.2	110	160
DBP (mmHg)	81.5 ± 8.9	60	100
Heart rate (bpm)	69.4 ± 6.3	52	80
FPG (mg/dL)	95.1 ± 15.6	73	155
HbA1c (mmol/mol)	36.98 ± 5.19	30.05	58.40
Insulin (mU/L)	12.65 ± 7.31	3.00	39.60
HOMA-IR	3.17 ± 2.67	0.64	15.16
Total cholesterol (mg/dL)	223.0 ± 37.7	159	339
HDL-cholesterol (mg/dL)	58.0 ± 12.9	37.3	82.7
Non HDL-cholesterol (mg/mL)	164.9 ± 35.7	101.3	289.3
Triglycerides (mg/dL)	104.7 ± 41.4	44	208
LDL-cholesterol (mg/dL) (*)	144.0 ± 33.6	80.1	248.3
Uric acid (mg/dL)	4.6 ± 1.0	3.1	6.6
AST (mg/dL)	18.5 ± 4.6	12	32
ALT (mg/dL)	20.5 ± 12.2	8	63
γ-GT (mg/dL)	21.0 ± 8.6	10	46
Creatinine (mg/dL)	0.74 ± 0.13	0.44	0.98
Creatinine clearance (mL/min)	122.40 ± 33.09	73.65	221.49
BUN (mg/dL)	33.39 ± 8.62	22.40	51.00
TSH (mUI/L)	2.38 ± 0.80	1.01	3.70

BMI, body mass index; SBP, systolic blood pressure; DBP, diastolic blood pressure; FPG, fasting plasma glucose; HbA1c, glycosylated hemoglobin; HOMA IR, homeostatic model assessment for insulin resistance; AST, aspartate transaminase; ALT, alanine transaminase; γ-GT, gamma-glutamyl transferase; BUN: blood urea nitrogen; TSH, thyroid-stimulating hormone. (*) calculated by the Friedewald formula.

3.2. Analysis of the Ketogenetic Effect of VLCKD

The determination of urinary ketones, an indirect index of carbohydrate restriction and adherence to the proposed dietary plan based on a VLCKD approach, was performed in order to evaluate the actual presence of ketogenesis produced by dietary carbohydrate restriction during the first 2 phases of the protocol. As expected, urinary ketones were not detectable at baseline. The occurrence of dietary-induced ketogenesis, detected by the presence of urinary ketones, was observed in 78% of the patients after phase 1 and in 50% of the patients after phase 2.

3.3. Effect of VLCKD on Anthropometric Parameters

Over the entire VLCKD program, which lasted 24 weeks, all anthropometric parameters were progressively improved, with a total significant decrease of 14.6% in body weight and BMI (Figure 3A), 12.4% in WC (Figure 3B) and 10.0% in HC, resulting in a lower (−2.7%) Waist-to-Hip ratio (WHR) (Figure 3C). It should be highlighted that the reduction of BMI and WC in a single-phase, although significant after each of them compared to the start value, was greater during phases 1–2 (BMI: −6.2% and −4.9%, WC: −4.7% and −4.6%, respectively) (Figure 3A,B), although some contribution to total weight loss was observed in all subsequent phases, leading to a cumulative 11.5 kg weight loss, on average. At the end of the VLCKD protocol, 33.3% of the participants reached a normal weight and the obesity prevalence was reduced from 70% to 16.7%. As a consequence, the overweight group rose from 30 to 50% (Figure 4).

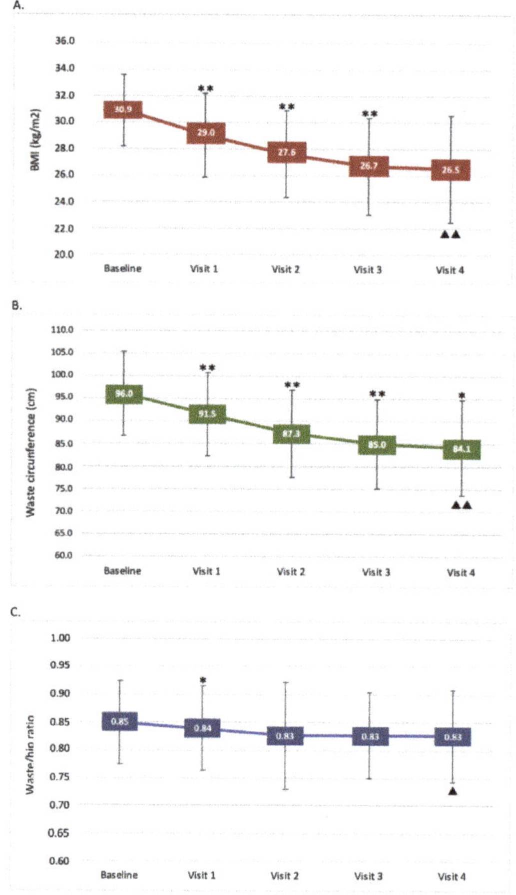

Figure 3. Effect of the VLCKD program on BMI, waist circumference and waist/hip ratio. (**A**) BMI changes during the 24-week program; (**B**) waist circumference during the 24-week program; (**C**) waist/hip ratio during the 24-week program. Data are mean ± SD. (*) $p < 0.05$ and (**) $p < 0.001$: *p*-value across consecutive visits. (▲) $p < 0.05$ and (▲▲) $p < 0.001$: *p*-value for trend.

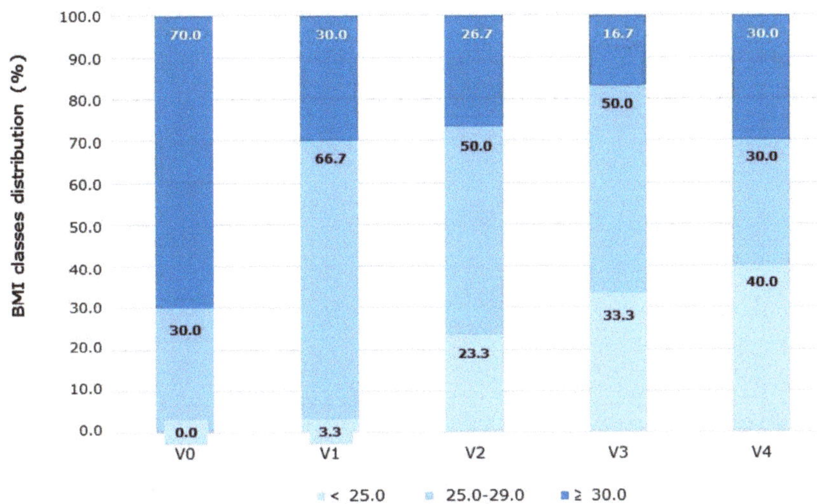

Figure 4. Effect of the VLCKD program on the BMI classes distribution. The relative % distribution of patients in the normal weight (<25.0 kg/m^2), overweight (<25.0–29.9 kg/m^2) and obese (>30 kg/m^2) BMI classes, over the 24-week VLCKD program, is reported. V, visit.

3.4. Effect of VLCKD on Glucometabolic and Cardiovascular Parameters

At baseline, patients enrolled in the study displayed a moderate rate of insulin resistance (HOMA-IR: 3.17 ± 2.67) (Table 1. The VLCKD program showed a specific effect on this parameter, as it was significantly reduced to 1.73 ± 1.23 (−38.0%; p = 0.003) at the end of phase 2, due to reduction of both plasma insulin (−35.0%; p < 0.001) and FPG (−8.7%; p = 0.002), in association with reduced HbA1c (−5.6%; p = 0.008), and then remained unchanged after phase 3 (Table 2). The TyG index was also significantly improved (p < 0.001) (Table 2). No changes in uric acid levels were observed (Table 3). SBP decreased after phase 1 (−3.5 mmHg; −2.5%; p = 0.006) and then remained unchanged until the end of the program. As reported above, the study subjects showed moderate baseline hypercholesterolemia (TC 223.0 ± 37.7) mg/dL). TC, TG and LDL-C were significantly reduced by the VLCKD program after phase 3, along with a significant increase of HDL-C (p = 0.027), resulting in reduced non-HDL-C (Table 3). The individual change of LDL-C level showed some variability since 6/30 patients displayed no changes and 6/30 had moderately increased concentrations (maximum 158 mg/dL in one case). Moreover, hsCRP (always below 0.1 mg/L; not shown) and uric acid (Table 3) concentrations did not significantly change during the intervention. At baseline, the study subjects were almost entirely at very low/low CVD risk, according to FRS lipids, FRS BMI and EAS/ESC SCORE algorithms. Interestingly, however, the BMI and lipid improvements driven by VLCKD resulted in a mean absolute reduction of these scores: FRS lipids (from 1.99 ± 1.57 to 1.53 ± 1.20), FRS BMI (from 6.23 ± 4.13 to 5.05 ± 3.12) and EAS/ESC SCORE (from 0.42 ± 0.34 to 0.36 ± 0.30), due to the specific reduction in the few with higher CVD risk.

Table 2. Effect of VLCKD on glucometabolic parameters.

		Mean $\pm$ SD	Absolute Change (% Change)	*p*-Value *
FPG (mg/dL)	Baseline	95.1 $\pm$ 15.6	−9.3 (−9.8)	0.001
	Visit 2	85.9 $\pm$ 12.1		
	Visit 3	85.8 $\pm$ 11.9		
HbA1c (mmol/mol)	Baseline	36.98 $\pm$ 5.19	−2.47 (−6.0)	0.001
	Visit 2	34.75 $\pm$ 2.82		
	Visit 3	34.51 $\pm$ 3.14		
Insulin (μU/mL)	Baseline	12.65 $\pm$ 7.31	−4.72 (−37.3)	0.001
	Visit 2	7.73 $\pm$ 4.92		
	Visit 3	7.93 $\pm$ 6.10		
HOMA-IR	Baseline	3.17 $\pm$ 2.67	−1.39 (−43.8)	0.001
	Visit 2	1.73 $\pm$ 1.23		
	Visit 3	1.78 $\pm$ 1.82		
TyG index	Baseline	8.43 $\pm$ 0.45	−0.41 (−4.9)	0.001
	Visit 2	8.05 $\pm$ 0.38		
	Visit 3	8.02 $\pm$ 0.49		

FPG, fasting plasma glucose; HbA1c: glycosylated hemoglobin; HOMA-IR: Homeostatic Model Assessment for Insulin Resistance; TyG index, triglyceride-glucose index; * Friedman test for k mutually related samples.

Table 3. Effect of VLCKD on cardiovascular, lipid and safety parameters.

		Mean $\pm$ SD	Absolute Change (% Change)	*p*-Value *
SBP (mmHg)	Baseline	127.2 $\pm$ 10.2	−3.5 (−2.8)	0.006
	Visit 1	123.3 $\pm$ 9.8		
	Visit 2	123.2 $\pm$ 9.6		
	Visit 3	121.4 $\pm$ 8.4		
	Visit 4	123.7 $\pm$ 9.6		
DBP (mmHg)	Baseline	81.5 $\pm$ 8.9	−3.5 (−4.3)	0.211
	Visit 1	80.1 $\pm$ 8.2		
	Visit 2	80.0 $\pm$ 9.3		
	Visit 3	78.6 $\pm$ 9.0		
	Visit 4	78.0 $\pm$ 8.6		
HR (bpm)	Baseline	69.4 $\pm$ 6.3	−0.3 (−0.4)	0.021
	Visit 1	72.0 $\pm$ 6.7		
	Visit 2	69.7 $\pm$ 5.0		
	Visit 3	70.4 $\pm$ 6.8		
	Visit 4	69.1 $\pm$ 12.9		
TC (mg/dL)	Baseline	223.0 $\pm$ 37.7	−13.2 (−5.9)	0.000
	Visit 2	194.8 $\pm$ 30.7		
	Visit 3	209.7 $\pm$ 28.4		
HDL-C (mg/dL)	Baseline	58.0 $\pm$ 12.9	3.3 (5.7)	0.000
	Visit 2	52.7 $\pm$ 12.7		
	Visit 3	61.7 $\pm$ 13.0		
TG (mg/dL)	Baseline	104.7 $\pm$ 41.4	−27.1 (−25.9)	0.000
	Visit 2	78.4 $\pm$ 29.1		
	Visit 3	77.6 $\pm$ 31.1		
LDL-C (mg/dL) (°)	Baseline	144.0 $\pm$ 33.6	−11.2 (−7.8)	0.000
	Visit 2	126.4 $\pm$ 23.4		
	Visit 3	132.8 $\pm$ 23.7		

Table 3. *Cont.*

		Mean ± SD	Absolute Change (% Change)	*p*-Value *
non HDL-C (mg/dL)	Baseline	164.9 ± 35.7	−16.5 (−10.1)	0.000
	Visit 2	142.1 ± 23.4		
	Visit 3	148.4 ± 24.4		
TG/HDL-C	Baseline	1.97 ± 1.14	−0.6 (−30.5)	0.001
	Visit 2	1.62 ± 1.05		
	Visit 3	1.35 ± 0.73		
Uric acid (mg/dL)	Baseline	4.6 ± 1.0	−0.3 (−6.5)	0.093
	Visit 2	4.5 ± 1.1		
	Visit 3	4.3 ± 1.1		
AST (UI/L)	Baseline	18.5 ± 4.6	−0.3 (−1.6)	0.246
	Visit 2	19.5 ± 6.0		
	Visit 3	18.2 ± 5.3		
ALT (UI/L)	Baseline	20.5 ± 12.2	−1.5 (−7.3)	0.899
	Visit 2	21.4 ± 13.5		
	Visit 3	19.0 ± 9.2		
γ-GT (UI/L)	Baseline	21.0 ± 8.6	−5.1 (−24.3)	0.000
	Visit 2	16.0 ± 8.3		
	Visit 3	15.9 ± 9.1		
Creatinine (mg/dL)	Baseline	0.74 ± 0.13	−0.09 (−12.2)	0.004
	Visit 2	0.73 ± 0.13		
	Visit 3	0.65 ± 0.11		
CC (mL/min)	Baseline	122.39 ± 33.09	−4.40 (−3.6)	0.026
	Visit 2	110.32 ± 26.15		
	Visit 3	117.99 ± 25.59		
BUN (mg/dL)	Baseline	33.39 ± 8.62	2.46 (7.4)	0.092
	Visit 2	35.35 ± 6.22		
	Visit 3	35.85 ± 8.94		
TSH (mUI/L)	Baseline	2.40 ± 0.77	−0.09 (−3.8)	0.629
	Visit 2	2.21 ± 0.88		
	Visit 3	2.31 ± 0.86		

SBP, systolic blood pressure; DBP, diastolic blood pressure; HR, heart rate; TC, total cholesterol; HDL-C, HDL-cholesterol; TG, triglycerides; LDL-C, LDL-cholesterol; non-HDL-C, non-HDL-cholesterol; AST, aspartate transaminase; ALT, alanine transaminase; γ-GT, gamma-glutamyl transferase; CC, creatinine clearance; BUN, blood urea nitrogen; TSH, thyroid-stimulating hormone; (°) calculated by the Friedewald formula; * Friedman test for k mutually related samples.

3.5. Effect of VLCKD on Markers of Liver, Kidney and Thyroid Function

At baseline, the markers of liver, kidney and thyroid function were within the reference range and remained within it over the entire duration of the VLCKD program (Table 3). A significant but moderate decrease was observed for γ-GT, creatinine and creatinine clearance (Table 3).

4. Discussion

This study aimed at evaluating the efficacy and the safety of a multiphasic VLCKD program, conducted in a multi-center real-world setting, in women with overweight or obesity. The main objective was to assess the actual health benefits of such approach in the context of the day-by-day management of these clinical conditions. The proposed multiphasic VLCKD program turned out to be safe, according to liver, kidney and thyroid biomarkers. Moreover, in the patients who completed the program, a set of important improvements related to cardiovascular function and cardiometabolic disease risk has been accomplished.

The efficacy data obtained show that the VLCKD program resulted in a significant reduction (−14.6%) of body weight and BMI, which is also greater than the 10% threshold proposed by the obesity guidelines [15]. The average absolute reduction of BMI (−4.4 kg/m^2) is similar to that obtained in hospital-based studies with a ketogenic phase up to 4 weeks (−4.2 kg/m^2) or at least 4 weeks (6.2 kg/m^2) [31]. Notably, the mean BMI value at the end of the protocol (26.5 kg/m^2) is just above the upper end (25 kg/m^2) of the normal range, with a reduction of subjects in the obese range from 70 to 16.7% at visit 3, but some weight regain at visit 4, leading to a final 30% obese subjects. On the other side, the percentage of subjects in the normal BMI range stably increased from 0% at baseline to 40% at the end of the VLCKD program. These findings may suggest that the health professional input is relevant not only in the initial phase of the VLCKD program but also in the last phase and the subsequent follow-up over the months and the years, in order to promote the longest time free of disease. Follow-up visits are important since, according to the obesity guidelines [15], once achieved, the body weight reduction of at least −10% or more should be maintained at least for 5 years to obtain an optimal benefit. Unfortunately, data from follow-up visits, after completion of the 6-month VLCKD program, could not be collected in this study, highlighting the relevant lack of long-term follow-up control visits in the real-world context. Possible reasons are lack of motivation, reduced synergy with the physician or the team and additional costs. In any case, this may clearly result in a long-term reduced benefit of the initial weight loss, since only one recommendation of the guidelines (weight reduction by at least −10%, but not 5 years maintenance) is fulfilled.

Interestingly, an additional important advantage of this VLCKD protocol was the marked decrease of WC, which was reduced by 11.9 cm to an average of 84.1 cm, which is even below the cut-off proposed by the harmonized criteria for metabolic syndrome [25] and in line with previous meta-analysis data [31]. Needless to say, this was a major benefit [8,32], which is reflected by the improvements of a several cardiometabolic biomarkers. In our study, we observed a reduction of SBP, TC, TG, LDL-C and a small but significant increase of HDL-C. The impact of VLCKD on LDL-C is still controversial in some instances, since it has been reported either unchanged [31] or reduced, such as, on average, in our study and in other recent studies conducted in men [33], or increased in a subset of patients (1 out of 4 patients) undergoing VLCKD [34], probably due to the impact of some gene variants [22]. These observations suggest that several factors, such as sex (our study included only women), the presence of selected gene variants, etc., may influence the individual LDL-C response to VLCKD and, indeed, also in our study we found some patients with no LDL-C changes and a few with a moderate increase of this marker. These findings then highlight the importance to evaluate LDL-C levels before and during/after a VLCKD program, making sure, when appropriate, to implement a specific diagnostic and therapeutic evaluation to assess ASCVD risk [35].

The overall reduction of CVD risk scores appears to be an important achievement of the VLCKD treatment evaluated in this study. Although the selected study cohort was already at low CVD risk at baseline, due to the female sex, no smoking, and the low-risk area (Italy) of their origin, the VLCKD program resulted in a further reduction (due to LDL-C and SBP reduction) of the SCORE CVD risk and of the FRS BMI and FRS lipids. Therefore, the VLCKD-driven improvement of several variables, either included or not in these risk algorithms, plays a role in reducing the global CVD risk.

A relevant reduction of insulin resistance, according to HOMA-IR reduction from 3.17 to 1.78 on average, represented another benefit, in line with other hospital-based studies [36]. Interestingly, subjects with HOMA-IR values above the threshold of 2, which indicates the presence of insulin resistance, were 66.7% at baseline but only 30% at the end of the protocol, suggesting that some participants did not fully improve their insulin resistance status.

These results obtained in a real-world setting thus appear comparable with those obtained in hospital-based studies and are relevant not only for body weight reduction per se but also of advantage in the overall reduction of primary CV and metabolic risk.

VLCKD may be a challenging approach for patients, especially in the first 2 phases, and requires a series of social and psychological features that may not be available to all subjects candidate to such treatment. This is reflected by a rather relevant rate of drop-out or non-compliance associated with VLCKD. Overall, 11/44 subjects either did not start our protocol and additional 3/33 (9%) dropped out within the first week of treatment, due to family reasons or lack of motivation to implement such a specific dietplan. Such drop-out rate is similar to that (7.5%) previously reported [31], suggesting that, since a VLCKD is obviously conducted as outpatients, the quality of the health personnel in our 5 clinical facilities was not substantially different from that present in research hospitals. It is important to emphasize that the maximum reduction of body weight/BMI and of WC as well as cardiometabolic improvements were achieved after completion of the entire VLCKD. Thus, it is important to avoid, especially in the real-world setting, the earlier interruption of such program after phases 1, 2 and 3, which sometimes happens due to excessively fast expectations by patients or quicker access to subsequent plastic surgery. Interestingly, some strategies to improve adherence to VLCKD in the real-world setting have been recently published [37,38]

This study has some limitations. A control group undergoing standard of care treatment (i.e., a low-calorie balanced diet) was not included, which does not allow one to compare this approach to the VLCKD one, when referring, for example, to CVD risk reduction. In this regard, a study reporting the comparison between VLCKD and standard low-calorie diet in the treatment of obesity in a hospital setting [39] showed that, over a 12-month timeframe, the VLCKD intervention was associated with much greater improvement of anthropometric parameters.

Moreover, no body composition assessment or indirect calorimetry could be conducted and no blinding was possible, nor was the compilation of a food diary was achievable. In addition, only three blood samplings were performed, along with the five visits, without the possibility to collect and store additional serum samples for additional experimental determinations (i.e., adipokines and pro-inflammatory cytokines). This precluded the opportunity for a more detailed cardiometabolic study, for example, evaluating the leptin: adiponectin ratio, which is markedly reduced by loss of adipose mass and has been shown to predict carotid intima-media thickness in males [40] or of circulating ghrelin levels [19,41]. Importantly, men and non-Caucasian subjects could not be included in this study since both are not referring in a relevant way to clinical practice for VLCKD in Italy.

The findings of this study on a multi-center VLCKD program conducted in a real-world setting in a cohort of women with overweight or obesity indicate that it is safe and effective since it results in a major improvement of cardiometabolic parameters, thus leading to benefits that span well beyond the mere body weight/abdominal adiposity reduction, as they lead to a decreased primary CVD and metabolic risk. Our data cannot however be directly extended to women with severe obesity (BMI > 37 kg/m^2) and relevant organ complication or failure, or to the male sex, which should be the focus of specific studies. Future developments in the practical application of VLCKD, especially in real-world clinics, may include the evaluation of genomic determinants of responsiveness to VLCKD and their clinical implementation following rigorous frameworks for gene variant interpretation [34].

Supplementary Materials: The following are available online at https://www.mdpi.com/article/10.3390/nu13061804/s1, Table S1: Concomitant medications; Table S2A: Composition of the PentaCal supplement; Table S2B: Composition of the Protiligne meal replacement (range of content per portion).

Author Contributions: Conceptualization, P.M.; methodology, L.V., M.R.; software, E.T.; validation, P.M., E.T. and A.L.C.; formal analysis, M.C., E.T.; investigation, L.V., A.S. and C.M.; resources, P.M., A.L.C.; data curation, P.M., M.C.; writing—original draft preparation, P.M., E.T.; writing, review and editing, P.M., A.L.C.; project administration, P.M.; funding acquisition, P.M. All authors have read and agreed to the published version of the manuscript.

Funding: The study was supported by a non-conditioning grant by New Penta srl (Milan, Italy) to P.M. The funding sponsors had no role in the design of the study; in collection, analyses, or interpretation of data; in the manuscript writing; or in the decision to publish the results. The work of A.L.C., P.M. and M.C. at MultiMedica was supported by Ministry of Health—Ricerca Corrente—IRCCS MultiMedica).

Institutional Review Board Statement: The study was conducted according to the guidelines of the Declaration of Helsinki and approved by the Institutional Review Board of IRCCS Ca' Granda Ospedale Maggiore Policlinico, Milan (Italy) (approval N°441/2011).

Informed Consent Statement: Informed consent was obtained from all subjects involved in the study.

Data Availability Statement: The data presented in this study are available on request from the corresponding author. The data are not publicly available due to privacy reasons.

Acknowledgments: The Authors gratefully acknowledge the expert support of Valter Rota and Mauro Giulietti and the clinical collaboration of Simona Ferrero, Raffaella Elena Monti, Simona Nichetti and Valeria Rizzi.

Conflicts of Interest: P.M., E.T., L.V., M.R., C.M., M.C., and A.S. have no conflict of interests. A.L.C. has received honoraria, lecture fees, or research grants from: Akcea, Amgen, Astrazeneca, Eli Lilly, Genzyme, Kowa, Mediolanum, Menarini, Merck, Pfizer, Recordati, Sanofi, Sigma Tau, Amryt, and Sandoz.

References

1. Casanueva, F.F.; Moreno, B.; Rodríguez-Azeredo, R.; Massien, C.; Conthe, P.; Formiguera, X.; Barrios, V.; Balkau, B. Relationship of abdominal obesity with cardiovascular disease, diabetes and hyperlipidaemia in Spain. *Clin. Endocrinol.* **2010**, *73*, 35–40. [CrossRef]
2. Flegal, K.M.; Carroll, M.D.; Kit, B.K.; Ogden, C.L. Prevalence of obesity and trends in the distribution of body mass index among US adults, 1999–2010. *JAMA* **2012**, *307*, 491–497. [CrossRef]
3. Afshin, A.; Forouzanfar, M.H.; Reitsma, M.B.; Sur, P.; Estep, K.; Lee, A.; Marczak, L.; Mokdad, A.H.; Moradi-Lakeh, M.; Naghavi, M.; et al. Health Effects of Overweight and Obesity in 195 Countries over 25 Years. *N. Engl. J. Med.* **2017**, *377*, 13–27. [CrossRef]
4. Blüher, M. Obesity: Global epidemiology and pathogenesis. *Nat. Rev. Endocrinol.* **2019**, *15*, 288–298. [CrossRef]
5. Magni, P.; Bier, D.M.; Pecorelli, S.; Agostoni, C.; Astrup, A.; Brighenti, F.; Cook, R.; Folco, E.; Fontana, L.; Gibson, R.A.; et al. Perspective: Improving Nutritional Guidelines for Sustainable Health Policies: Current Status and Perspectives. *Adv. Nutr.* **2017**, *8*, 532–545. [CrossRef] [PubMed]
6. Yumuk, V.; Tsigos, C.; Fried, M.; Schindler, K.; Busetto, L.; Micic, D.; Toplak, H. European Guidelines for Obesity Management in Adults. *Obes. Facts* **2015**, *8*. [CrossRef]
7. Riaz, H.; Khan, M.S.; Siddiqi, T.J.; Usman, M.S.; Shah, N.; Goyal, A.; Khan, S.S.; Mookadam, F.; Krasuski, R.A.; Ahmed, H. Association between Obesity and Cardiovascular Outcomes: A Systematic Review and Meta-analysis of Mendelian Randomization Studies. *JAMA Netw. Open* **2018**, *1*, e183788. [CrossRef] [PubMed]
8. Neeland, I.J.; Ross, R.; Després, J.P.; Matsuzawa, Y.; Yamashita, S.; Shai, I.; Seidell, J.; Magni, P.; Santos, R.D.; Arsenault, B.; et al. Visceral and ectopic fat, atherosclerosis, and cardiometabolic disease: A position statement. *Lancet Diabetes Endocrinol.* **2019**, *7*. [CrossRef]
9. Yumuk, P.F.; Dane, F.; Yumuk, V.D.; Yazici, D.; Ege, B.; Bekiroglu, N.; Toprak, A.; Iyikesici, S.; Basaran, G.; Turhal, N.S. Impact of body mass index on cancer development. *J. BUON* **2008**, *13*, 55–59. [PubMed]
10. De Backer, G.G. Prevention of cardiovascular disease: Much more is needed. *Eur. J. Prev. Cardiol.* **2018**, *25*, 1083–1086. [CrossRef]
11. Joseph, P.; Leong, D.; McKee, M.; Anand, S.S.; Schwalm, J.D.; Teo, K.; Mente, A.; Yusuf, S. Reducing the Global Burden of Cardiovascular Disease, Part 1: The Epidemiology and Risk Factors. *Circ. Res.* **2017**, *121*, 677–694. [CrossRef] [PubMed]
12. Kraak, V.I.; Swinburn, B.; Lawrence, M.; Harrison, P. An accountability framework to promote healthy food environments. *Public Health Nutr.* **2014**, *17*, 2467–2483. [CrossRef]
13. Donini, L.M.; Cuzzolaro, M.; Gnessi, L.; Lubrano, C.; Migliaccio, S.; Aversa, A.; Pinto, A.; Lenzi, A. Obesity treatment: Results after 4 years of a Nutritional and Psycho-Physical Rehabilitation Program in an outpatient setting. *Eat. Weight. Disord.* **2014**, *19*, 249–260. [CrossRef] [PubMed]
14. Sacks, F.M.; Bray, G.A.; Carey, V.J.; Smith, S.R.; Ryan, D.H.; Anton, S.D.; McManus, K.; Champagne, C.M.; Bishop, L.M.; Laranjo, N.; et al. Comparison of weight-loss diets with different compositions of fat, protein, and carbohydrates. *N. Engl. J. Med.* **2009**, *360*, 859–873. [CrossRef]
15. Bray, G.A.; Heisel, W.E.; Afshin, A.; Jensen, M.D.; Dietz, W.H.; Long, M.; Kushner, R.F.; Daniels, S.R.; Wadden, T.A.; Tsai, A.G.; et al. The Science of Obesity Management: An Endocrine Society Scientific Statement. *Endocr. Rev.* **2018**, *39*, 79–132. [CrossRef]
16. Gadde, K.M.; Martin, C.K.; Berthoud, H.R.; Heymsfield, S.B. Obesity: Pathophysiology and Management. *J. Am. Coll. Cardiol.* **2018**, *71*, 69–84. [CrossRef]

17. Paddon-Jones, D.; Westman, E.; Mattes, R.D.; Wolfe, R.R.; Astrup, A.; Westerterp-Plantenga, M. Protein, weight management, and satiety. *Am. J. Clin. Nutr.* **2008**, *87*, 1558S–1561S. [CrossRef]

18. Spadafranca, A.; Rinelli, S.; Riva, A.; Morazzoni, P.; Magni, P.; Bertoli, S.; Battezzati, A. Phaseolus vulgaris extract affects glycometabolic and appetite control in healthy human subjects. *Br. J. Nutr.* **2013**, *109*. [CrossRef]

19. Crujeiras, A.B.; Goyenechea, E.; Abete, I.; Lage, M.; Carreira, M.C.; Martínez, J.A.; Casanueva, F.F. Weight regain after a diet-induced loss is predicted by higher baseline leptin and lower ghrelin plasma levels. *J. Clin. Endocrinol. Metab.* **2010**, *95*, 5037–5044. [CrossRef] [PubMed]

20. Hemmingsson, E.; Johansson, K.; Eriksson, J.; Sundström, J.; Neovius, M.; Marcus, C. Weight loss and dropout during a commercial weight-loss program including a very-low-calorie diet, a low-calorie diet, or restricted normal food: Observational cohort study. *Am. J. Clin. Nutr.* **2012**, *96*, 953–961. [CrossRef] [PubMed]

21. Basciani, S.; Costantini, D.; Contini, S.; Persichetti, A.; Watanabe, M.; Mariani, S.; Lubrano, C.; Spera, G.; Lenzi, A.; Gnessi, L. Safety and efficacy of a multiphase dietetic protocol with meal replacements including a step with very low calorie diet. *Endocrine* **2015**, *48*, 863–870. [CrossRef]

22. Caprio, M.; Infante, M.; Moriconi, E.; Armani, A.; Fabbri, A.; Mantovani, G.; Mariani, S.; Lubrano, C.; Poggiogalle, E.; Migliaccio, S.; et al. Very-low-calorie ketogenic diet (VLCKD) in the management of metabolic diseases: Systematic review and consensus statement from the Italian Society of Endocrinology (SIE). *J. Endocrinol. Investig.* **2019**, *42*, 1365–1386. [CrossRef]

23. Rolland, C.; Johnston, K.L.; Lula, S.; Macdonald, I.; Broom, J. Long-term weight loss maintenance and management following a VLCD: A 3-year outcome. *Int. J. Clin. Pract.* **2014**, *68*, 379–387. [CrossRef]

24. Moriconi, E.; Camajani, E.; Fabbri, A.; Lenzi, A.; Caprio, M. Very-Low-Calorie Ketogenic Diet as a Safe and Valuable Tool for Long-Term Glycemic Management in Patients with Obesity and Type 2 Diabetes. *Nutrients* **2021**, *13*, 758. [CrossRef] [PubMed]

25. Alberti, K.G.; Eckel, R.H.; Grundy, S.M.; Zimmet, P.Z.; Cleeman, J.I.; Donato, K.A.; Fruchart, J.C.; James, W.P.; Loria, C.M.; Smith, S.C.; et al. Harmonizing the metabolic syndrome: A joint interim statement of the International Diabetes Federation Task Force on Epidemiology and Prevention; National Heart, Lung, and Blood Institute; American Heart Association; World Heart Federation; International Atherosclerosis Society; and International Association for the Study of Obesity. *Circulation* **2009**, *120*, 1640–1645. [CrossRef]

26. Friedewald, W.T.; Levy, R.I.; Fredrickson, D.S. Estimation of the concentration of low-density lipoprotein cholesterol in plasma, without use of the preparative ultracentrifuge. *Clin. Chem.* **1972**, *18*, 499–502. [CrossRef]

27. Matthews, D.R.; Hosker, J.P.; Rudenski, A.S.; Naylor, B.A.; Treacher, D.F.; Turner, R.C. Homeostasis model assessment: Insulin resistance and beta-cell function from fasting plasma glucose and insulin concentrations in man. *Diabetologia* **1985**, *28*, 412–419. [CrossRef] [PubMed]

28. Cockcroft, D.W.; Gault, M.H. Prediction of creatinine clearance from serum creatinine. *Nephron* **1976**, *16*, 31–41. [CrossRef] [PubMed]

29. Andersson, C.; Johnson, A.D.; Benjamin, E.J.; Levy, D.; Vasan, R.S. 70-year Legacy of the Framingham Heart Study. *Nat. Rev. Cardiol.* **2019**, *16*. [CrossRef] [PubMed]

30. HeartScore. Access HeartScore—Full Version. Available online: https://www.heartscore.org/en_GB/access (accessed on 2 April 2021).

31. Castellana, M.; Conte, E.; Cignarelli, A.; Perrini, S.; Giustina, A.; Giovanella, L.; Giorgino, F.; Trimboli, P. Efficacy and safety of very low calorie ketogenic diet (VLCKD) in patients with overweight and obesity: A systematic review and meta-analysis. *Rev. Endocr. Metab. Disord.* **2020**, *21*, 5–16. [CrossRef] [PubMed]

32. Ross, R.; Neeland, I.J.; Yamashita, S.; Shai, I.; Seidell, J.; Magni, P.; Santos, R.D.; Arsenault, B.; Cuevas, A.; Hu, F.B.; et al. Waist circumference as a vital sign in clinical practice: A Consensus Statement from the IAS and ICCR Working Group on Visceral Obesity. *Nat. Rev. Endocrinol.* **2020**, *16*. [CrossRef]

33. Mongioì, L.M.; Cimino, L.; Condorelli, R.A.; Magagnini, M.C.; Barbagallo, F.; Cannarella, R.; La Vignera, S.; Calogero, A.E. Effectiveness of a Very Low Calorie Ketogenic Diet on Testicular Function in Overweight/Obese Men. *Nutrients* **2020**, *12*, 2967. [CrossRef] [PubMed]

34. Aronica, L.; Volek, J.; Poff, A.; D'agostino, D.P. Genetic variants for personalised management of very low carbohydrate ketogenic diets. *BMJ Nutr. Prev. Health* **2020**, *3*, 363–373. [CrossRef] [PubMed]

35. Ference, B.A.; Graham, I.; Tokgozoglu, L.; Catapano, A.L. Impact of Lipids on Cardiovascular Health: JACC Health Promotion Series. *J. Am. Coll. Cardiol.* **2018**, *72*, 1141–1156. [CrossRef] [PubMed]

36. Watanabe, M.; Risi, R.; Camajani, E.; Contini, S.; Persichetti, A.; Tuccinardi, D.; Ernesti, I.; Mariani, S.; Lubrano, C.; Genco, A.; et al. Baseline HOMA IR and Circulating FGF21 Levels Predict NAFLD Improvement in Patients Undergoing a Low Carbohydrate Dietary Intervention for Weight Loss: A Prospective Observational Pilot Study. *Nutrients* **2020**, *12*, 2141. [CrossRef] [PubMed]

37. Gibson, A.A.; Sainsbury, A. Strategies to Improve Adherence to Dietary Weight Loss Interventions in Research and Real-World Settings. *Behav. Sci.* **2017**, *7*, 44. [CrossRef]

38. Muscogiuri, G.; Barrea, L.; Laudisio, D.; Pugliese, G.; Salzano, C.; Savastano, S.; Colao, A. The management of very low-calorie ketogenic diet in obesity outpatient clinic: A practical guide. *J. Transl. Med.* **2019**, *17*, 356. [CrossRef] [PubMed]

39. Moreno, B.; Bellido, D.; Sajoux, I.; Goday, A.; Saavedra, D.; Crujeiras, A.B.; Casanueva, F.F. Comparison of a very low-calorie-ketogenic diet with a standard low-calorie diet in the treatment of obesity. *Endocrine* **2014**, *47*, 793–805. [CrossRef]

40. Norata, G.D.; Raselli, S.; Grigore, L.; Garlaschelli, K.; Dozio, E.; Magni, P.; Catapano, A.L. Leptin:adiponectin ratio is an independent predictor of intima media thickness of the common carotid artery. *Stroke* **2007**, *38*. [CrossRef] [PubMed]
41. Bertoli, S.; Magni, P.; Krogh, V.; Ruscica, M.; Dozio, E.; Testolin, G.; Battezzati, A. Is ghrelin a signal of decreased fat-free mass in elderly subjects? *Eur. J. Endocrinol.* **2006**, *155*, 321–330. [CrossRef] [PubMed]

nutrients

Article

Gut Microbiota Functional Dysbiosis Relates to Individual Diet in Subclinical Carotid Atherosclerosis

Andrea Baragetti [1,†], Marco Severgnini [2,†], Elena Olmastroni [3], Carola Conca Dioguardi [4], Elisa Mattavelli [1], Andrea Angius [5], Luca Rotta [6], Javier Cibella [7], Giada Caredda [1], Clarissa Consolandi [2], Liliana Grigore [8,9], Fabio Pellegatta [8,9], Flavio Giavarini [1], Donatella Caruso [1], Giuseppe Danilo Norata [1], Alberico Luigi Catapano [1,9,*] and Clelia Peano [4,7]

1   Department of Pharmacological and Biomolecular Sciences, University of Milan, 20133 Milan, Italy; andrea.baragetti@unimi.it (A.B.); elisa.mattavelli@unimi.it (E.M.); giada.caredda@unimi.it (G.C.); flavio.giavarini@unimi.it (F.G.); donatella.caruso@unimi.it (D.C.); danilo.norata@unimi.it (G.D.N.)
2   Institute of Biomedical Technologies, National Research Council, 20090 Segrate, Milan, Italy; marco.severgnini@itb.cnr.it (M.S.); clarissa.consolandi@itb.cnr.it (C.C.)
3   Epidemiology and Preventive Pharmacology Service (SEFAP), Department of Pharmacological and Biomolecular Sciences, University of Milan, 20133 Milan, Italy; olmastronielena92@gmail.com
4   Institute of Genetic and Biomedical Research, UoS Milan, National Research Council, Rozzano, 20125 Milan, Italy; carola.conca_dioguardi@humanitas.it (C.C.D.); clelia.peano@humanitasresearch.it (C.P.)
5   Institute of Genetic and Biomedical Research, National Research Council, 09042 Cagliari, Italy; andrea.angius@irgb.cnr.it
6   Department of Experimental Oncology, IEO, European Institute of Oncology, IRCCS, 20141 Milan, Italy; luca.rotta@ieo.it
7   Genomic Unit, IRCCS, Humanitas Clinical and Research Center, 20090 Rozzano, Milan, Italy; Huge@humanitasresearch.it
8   S.I.S.A. Center for the Study of Atherosclerosis, Bassini Hospital, 20092 Cinisello Balsamo, Milan, Italy; grigore.centroatero@gmail.com (L.G.); fabio.pellegatta@guest.unimi.it (F.P.)
9   MultiMedica IRCCS, 20092 Cinisello Balsamo, Milan, Italy
*   Correspondence: alberico.catapano@unimi.it; Tel.: +39-0250-318-302
†   These authors equally contributed.

Citation: Baragetti, A.; Severgnini, M.; Olmastroni, E.; Dioguardi, C.C.; Mattavelli, E.; Angius, A.; Rotta, L.; Cibella, J.; Caredda, G.; Consolandi, C.; et al. Gut Microbiota Functional Dysbiosis Relates to Individual Diet in Subclinical Carotid Atherosclerosis. *Nutrients* **2021**, *13*, 304. https://doi.org/10.3390/nu13020304

Received: 9 December 2020
Accepted: 15 January 2021
Published: 21 January 2021

Publisher's Note: MDPI stays neutral with regard to jurisdictional claims in published maps and institutional affiliations.

Abstract: Gut Microbiota (GM) dysbiosis associates with Atherosclerotic Cardiovascular Diseases (ACVD), but whether this also holds true in subjects without clinically manifest ACVD represents a challenge of personalized prevention. We connected exposure to diet (self-reported by food diaries) and markers of Subclinical Carotid Atherosclerosis (SCA) with individual taxonomic and functional GM profiles (from fecal metagenomic DNA) of 345 subjects without previous clinically manifest ACVD. Subjects without SCA reported consuming higher amounts of cereals, starchy vegetables, milky products, yoghurts and bakery products versus those with SCA (who reported to consume more mechanically separated meats). The variety of dietary sources significantly overlapped with the separations in GM composition between subjects without SCA and those with SCA (RV coefficient between nutrients quantities and microbial relative abundances at genus level = 0.65, *p*-value = 0.047). Additionally, specific bacterial species (*Faecalibacterium prausnitzii* in the absence of SCA and *Escherichia coli* in the presence of SCA) are directly related to over-representation of metagenomic pathways linked to different dietary sources (sulfur oxidation and starch degradation in absence of SCA, and metabolism of amino acids, syntheses of palmitate, choline, carnitines and Trimethylamine *n*-oxide in presence of SCA). These findings might contribute to hypothesize future strategies of personalized dietary intervention for primary CVD prevention setting.

Keywords: Atherosclerotic Cardiovascular Diseases; Gut Microbiota; next generation sequencing

1. Introduction

Atherosclerotic Cardiovascular Diseases (ACVD) still contribute significantly to excessive mortality, despite pharmacological weapons substantially improving their treatment [1]. Moreover, the preventive perspectives are complicated yet at early stages because the approaches to identify high risk subjects are complex. First, the presence of focal atherosclerotic lesions, detected by currently used techniques (such as ultrasound) identify subjects at increased risk of ACVD [1], although tracking Subclinical Carotid Atherosclerosis (SCA) progression by preclinical markers (including carotid Intima-Media Thickness—IMT) remains an important step to further identify subjects in primary prevention [2]. Second, IMT and the presence of focal carotid vascular lesions (a more robust indicator of ACVD risk [1]) are both predicted by classical cardiovascular risk factors (CVRFs) (e.g., Type 2 Diabetes (T2D), Metabolic Syndrome (MetS), dyslipidemia), and patterns of individual predisposition to other emerging factors such as low-grade inflammation [3,4]. Thereby, early approaches to reduce the onset and burden of these factors in a personalized fashion are surmised [5] and become feasible with the understanding of modifiable factors, like changes in lifestyle and exposure to environmental factors that differently affect risk of ACVD [6]. Under this vision, diet represents the first target for personalized approaches since digestion, the metabolism of nutrients and the absorption of potentially bioactive compounds shaping immune-metabolic functions of the host are demanded to Gut Microbiota (GM), which reflects the individual interaction with the environment [7]. Actually, the physiological cross-talk between diet with GM richness and variety (namely "eubiosis") is well proven. On one hand, acute changes in dietary habits rapidly re-shape GM composition [8] while, on the other, GM compositions appears to be a superior factor (even beyond the genetic and clinical background of the host) in determining individual metabolic and inflammatory postprandial response to different foods [9]. This cross-talk is supported by the exposure to different dietary patterns, contributing to the absorption of a multitude of dietary metabolites that, in turn, critically foster well-described anti- or pro-inflammatory metagenomic molecular pathways. For example, some short-chain fatty acids are beneficial (e.g., butyrate and propionate activate intestinal gluconeogenesis), whereas others promote pathogenic mechanisms leading metabolic impairments (e.g., acetate promotes hepatic gluconeogenesis predisposing to glucose intolerance [10]). Bile salts contribute to preserving the intestinal barrier while favoring the proliferation of inflammatory bacteria like Clostridia by the Aryl hydrocarbon-receptor system [11,12]. Additionally, branched-chain amino acids [13] in protein-based foods and products of tyrosine/tryptophan metabolism (*p*-Cresol and indoles) exert inflammatory potential and promote insulin resistance [14]. Finally, the atherogenic properties of Trimethylamine *n*-oxide (TMAO), metabolized in the liver starting from dietary choline, have been deeply described [15–20]. Despite these robust data, whether the individual exposure to different dietary sources relates with GM taxonomic alterations (namely, "dysbiosis") even during initial stages of atherosclerosis and before clinical establishment of ACVD is still to be understood. Whether this relation is explained by the activation of bacterial cellular pathways involved in the metabolism of dietary sources towards potentially active metagenomic compounds represents a significant add-on the potential causal effect of GM in atherosclerosis [21–23]. This scientific question is of current clinical concern given the continuous changes in dietary sources nowadays among societies [6]. However, methodological criticisms affect GM composition analysis [24], the design of interventional dietary trials is scarce up to now, small-sized trials gave contrasting data about the effect of dietary intervention on changes of GM composition and subsequent effect on markers of ACVD risk [25]. This sets the stage for an immediate clinical value, since the clustering of taxonomic and metagenomic signatures with individual dietary lifestyle might represent a pioneering approach of primary prevention, identifying patients among the population at increased risk of future occurrence of CVD. We here address the relation between functional metagenomic signatures and individual exposure to diet during subclinical manifestation of CVD, studying people from a general population-based study, in primary prevention, with low prevalence of CVRFs and characterized by different stages of SCA.

2. Materials and Methods

2.1. Study Population

For the purposes of this study, we collected fecal samples from 345 subjects in primary prevention for CVD from the population-based study representative of the general population of the northern area of Milan (Progressione delle Lesioni Intimali Carotidee—PLIC), which has been extensively described elsewhere [26,27]. Subjects of this study were screened at the Center for the Study of Atherosclerosis at E. Bassini Hospital (Cinisello Balsamo, Milan, Italy) for personal and familial clinical history and for absence of previous CVDs and personal history of T2D or MetS (defined according to validated criteria). Additionally, we excluded: (i) subjects reporting use of glucose-lowering drugs, (ii) with positive personal history of CVD (either ischemic heart disease, ST segment elevation or non-ST elevation myocardial infarction, aortic-coronary by-pass grafting, angioplasty, transient ischemic attack, stroke, heart failure from Class II to IV (according to New York Heart Association (NYHA) definition or documented peripheral arteriopathy), (iii) with MetS (defined according to harmonized criteria of the American Heart Association [28]), (iv) chronic kidney disease (Glomerular Filtration Rate, GFR, <60 mL/min or documented albuminuria > 30 mg/g), (iv) pregnancy and (v) reported malignancies. Data management and statistical analyses were performed with the coordination of the Epidemiology and Preventive Pharmacology Centre (SEFAP) of the University of Milan. The study was approved by the Scientific Committee of the University of Milan (SEFAP/Pr.0003). Informed consent was obtained from subjects (all over 18 years-old), in accordance with the Declaration of Helsinki. Systolic and diastolic blood pressure and Body Mass Index (BMI), waist and waist/hip ratio were measured. Information on the presence of hepatic steatosis, available on a subgroup of 133 subjects, was defined via ultrasound, as per already published protocols [26]. Blood samples were collected from the antecubital vein after 12 h fasting on NaEDTA tubes (BD Vacuette®, Franklin Lakes, NJ, USA) and then centrifuged at 3000 rpm for 12 min (Eppendorf 580r, Eppendorf, Hamburg, Germany) for biochemical parameters profiling including: total cholesterol, HDL-C, triglycerides, ApoB, ApoA-I, glucose, liver enzymes, creatinine and creatinine-phospho kinase (CPK). Measurements were performed using immuno-turbidimetric and enzymatic methods thorough automatic analyzers (Randox, Crumlin, UK). LDL-C was derived from Friedewald formula. Separately, whole blood in NaEDTA tubes was used for hematocrit analysis to derive a total count of leukocytes and their fractions (neutrophils, lymphocytes, monocytes, eosinophils and basophils, indicated as cells*1000/microliter).

Fecal samples of 345 subjects were collected and used for the analysis of GM taxonomic composition. SCA was defined by ultrasound-based analysis of bilateral carotid arteries as previously described [27]. In detail, common carotid IMT (one centimeter from the bulb) was measured in longitudinal view, far wall, by a high resolution B-mode ultrasound-based system (Vivid S5—GE Healthcare, Wauwatosa, WI, USA) connected to linear probe—4.0 × 13.0 MHz frequency; 14 × 48 mm footprint, 38 mm field of view). A mean value for both sides was averaged. "+ IMT" was determined as the presence of IMT above the 75th percentile of the median IMT for a Caucasian population according to ASE guidelines [29].

SCA was defined when mean IMT was ≥1.3 mm or in presence of focal atherosclerotic lesions larger than 1.3 mm using a manual caliper in longitudinal view either in far or near wall and over every carotid tract (common, bulb section, bifurcation, internal or external branches). In two scans performed by the same operator in 75 subjects, the mean difference in IMT was 0.005 ± 0.002 mm and the coefficient of variation (CV) was 1.93%. The correlation between two scans was significant ($r = 0.96$; $p < 0.0001$). The combination of information from IMT measurement and from presence/absence of SCA allowed four different SCA stages to be identified: subjects without intimal thickening and without SCA ("−IMT/−SCA", $n = 23$); subjects with intimal thickening but without SCA ("+IMT/−SCA", $n = 173$); subjects without intimal thickening but with SCA ("−IMT/+SCA", $n = 121$); subjects with both intimal thickening and SCA ("+IMT/+SCA", $n = 23$).

Whole shotgun metagenomic sequencing analyses were performed on the same fecal samples of 23 "−IMT/−SCA" and 23 "+IMT/+SCA", whose clinical characteristics are reported in (Supplementary Table S1). Further vascular characterization according to validated criteria [30] allowed, among the +IMT/+SCA group, subjects with "no advanced SCA" (stenosis < 30% and p/S < 125 cm/s) vs. the "advanced SCA" (stenosis 30% and elevation of the p/S wave in the bilateral internal carotid branches) to be distinguished. The advanced SCA was then divided by further characterization identifying: (a) SCA causing stenosis between 30 and 50% with p/S < 125 cm/s; (b) SCA causing stenosis between 50 and 70% and p/S between 125 and 250 cm/s; (c) SCA causing stenosis over 70% and p/S over 250 cm/s. We evaluated echolucencies of the atherosclerotic lesions among all subjects from the −IMT/+SCA and from the +IMT/+SCA group, using grey-scale definition and parameters of the QuickScan® and autoIMT® software included in the ultrasound machinery (Samsung HM70a®, Samsung®, Seoul, South Korea).

Additional information, clinical criteria, and determination of biochemical parameters are reported in Supplementary Materials.

2.2. Lifestyle Data, Collection and Analysis of Dietary Habits

Subjects self-reported their level and type of physical activity and smoking habit and information about individual diet were collected in the PLIC study as previously reported [31]. In detail, all subjects were requested to complete a semi-quantitative daily food diary representative of seven days before the clinical evaluation and collection of the fecal sample. The food diary was administered to subjects following instructions about the reporting of quali/quantitative dietary information by two dieticians (blinded on subject's clinical history). In the food diary, subjects reported for each meal (breakfast, lunch, dinner and snacks) the foods, the brand names of foods (where applicable), the methods of preparation and dressings. During the seven days, dieticians were available for help and to provide more instructions to subjects by phone or by email. A portion reference from validated color photographs (the "Atlante Fotografico delle Porzioni degli Alimenti"; [32]) was also given to subjects, for further help in the interpretation of food quantities. Then, after seven days, the filled-out food diary was analyzed by dieticians during the outpatient evaluation in front of the subject: (i) to clarify details and improper indications and (ii) to derive individual daily energy and the seven-day dietary averaged nutrient intakes (as g/week), referring to the Italian BDA database [33]. BDA- Food Composition Database for Epidemiological Studies in Italy—2015). Additionally, BDA and the reference values of the Italian Society of Nutrition ("LARN", Livelli di Assunzione di Riferimento di Nutrienti ed Energia) were used to exclude outlier data about energy intake, deriving from the improper self-reporting of the subject.

2.3. DNA Extraction from Fecal Samples

Total microbial DNA from all the fecal samples collected has been extracted as previously described [34], since the protocol herein described was specifically modified to allow an efficient and unbiased bacterial DNA extraction from human fecal samples. Genomic DNA quality was assessed by using the TapeStation 2200 system (Agilent, Santa Clara, CA, USA); only samples having a DNA Integrity Number (DIN) > 4 were used for successive analyses.

2.4. Libraries Construction and Sequencing
Microbiome 16S Analysis

For each sample, the V3–V4 region of the 16S rRNA gene was PCR-amplified by using primers carrying overhanging adapter sequences (primer selection originally described in [35]), following the Illumina 16S Metagenomic Sequencing Library Preparation protocol [36] (Illumina, San Diego, CA, USA), and libraries were barcoded using dual Nextera® XT indexes (Illumina). Indexed libraries were pooled at equimolar concentrations and sequenced on a HiSeq 2500 Illumina sequencing platform generating 2×250 bp paired-end reads, according to manufacturer's instructions (Illumina).

2.5. Metagenome Analysis

A total of 46 Metagenomic shotgun libraries were prepared from the DNA extracted from fecal samples from 23 "−IMT/−SCA" and 23 "+IMT/+SCA; dual indexed libraries were prepared following the Nextera® DNA Flex Library Prep Kit (Illumina); then, they were pooled at equimolar concentrations and sequenced on a Novaseq 6000 Illumina sequencing platform, generating 2×100 bp paired-end reads.

2.6. Statistical and Bioinformatic Data Analysis

Microbiome 16S Data Analysis

The 16S rRNA raw sequences were processed through a bioinformatic pipeline composed of PANDAseq [37] and QIIME (release 1.8.0 [38]); Operational Taxonomic Units (OTUs) were assigned at 97% similarity level and classified against the Greengenes database (release 13.8; [39]). Biodiversity and distribution of the microorganisms were characterized via alpha- and beta-diversity analysis evaluating specific metrics and distances. Statistical evaluation was performed by non-parametric Monte Carlo-based tests and by analysis of variance with partitioning among sources of variation ("adonis" function) in the R package "vegan" (version 2.0–10; [40]) for alpha- and beta-diversity, respectively. Differences in abundances of bacterial taxa were analyzed by non-parametric Mann-Whitney U-test using MATLAB software (Natick, MA, USA). Unless otherwise stated, $p < 0.05$ were considered as significant for each statistical analysis. Detailed procedures are available as Supplementary Materials.

2.7. Metagenome Data Analysis

Metagenomic reads were quality filtered using the recommended pipelines from the Human Microbiome Project [41,42]. Resulting reads were then processed by HUMAnN2 pipeline (v. 0.11.2 [43]). In order to compensate for different sequencing depths, all measures were expressed as copies-per-million (CPM).

Alpha- and beta-diversity analyses were performed on species-level taxonomic classification and MetaCyc reaction-level functional classification [44], using non-phylogenetic indexes and distances in QIIME. Statistical evaluation was performed as described above. Pathways were grouped to upper levels thanks to their lineage association in MetaCyc. Detailed procedures are available as Supplementary Materials.

2.8. Dietary Data Analysis

Statistical data of nutrients composition for individuals with and without SCA was performed by employing the non-parametric Mann-Whitney U-test. Overall separation between patients was assessed calculating Bray-Curtis distances among patients on the basis of the nutrients table and "adonis" function in the R package "vegan" was used. In order to assess the correlation between dietary and microbial composition data, the RV coefficient [45] was calculated; coefficient statistical significance of the coefficient was calculated by 99,999 random permutations [46].

2.9. Data Availability

Sequencing data of 16S rRNA amplicons (raw reads, $n = 345$) and metagenomes (after removal of human sequences and duplicates, $n = 46$) have been deposited in NCBI Short-Read Archive (SRA) under accession number PRJNA615842 [47].

3. Results

3.1. Gut Microbiota Dysbiosis Associates with Cubclinical Carotid Atherosclerosis

We identified 144 subjects with SCA by carotid ultrasound examination and 201 gender-matched subjects without SCA (clinical characteristics of both groups reported in Table 1).

Table 1. Descriptives of the population according to SCA.

	Total	SCA		
Variable	*n* = 345	No (*n* = 201)	Yes (*n* = 144)	*p*-Value
Men gender, *n* (%)	158 (45.80)	93 (46.50)	65 (44.83)	0.760
Age (years)	67.32 (11.0)	63.83 (11.52)	72.04 (8.18)	<0.001 ***
<= 60 years-old	86	73	13	
60–70 years-old	105	64	41	
70–80 years-old	130	57	73	
>80 years-old	24	6	18	
Alcohol consumption, *n* (%)	231 (67.74)	129 (65.82)	102 (70.34)	0.380
Smoking, *n* (%)	37 (10.85)	22 (11.22)	15 (10.34)	0.800
Physical Activity, *n* (%)	175 (51.47)	115 (58.97)	60 (41.38)	0.001 ***
BMI (Kg/m^2)	26.44 (3.90)	26.22 (3.85)	26.75 (4.00)	0.230
Lean, *n* (%)	114	76	38	
Overweight, *n* (%)	165	86	79	
Obese, *n* (%)	53	30	23	
Waist-hip Ratio	0.89 (0.08)	0.87 (0.08)	0.91 (0.08)	<0.001 ***
Systolic pressure (mmHg)	126.91 (14.41)	125.20 (13.74)	129.30 (15.01)	0.008 **
Diastolic pressure (mmHg)	75.75 (9.06)	75.06 (8.74)	76.71 (9.44)	0.100
Antihypertensive drugs, *n* (%)	155 (45.45)	73 (37.24)	82 (56.55)	0.004 ***
Total Cholesterol (mg/dL)	199.0 (32.71)	202.90 (32.13)	193.60 (32.86)	0.009 **
HDL-C (mg/dL)	60.57 (13.60)	62.23 (13.59)	58.29 (13.32)	0.008 **
LDL-C (mg/dL)	118.53 (27.62)	121.20 (27.23)	114.90 (27.83)	0.037 *
Triglycerides (mg/dL)	99.46 (37.75)	97.42 (36.76)	102.30 (39.03)	0.240
Apolipoprotein A1 (mg/dL)	157.91 (19.87)	160.30 (20.24)	154.60 (18.92)	0.008 **
Apolipoprotein B (mg/dL)	101.55 (24.70)	103.60 (25.35)	98.65 (23.57)	0.060
Lipid lowering drugs, *n* (%)	164 (48.09)	79 (40.31)	85 (58.62)	0.008 **
Fasting glucose (mg/dL)	100.36 (10.49)	100.30 (9.68)	100.40 (11.56)	0.910
Uric acid (mg/dL)	5.17 (1.44)	4.96 (1.35)	5.45 (1.52)	0.002 **
Creatinine (mg/dL)	0.84 (0.19)	0.82 (0.18)	0.86 (0.20)	0.030 *
ALT (UI/L)	21.13 (14.59)	21.63 (16.68)	20.43 (11.10)	0.420
AST (UI/L)	23.64 (6.03)	23.68 (6.39)	23.57 (5.52)	0.870
GGT (UI/L)	27.73 (36.43)	27.62 (37.85)	27.87 (34.50)	0.950
Liver steatosis, *n* (%)	33 (24.80)	11 (5.50)	20 (13.80)	0.061
CPK (mg/dL)	120.79 (62.65)	121.00 (66.09)	120.50 (57.78)	0.940
Hs-CRP (mg/dL)	0.11 (0.06–0.21)	0.10 (0.05–0.19)	0.11 (0.06–0.21)	0.176
Neutrophils (cells*10^3/μL)	3.50 (1.25)	3.42 (1.22)	3.61 (1.28)	0.180
Leucocytes (cells*10^3/μL)	6.26 (1.68)	6.12 (1.49)	6.46 (1.89)	0.080
Lymphocytes (cells*10^3/μL)	2.05 (0.92)	2.00 (0.52)	2.11 (1.28)	0.370
Monocytes (cells*10^3/μL)	0.51 (0.14)	0.49 (0.14)	0.53 (0.15)	0.019 *
Eosinophils (cells*10^3/μL)	0.16 (0.11)	0.16 (0.10)	0.17 (0.11)	0.460
Basophils (cells*10^3/μL)	0.04 (0.02)	0.04 (0.02)	0.04 (0.02)	0.170
C-IMT (mm)	0.77 (0.18)	0.71 (0.13)	0.85 (0.19)	<0.001 ***

Clinical parameters for the 345 selected subjects from the entire PLIC cohort (second column from left) and when divided for subjects without Subclinical Carotid Atherosclerosis (SCA) (third column from left) vs. those with SCA (fourth column from left). "***" indicates $p < 0.005$; "**" indicates $p < 0.01$; "*" indicates $p < 0.05$. p refers to that of the two-sided Mann-Whitney U-test between subjects with and without SCA. Data are presented as mean (standard deviation) if normally distributed or as median (Inter-Quartile Range) if not normally distributed (Shapiro-Wilk test). BMI: "Body Mass Index"; HDL-C: "High Density Lipoprotein cholesterol"; LDL-C "Low Density Lipoprotein cholesterol"; ALT: "Alanine aminotransferase"; AST: "Aspartate aminotransferase"; GGT: "Gamma-glutamyl transpeptidase"; CPK: "Creatine phosphokinase"; c-IMT: "carotid Intima Media Thickness" (as averaged value of IMT values of right and left carotid arteries at the common tract). Information on ultrasound hepatic steatosis was available on 133 subjects out of total studied cohort.

Subjects with SCA were older than those without SCA, showed higher waist/hip ratio (0.91 ± 0.08 vs. 0.87 ± 0.08, $p < 0.001$) and they were more hypertensive (on antihypertensive 56.55% vs. 37.24% respectively, $p = 0.004$). LDL-C was 7 mg/dL lower in subjects with SCA vs. those without (114.90 ± 27.83 vs. 121.20 ± 27.23 mg/dL, $p = 0.037$) because of the higher prevalence of lipid lowering treatments (58.62% vs. 40.31% respectively, $p = 0.008$). Subjects with SCA presented with comparable plasma high-sensitivity C-Reactive Protein (hs-CRP) but with higher blood monocytes counts versus those without

SCA (0.11 (0.06–0.21) mg/dL vs. 0.10 (0.05–0.19) mg/dL, $p = 0.179$ for hs-CRP; 0.53 ± 0.15 vs. $0.49 \pm 0.14 * 1000$ cells/μL, $p = 0.019$ for monocytes).

We firstly investigated whether changes in GM composition still occur over SCA progression, thereby analyzing taxonomic GM composition of the entire cohort ($n = 345$) by 16S rRNA-based sequencing (Supplementary Figure S1), and we then performed a metagenome shotgun sequencing on a subset of subjects with +IMT/ + SCA and of those with –IMT/-SCA phenotype ($n = 46$), selected according to SCA presence and IMT measurements as described above. This dual strategy allowed to gather, with a high degree of consistency (Supplementary Figure S2), information about different relative abundances of bacterial genera and species in the presence or absence of SCA.

The GM taxonomic composition significantly differed between subjects with SCA ($n = 144$) and those without ($n = 201$) ($p = 0.016$, unweighted Unifrac distance, Figure 1A), although no changes in GM richness were found ($p > 0.05$ in all alpha-diversity metrics). The 16S rRNA-based analysis highlighted increased relative abundance of members of Escherichia (2.8% vs. 1.4%, $p = 0.008$ in SCA and no SCA subjects, respectively) and Oscillospira (6.5% vs. 5.7%, $p = 0.013$ in SCA and no SCA subjects, respectively) genera (Figure 1B,C, Supplementary Table S2) in subjects with SCA.

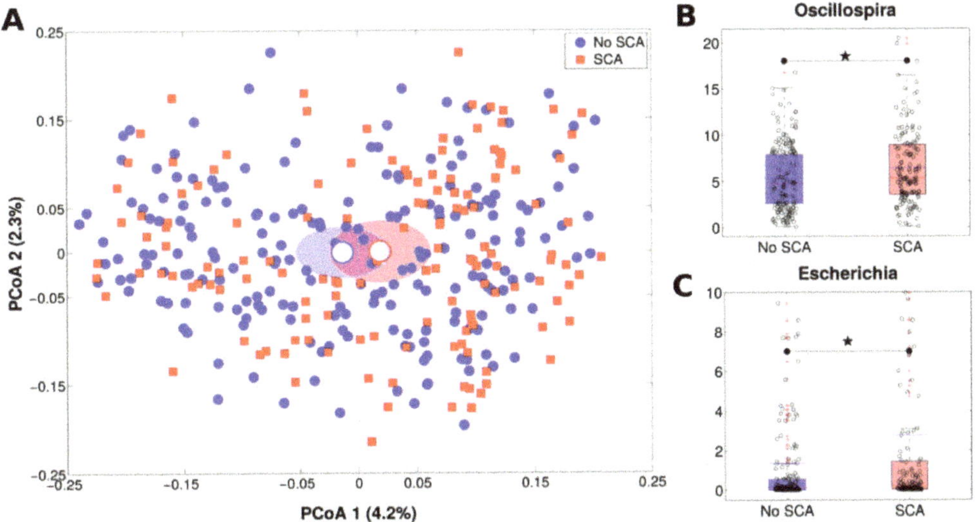

Figure 1. Taxonomic Gut Microbiota (GM) differences over SCA progression. (**A**) Principal Coordinate Analysis (PCoA) plot of the unweighted Unifrac distances; data were divided according to experimental category (SCA vs. no SCA); each point represents a sample; centroids are calculated as the mean coordinate of all samples per experimental category; ellipses represent the standard error of the mean (SEM)-based estimation of the variance. The first and second components of the variance are shown. (**B,C**) Boxplots of relative abundances of (**B**) Oscillospira and (**C**) Escherichia genera, according to SCA vs. no SCA experimental categories. Red lines indicate median and blue lines indicate mean values. Star indicates a significant difference ($p < 0.05$, Mann-Whitney U-test).

Notably, these differences in GM diversity in the presence of SCA were significantly explained by those from the subset of 23 subjects with +IMT/ +SCA phenotype, analyzed by metagenome shotgun sequencing, as compared to all the other three groups ($p \leq 0.030$ and $p \leq 0.004$ for all pair-wise comparisons on unweighted and weighted Unifrac-based PCoA, respectively, Supplementary Figure S3A). Moreover, we found few genera whose taxonomic abundances were significantly different in fecal samples from the subset of subjects with +IMT/+SCA ($n = 23$). In particular, Escherichia, Shigella and Streptococcus were those mostly significantly increased while Bacteroides were reduced in subjects with the most advanced SCA stage (Supplementary Figure S3B).

On top of significantly different metagenomic profiles in subjects with −IMT/−SCA vs. +IMT/SCA, both on alpha (p = 0.001, permutation-based t-test on observed species metrics) and beta-diversity (p = 0.002, adonis test on Bray-Curtis distance) measures, this strategy allowed the identification of increased abundance of *E. coli*, as well as of members of the Streptococcus genus (i.e., *S. salivarius*, *S. parasanguinis*, *S. anginosus*) in metagenomes of subjects with +IMT/+SCA. Vice versa, we found increased abundance of members of Bacteroides genus (i.e., *B. uniformis* and *B. thetaiotaomicron*) in the metagenomes of subjects with −IMT/−SCA (Table 2). Together these data allowed to conclude that specific taxonomic and metagenomic markers can be found still during early stages of carotid intimal thickening and SCA.

Table 2. Reduced and increased abundances of genera and species in +IMT/ + SCA.

Genus	Species	−IMT/−SCA	+IMT/+SCA
Escherichia	*E.coli*	0.83	7.50 (**)
	Uncl. Escherichia	0.18	1.93 (**)
Streptococcus	*S. salivarius*	0.23	0.50 (*)
	S. parasanguinis	0.06	0.44 (**)
	S. anginosus	0.00	0.04 (**)
Ruminococcus	*R. obeum*	0.25	0.49 (*)
Lactobacillus	*L. gasseri*	0.00	0.11 (*)
	L. fermentum	0.00	0.02 (*)
Dorea	*D. longicatena*	0.16	0.50 (**)
-	*C2likevirus*	0.00	0.04 (*)
Coprococcus	*Co. comes*	0.18	0.40 (**)
Clostridium	*C. leptum*	0.05	0.31 (*)
Parabacteroides	*Pa. goldsteinii*	0.06	0.30 (*)
Eubacterium	*Eu. ramulus*	0.08	0.26 (*)
Bifidobacterium	*B. dentium*	0.05	0.16 (*)
Bacteroides	*B. uniformis*	5.19	1.69 (*)
	B. thetaiotaomicron	0.86	0.25 (*)
Ruminococcus	*R. bromii*	2.08	1.10 (*)

List of bacterial species whose relative abundance was statistically different between +IMT/+SCA and −IMT/−SCA individuals (n = 23, each). "**" indicates $p < 0.01$; "*" indicates $p < 0.05$. p values refer to that of the two-sided Mann-Whitney U-test. For clarity, bacteria are grouped according to increase/decrease status and genus.

3.2. Functional Relevance of GM Dysbiosis over Subclinical Carotid Atherosclerosis

In order to highlight whether different shapes in GM composition over SCA stages have a functional relevance, we harnessed data from metagenome shotgun sequencing to predict the MetaCyc reactions (see Supplementary Materials) mostly encoded in GM of subjects with +IMT/+SCA (n = 23).

Notably, in the metagenomes of this group of subjects, we predicted a higher number of MetaCyc reactions as compared to those without subclinical atherosclerosis (−IMT/−SCA, n = 23) (p = 0.009 and p = 0.003 for observed species and Simpson's index, respectively) (Figure 2A); moreover, considering MetaCyc pathway abundances, significantly different functional profiles were evidenced (p = 0.001, adonis test on Bray-Curtis distances) (Figure 2B).

Based on these findings we then sought to identify which were the over- or down-represented bacterial metagenomic pathways in GM of subjects with +IMT/+SCA versus those without subclinical atherosclerosis (−IMT/−SCA). Moreover, the contribution of each species to MetaCyc reaction pathways was elucidated, highlighting the bacterial species in GM related to over or down-represented pathways (Figure 3).

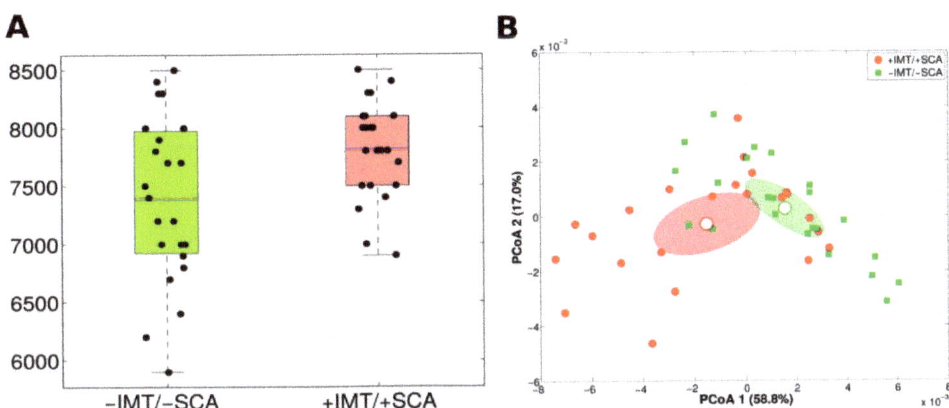

Figure 2. Hit functional markers associated with +IMT/+SCA. (**A**) Boxplots deriving from level L4 of the MetaCyc pathway hierarchy based on shotgun metagenome sequencing of +IMT/+SCA (n = 23) and −IMT/−SCA (n = 23) individuals as determined by "observed_species" metric. (**B**) Principal Coordinates Analysis (PCoA) plots of Bray-Curtis distances among +IMT/+SCA and −IMT/−SCA samples calculated on level L4 functional MetaCyc pathway classification. Each point represents a sample; centroids are calculated as the mean coordinate of all samples per experimental category; ellipses represent the standard error of the mean (SEM)-based estimation of the variance. The first and second components of the variance are shown.

We found that *E. coli* was the bacterial species associated with altered microbial pathways in samples from +IMT/+SCA individuals, including an increase in those related to the biosynthesis of palmitate, arginine, glutamine, biotin, phylloquinone, ubiquinone, menaquinone and phosphatidylethanolamine (PE). Notably, these pathways all clustered together in metagenomes from +IMT/+SCA subjects, characterized by advanced SCA stage.

By contrast, *Faecalibacterium prausnitzii* was found to contribute to thirteen significantly different MetaCyc reactions (with sulfur oxidation, starch degradation and multiple biosynthetic routes of purine and pyrimidines as the most over-represented) in samples from −IMT/−SCA subjects.

Together, these observations allowed the conclusion that taxonomic changes in GM composition, occurring during first stages of subclinical atherosclerosis and coinciding with individual exposure to dietary sources, highlight functional metagenomic relevance.

3.3. Individual Diet Clusters with Changes in Taxonomic GM Composition and SCA

These observations supported that early re-shape in taxonomic GM composition occurs during the first SCA stages and prompted us to explore whether this might be associated with different individual exposure to dietary sources.

Daily alimentary habits were profiled by the analysis of daily food diaries, filled in by subjects one week before the collection of the fecal sample (see Methods and Supplementary Table S3).

Subjects with SCA reported to consume higher amounts of mechanically separated meats (p = 0.013) although a similar amount of ham, salami, sausages (p = 0.414), meat products and substitutes (p = 0.470) and a similar amount of not preserved meat (beef, veal, poultry and pork) (p = 0.683) as compared to subjects without SCA. Moreover, they reported consuming more dried fruits (p = 0.02), also with a trend towards higher quantities of fruits (both processed and fresh) and eggs. By contrast, subjects without SCA reported to consume a higher amount of cereals (p = 0.009), starchy vegetables (p = 0.027), milky products and beverages (p = 0.004 and p = 0.016, respectively), yoghurts (p = 0.047) and bakery (p < 0.001) as compared to those with SCA (Figure 4A).

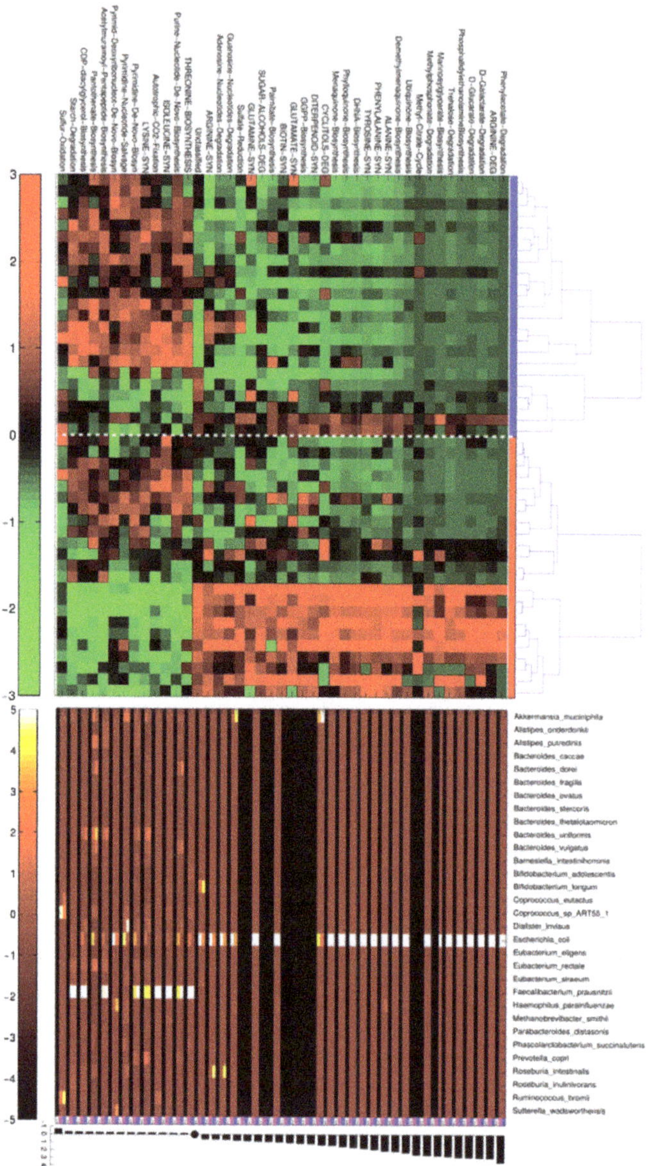

Figure 3. Metagenomic pathways and bacterial genera more expressed with +IMT/+SCA. Heatmap on top shows the MetaCyc L4-pathways statistically different between +IMT/+SCA (*n* = 23, red bars at right) and −IMT/−SCA (*n* = 23, blue bar at right) individuals. Normalized and scaled read counts (CPM) per pathway are standardized along rows. Within each experimental group, samples are clustered for similarity using Pearson's correlation metric and average linkage; the plot in the middle represents the relative contribution of bacterial species to each differential pathway. Values are average CPM calculated for −IMT/−SCA samples (indicated by the blue square below) and +IMT/+SCA (indicated by the magenta square below). Average CPM values are standardized on a per-column basis (i.e., on pathways), in order to highlight, for each pathway which bacterial taxa contributes most. On the bottom, of the figure, the barplot indicates log2 fold-change between +IMT/+SCA and −IMT/−SCA average CPM.

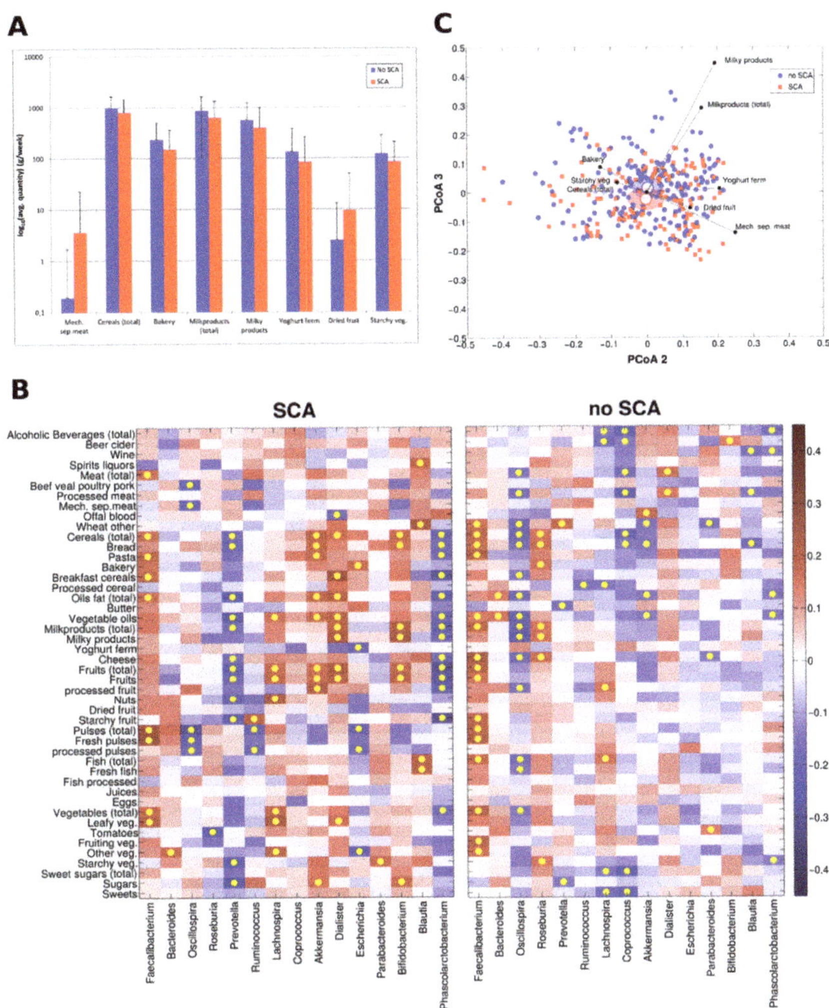

Figure 4. (**A**) Barplots of significantly different ($p < 0.05$, Mann-Whitney U-test) nutrients based on the analysis of daily food diaries for individuals with ($n = 144$) and without ($n = 201$) SCA. Bars represent average intake and standard deviations are represented as error bars. Due to graphical reasons, intakes were represented as $\log_{10}$ (**B**) Heatmap of the Spearman's correlation coefficients between bacterial genera relative abundances and nutrients intake. Correlations were calculated for individuals belonging to "SCA" ($n = 144$) and "no SCA" ($n = 201$), respectively. Yellow dots correspond to significant correlation (p-value of the linear model < 0.05) and bacteria with an average relative abundance $\geq 1\%$ in either experimental group were represented. (**C**) Biplot representing samples according to dietary patterns for individuals with ($n = 144$) and without ($n = 201$) SCA. Principal Coordinate Analysis (PCoA) was based on Bray-Curtis distances. Each point represents a sample, centroids represent the average coordinate for the data points in each category and ellipses represent the 95% Standard error of the mean (SEM)-based confidence interval of the data points. The second and third principal coordinates are represented. The average coordinates of the statistically different dietary component ($p < 0.05$) weighted by the corresponding abundance per sample was superimposed on the PCoA plot to identify those mainly contributing to the ordination space (black arrows).

Moreover, these dietary intakes showed more significant correlations to bacterial genera in subjects with SCA as compared to those without SCA (Figure 4B; red square represent positive correlation while blue square indicates negative correlation; yellow dots indicate significant correlations, $p < 0.05$). Of note, some of these correlation recapitulated data from metagenomic analysis. For example, increased relative abundance of *Escherichia* was inversely related with intakes of yoghurts, pulses and fresh vegetables while positively correlated to bakery only in individuals with SCA.

We were overall able to cluster intakes of specific food patterns (including milk-derived products, yogurts, total and leafy vegetables, processed cereals, mechanically separated meats, fish, bread and bakery products) that significantly co-segregated with the taxonomic diversities between subjects with SCA as compared to subjects without SCA (RV coefficient between nutrients quantities and microbial relative abundances at genus level = 0.65, p-value = 0.047) (Figure 4C). In addition to this finding, we also found similarly significant changes ($p = 0.04$) in the individual diet in subjects with +IMT/+SCA versus those without subclinical atherosclerosis (−IMT/−SCA) (not shown), further supporting the interaction of dietary exposure on GM changes and early stages of SCA.

Together these data support that changes in taxonomic GM composition, occurring still at subclinical stages of SCA, cluster and occur together with differences in individual exposure to dietary sources.

4. Discussion

Despite effective strategies for the treatment of patients with clinically manifest ACVD [1], the identification of subjects at higher risk of future development of the disease is currently far from optimal, due to hardwired CVRFs [1] and individual predisposition to low-grade inflammation [3]. Therefore, personalized approaches are sought, targeting the host-derived CVRFs and those dependent on the individual exposure to environment, in a tailored way [6]. The microbiome appears the most relevant potential target under this perspective, due to its involvement in the metabolism of dietary sources and since it is particularly sensitive to rapid changes in dietary habits [7]. We hypothesized, via an innovative research design, that GM changes in taxonomic and functional signatures still occur during subclinical stages of ACVD, before the clinical manifestation of CVRFs.

Currently, our data add further knowledge about the relation between GM and ACVD since, in comparison to other studies involving patients with clinically established ACVD, (either as coronary ischemic atherosclerosis [23], cerebrovascular events [23,40], ST-elevation myocardial infarction (STEMI) [21], stable angina or coronary artery disease [22]), we here show changes in taxonomic GM composition in subclinical stages of ACVD progression, when the effect of high cardio-metabolic impairment is not yet clinically evident.

We analyzed the GM profile in 345 subjects (the majority of whom were aged between 60 and 80 years-old), prevalently lean/overweight, without clinically manifest ACVD, T2D and MetS. Taxonomic compositions taken from this cohort (one of the largest of this kind) were comparable to those found in other geographically distant Italian cohorts [48–50] although they did not confirm some correlations between relative abundances of genera/species and cardio-metabolic markers (Supplementary Materials Figure S4), that were previously reported in populations characterized by different clinical phenotypes. For example, we did not find correlation between relative abundances of *Clostridium* species with increased BMI and with higher fasting glucose, findings that were reported in severe obese post-menopausal women (34.5 Kg/m^2 as mean BMI) [51], in 70 year-old overweight subjects (28.0 Kg/m^2 as mean BMI) but with T2D [52] and in younger and prevalently lean subjects (43–63 y-old and 23.7 Kg/m^2 as mean BMI) with diabetes (possibly including type 1 forms) [53]. Additionally, hepatic steatosis (which was only ultrasound-determined and poorly prevalent in our cohort) as well as plasma hs-CRP, a marker of low-grade inflammation, did not correlate either with GM diversity (while recently reported in morbidly obese patients [54] and in liver steatosis [55]), or with a reduced relative abundance of specific bacterial species (e.g., *Akkermansia muciniphila* (less abundant in insulin resistance [56] and in obese individuals [57]). Although these data might exclude an impact of a gut-liver connection during early stages of SCA, in depth evaluation on larger cohorts with more advanced stages of liver disease should is warranted.

Vice versa, we found significant relations with taxonomic species which down- or over-represent metabolic pathways of multiple dietary sources, hereby supporting an early immune-inflammatory activation of GM dysbiosis engaged by different dietary sources over SCA stages, before the clinical manifestation of ACVD and CVRFs. Different lines of evidence sustain our scientific question. Firstly, in contrast to previous reports in patients with metabolic syndrome [54,58], we here found an inverse relation between reduced HDL-C and increased abundance of Escherichia only in subjects with +IMT/+SCA, who also showed reduced HDL-C and increased plasma levels of the atherogenic molecule TMAO (a coincidence with previous data [15–20,59,60], see Supplementary Materials). Whether this is a consequence of TMAO produced by Escherichia (which has been reported to inhibit both HDL-mediated reverse cholesterol transport and intestinal HDL lipoprotein maturation [61]) requires further investigation. Additionally, we observed that *E. coli* caiTABCDE operon genes (encoding for membrane transport and metabolism of L-Carnitine to γBB and TMA [15,19]) were overrepresented in subjects with +IMT/+SCA (see Supplementary Materials, Figure S5). Carnevale et al. [62] recently showed higher *E. coli* abundance in GM of patients with STEMI, correlating to increased systemic Lipopolysaccharide (LPS) absorption and infiltration in atheromas from endarterectomies leading to macrophage activation [48]. In our investigation, we did not find increased plasma levels of zonulin in +IMT/+SCA subjects (see Supplementary Figure S6), therefore prompting the exclusion of systemic LPS absorption through a "leaky-gut" [63] yet at initial stages of atherosclerosis. Vice versa, we found a more over-expressed biosynthetic pathway of PE (strictly linked to the hepatic conversion to atherogenic TMAO [15] mediated by the L-carnitine/γBB/TMA metabolic cascade [64]), as documented by the data presented in Supplementary Figure S5, together with an increased number of circulating monocytes and neutrophils in subjects with SCA. These subjects reported indeed to consume only higher amounts of mechanically separated meats but a similar amount of unprocessed meats (as more complex food matrices containing other nutrients, phospholipids and probiotics that have been not associated with higher ACVD risk [65]) versus subjects without SCA. Future analyses and interventional dietary approaches are requested to unveil whether these connections reflect a gut-bone marrow connection fostering an activation of the innate immune system.

Secondly, we found in metagenomes of +IMT/+SCA subjects a reduced contribution of pathways (such as starch degradation, sulfur oxidation and the biosynthetic routes of purine and pyrimidines) encoded by *Faecalibacterium prausnitzii*, previously reported to be actively involved in gut permeability through the production of anti-inflammatory butyrate [66–68]. This finding coincides with intakes of fibers, carbohydrates and proteins from higher amounts of starchy vegetables, milky products and beverages, yoghurts and bakery products that subjects without SCA reported consuming. In fact, inulin-type fructans, fructo-oligosaccharides, polydextrose or soluble corn fiber support the proliferation of *Faecalibacterium prausnitzii* which, in turn, mediates the metabolism of fibers [68], ensures gut physiological transit time [69] and attenuates the pro-inflammatory potential of specific dietary proteins [70].

We have to acknowledge several limitations in our study. Firstly, this epidemiological study, gathering self-reported dietary data, does not allow the unveiling of the actual relation of causality. These limitations pave the road to dissecting this aspect in the near future, a perspective that might be pursued: (i) by clustering a larger number of subjects on the basis of their exposure to different dietary patterns (daily collected using smartphone health apps or other health mobile technologies to improve self-reporting) or (ii) by longitudinally evaluating the actual effect of a single dietary pattern/habit on changes of GM composition/functionality in subjects with different SCA stages. Different hurdles undermining these approaches are to be accounted in the design of lifestyle and dietary intervention studies (e.g., technical criticisms, difficulties in the interpretation of their resulting data and the export to wider populations [24]). The cross-sectional design of this investigation cannot rule out the burden of CVRFs prior to the examination, although significant differences we observed in microbial composition comparing beta-diversity measures via Principal Coordinate Analysis (PCoA) still take into account potential confounders, since they are based on multivariate models, associated with data-reduction techniques that produce a set of uncorrelated (orthogonal) axes to summarize the variability in the data set. Although the effect of pharmacological interventions over time (especially anti-hypertensive and lipid lowering drugs) cannot be ruled out, we did not find differences in the GM alpha and beta diversities when comparing subjects with pharmacological therapy and those with any treatments [23,71]).

Secondly, subjects with SCA were older in our cohort and we acknowledge that age, the principal predictor of faster SCA progression over time [4], might act as a confounding factor in the relation between GM taxonomy and SCA. However, the majority of subjects in our study were aged between 60 and 80 y/old (Table 1), therefore prompting us to conclude that data from younger cohorts or longitudinal evaluation on this same cohort are needed in order to dissect how much age interacts in the relation between GM composition and SCA progression over time.

Thirdly, our knowledge of the actual differences in metagenome-encoded functions and their relation to the host health is limited by the fact that a relevant part of the metagenome still remains undescribed (e.g., in our data, about half of the metagenomic reads could not be mapped and a majority of the mapped ones could not be annotated to a known reaction or pathway), due to the lack of complete description of bacterial genes in databases.

Finally, we did not perform a complete analysis of circulating proteomic and metabolomics markers validating the metagenomics pathways that emerged from our data. Since multiple circulating proteins and metabolites (the majority of which are of lipid origin) have been recently associated with different types of dietary patterns [72], this aspect is of relevance and will be analyzed in the near future.

However, to the best of our knowledge, this is the first extensive taxonomic and metagenomic characterization connecting GM dysbiosis with individual diet and SCA. These cross-sectional data aim at setting the stage for future longitudinal studies and dietary interventions, testing if personalized modifications in dietary habits over time could affect GM composition contributing to the prevention of the onset of CVRF and the clinical manifestation of ACVD.

Supplementary Materials: The following are available online at https://www.mdpi.com/2072-6643/13/2/304/s1. Table S1: Descriptives of the entire cohort divided by SCA stages; Table S2: Reduced and increased abundances of genera in SCA; Table S3: Daily intakes of food patterns reported by subjects of the studied population divided according to SCA; Figure S1: GM Taxonomic relative abundances of phyla, families and genera in SCA; Figure S2: Accordance between 16s rRNA and shotgun metagenomic sequencing in the evaluation of relative abundances at the level of genera; Figure S3: Hit taxonomic markers associated with +IMT/+SCA Figure S4: Correlation between most abundant bacteria and cardio-metabolic markers according to SCA; Figure S5: *E.coli* metagenetic markers associated with advanced SCA stage. Figure S6: Zonulin plasma levels are not increased in subjects with +IMT/+SCA.

Author Contributions: All the authors approved the manuscript in its contents and details and agree to be personally accountable for the author's own contributions and for ensuring that questions related to the accuracy or integrity of any part of the work, even ones in which the author was not personally involved, are appropriately investigated, resolved, and documented in the literature. A.B., M.S., C.P. contributed to conceptualization, investigation, project administration, formal analyses and use of software, data Curation, writing, reviewing and editing the original draft. A.L.C. contributed to conceptualization, investigation, supervision, reviewing and editing the original draft. E.O., C.C.D., E.M., J.C., G.C., C.C., L.G., F.P., F.G. and D.C. contributed to methodology and formal analyses. G.D.N., A.A. and L.R. contributed to visualization and supervision. A.B., G.D.N., and A.L.C. contributed to resources and funding acquisition. All authors have read and agreed to the published version of the manuscript.

Funding: The authors research is supported by: "Cibo, Microbiota, Salute" by "Vini di Batasiolo S.p.A" AL_RIC19ABARA_01 (AB), "Post-Doctoral Fellowship 2020" by "Fondazione Umberto Veronesi" 2020-3318 (AB); Fondazione Cariplo 2016-0852 (GDN); EFSD/Lilly European Diabetes Research Programme 2018 (GDN), Fondazione Telethon GGP19146 (GDN), PRIN 2017K55HLC (GDN); H2020 REPROGRAM PHC-03-2015/667837-2 (ALC); Fondazione Cariplo 2015-0524 and 2015-0564 (ALC); ERANET ER-2017-2364981 (ALC); Ministry of Health-IRCCS MultiMedica GR-2011-02346974 (ALC); PRIN 2017H5F943 (ALC); PRIN 2017—LS 4 2017HTKLRF_003 (CP). This work has been supported by Ministry of Health—Ricerca Corrente—IRCCS MultiMedica.

Institutional Review Board Statement: The study was conducted according to the guidelines of the Declaration of Helsinki, and approved by the Ethics Committee of the University of Study of Milan (SEFAP protocol no 0003/2001).

Informed Consent Statement: Informed consent was obtained from all subjects involved in the study.

Data Availability Statement: Publicly available datasets were analyzed in this study. This data can be found in NCBI Short-Read Archive (SRA) under accession number PRJNA615842.

Conflicts of Interest: The authors declare no conflict of interest.

References

1. Mach, F.; Baigent, C.; Catapano, A.L.; Koskinas, K.C.; Casula, M.; Badimon, L.; Chapman, M.J.; De Backer, G.G.; Delgado, V.; Ference, B.A.; et al. 2019 ESC/EAS Guidelines for the management of dyslipidaemias: Lipid modification to reduce cardiovascular risk: The Task Force for the management of dyslipidaemias of the European Society of Cardiology (ESC) and European Atherosclerosis Society (EAS). *Eur. Heart J.* **2020**, *41*, 111–188. [CrossRef] [PubMed]
2. Lorenz, M.W.; Polak, J.F.; Kavousi, M.; Mathiesen, E.B.; Völzke, H.; Tuomainen, T.-P.; Sander, D.; Plichart, M.; Catapano, A.L.; Robertson, C.M.; et al. Carotid intima-media thickness progression to predict cardiovascular events in the general population (the PROG-IMT collaborative project): A meta-analysis of individual participant data. *Lancet* **2012**, *379*, 2053–2062. [CrossRef]
3. Hoogeveen, R.M.; Pereira, J.P.B.; Nurmohamed, N.S.; Zampoleri, V.; Bom, M.J.; Baragetti, A.; Boekholdt, S.M.; Knaapen, P.; Khaw, K.-T.; Wareham, N.J.; et al. Improved cardiovascular risk prediction using targeted plasma proteomics in primary prevention. *Eur. Hear. J.* **2020**, *41*, 3998–4007. [CrossRef] [PubMed]
4. Olmastroni, E.; Baragetti, A.; Casula, M.; Grigore, L.; Pellegatta, F.; Pirillo, A.; Tragni, E.; Catapano, A.L. Multilevel Models to Estimate Carotid Intima-Media Thickness Curves for Individual Cardiovascular Risk Evaluation. *Stroke* **2019**, *50*, 1758–1765. [CrossRef]
5. Leopold, J.A.; Loscalzo, J. Emerging Role of Precision Medicine in Cardiovascular Disease. *Circ. Res.* **2018**, *122*, 1302–1315. [CrossRef]
6. Yusuf, S.; Joseph, P.; Rangarajan, S.; Islam, S.; Mente, A.; Hystad, P.; Brauer, M.; Kutty, V.R.; Gupta, R.; Wielgosz, A.; et al. Modifiable risk factors, cardiovascular disease, and mortality in 155 722 individuals from 21 high-income, middle-income, and low-income countries (PURE): A prospective cohort study. *Lancet* **2020**, *395*, 795–808. [CrossRef]
7. Kolodziejczyk, A.A.; Zheng, D.; Elinav, E. Diet–microbiota interactions and personalized nutrition. *Nat. Rev. Microbiol.* **2019**, *17*, 742–753. [CrossRef]
8. David, L.A.; Maurice, C.F.; Carmody, R.N.; Gootenberg, D.B.; Button, J.E.; Wolfe, B.E.; Ling, A.V.; Devlin, A.S.; Varma, Y.; Fischbach, M.A.; et al. Diet rapidly and reproducibly alters the human gut microbiome. *Nat. Cell Biol.* **2014**, *505*, 559–563. [CrossRef]
9. Berry, S.E.; Valdes, A.M.; Drew, D.A.; Asnicar, F.; Mazidi, M.; Wolf, J.; Pujol, J.C.; Hadjigeorgiou, G.; Davies, R.; Al Khatib, H.; et al. Human postprandial responses to food and potential for precision nutrition. *Nat. Med.* **2020**, *26*, 964–973. [CrossRef]
10. Perry, R.J.; Peng, L.; Barry, N.A.; Cline, G.W.; Zhang, D.; Cardone, R.L.; Petersen, K.F.; Kibbey, R.G.; Goodman, N.A.B.A.L.; Shulman, G.I. Acetate mediates a microbiome–brain–β-cell axis to promote metabolic syndrome. *Nat. Cell Biol.* **2016**, *534*, 213–217. [CrossRef]
11. Zmora, N.; Bashiardes, S.; Levy, M.; Elinav, E. The Role of the Immune System in Metabolic Health and Disease. *Cell Metab.* **2017**, *25*, 506–521. [CrossRef] [PubMed]
12. Zhang, L.; Nichols, R.G.; Correll, J.; Murray, I.A.; Tanaka, N.; Smith, P.B.; Hubbard, T.D.; Sebastian, A.; Albert, I.; Hatzakis, E.; et al. Persistent Organic Pollutants Modify Gut Microbiota–Host Metabolic Homeostasis in Mice Through Aryl Hydrocarbon Receptor Activation. *Environ. Heal. Perspect.* **2015**, *123*, 679–688. [CrossRef] [PubMed]
13. Pedersen, H.K.; Gudmundsdottir, V.; Nielsen, H.B.; Hyotylainen, T.; Nielsen, T.; Jensen, B.A.H.; Forslund, K.; Hildebrand, F.; Prifti, E.; Falony, G.; et al. Human gut microbes impact host serum metabolome and insulin sensitivity. *Nat. Cell Biol.* **2016**, *535*, 376–381. [CrossRef] [PubMed]

14. Velmurugan, G.; Dinakaran, V.; Rajendhran, J.; Swaminathan, K. Blood Microbiota and Circulating Mi-crobial Metabolites in Diabetes and Cardiovascular Disease. *Trends Endocrinol. Metab.* **2020**, *31*, 835–847. [CrossRef] [PubMed]

15. Tang, W.W.; Wang, Z.; Fan, Y.; Levison, B.S.; Hazen, J.E.; Donahue, L.M.; Wu, Y.; Hazen, S.L. Prognostic Value of Elevated Levels of Intestinal Microbe-Generated Metabolite Trimethylamine-N-Oxide in Patients With Heart Failure. *J. Am. Coll. Cardiol.* **2014**, *64*, 1908–1914. [CrossRef]

16. Mente, A.; Chalcraft, K.R.; Ak, H.; Davis, A.D.; Lonn, E.M.; Miller, R.; Potter, M.A.; Yusuf, S.; Anand, S.S.; McQueen, M.J. The Relationship Between Trimethylamine-N-Oxide and Prevalent Cardiovascular Disease in a Multiethnic Population Living in Canada. *Can. J. Cardiol.* **2015**, *31*, 1189–1194. [CrossRef] [PubMed]

17. Trøseid, M.; Ueland, T.; Hov, J.R.; Svardal, A.; Gregersen, I.; Dahl, C.P.; Aakhus, S.; Gude, E.; Bjørndal, B.; Halvorsen, B.; et al. Microbiota-dependent metabolite trimethylamine-N-oxide is associated with disease severity and survival of patients with chronic heart failure. *J. Intern. Med.* **2015**, *277*, 717–726. [CrossRef] [PubMed]

18. Wang, Z.; Klipfell, E.; Bennett, B.J.; Koeth, R.A.; Levison, B.S.; Dugar, B.; Feldstein, A.E.; Britt, E.B.; Fu, X.; Chung, Y.-M.; et al. Gut flora metabolism of phosphatidylcholine promotes cardiovascular disease. *Nat. Cell Biol.* **2011**, *472*, 57–63. [CrossRef] [PubMed]

19. Koeth, R.A.; Lam-Galvez, B.R.; Kirsop, J.; Wang, Z.; Levison, B.S.; Gu, X.; Copeland, M.F.; Bartlett, D.; Cody, D.B.; Dai, H.J.; et al. l-Carnitine in omnivorous diets induces an atherogenic gut microbial pathway in humans. *J. Clin. Investig.* **2018**, *129*, 373–387. [CrossRef] [PubMed]

20. Skagen, K.; Trøseid, M.; Ueland, T.; Holm, S.; Abbas, A.; Gregersen, I.; Kummen, M.; Bjerkeli, V.; Reier-Nilsen, F.; Russell, D.; et al. The Carnitine-butyrobetaine-trimethylamine-N-oxide pathway and its association with cardiovascular mortality in patients with carotid atherosclerosis. *Atherosclerosis* **2016**, *247*, 64–69. [CrossRef]

21. Karlsson, F.H.; Fåk, F.; Nookaew, I.; Tremaroli, V.; Fagerberg, B.; Petranovic, D.; Bäckhed, F.; Nielsen, J. Symptomatic atherosclerosis is associated with an altered gut metagenome. *Nat. Commun.* **2012**, *3*, 1245. [CrossRef] [PubMed]

22. Emoto, T.; Yamashita, T.; Sasaki, N.; Hirota, Y.; Hayashi, T.; So, A.; Kasahara, K.; Yodoi, K.; Matsumoto, T.; Mizoguchi, T.; et al. Analysis of Gut Microbiota in Coronary Artery Disease Patients: A Possible Link between Gut Microbiota and Coronary Artery Disease. *J. Atheroscler. Thromb.* **2016**, *23*, 908–921. [CrossRef] [PubMed]

23. Jie, Z.; Xia, H.; Zhong, S.-L.; Feng, Q.; Li, S.; Liang, S.; Zhong, H.; Liu, Z.; Gao, Y.; Zhao, H.; et al. The gut microbiome in atherosclerotic cardiovascular disease. *Nat. Commun.* **2017**, *8*, 845. [CrossRef] [PubMed]

24. Swann, J.R.; Rajilic-Stojanovic, M.; Salonen, A.; Sakwinska, O.; Gill, C.; Meynier, A.; Fança-Berthon, P.; Schelkle, B.; Segata, N.; Shortt, C.; et al. Considerations for the design and conduct of human gut micro-biota intervention studies relating to foods. *Eur. J. Nutr.* **2020**, *59*, 3347–3368. [CrossRef] [PubMed]

25. Gerdes, V.; Gueimonde, M.; Pajunen, L.; Nieuwdorp, M.; Laitinen, K. How strong is the evidence that gut microbiota composition can be influenced by lifestyle interventions in a cardio-protective way? *Atheroscler.* **2020**, *311*, 124–142. [CrossRef]

26. Fracanzani, A.L.; Pisano, G.; Consonni, D.; Tiraboschi, S.; Baragetti, A.; Bertelli, C.; Norata, G.D.; Dongiovanni, P.; Valenti, L.; Grigore, L.; et al. Epicardial Adipose Tissue (EAT) Thickness Is Associated with Cardiovascular and Liver Damage in Nonalcoholic Fatty Liver Disease. *PLoS ONE* **2016**, *11*, e0162473. [CrossRef]

27. Baragetti, A.; Pisano, G.; Bertelli, C.; Garlaschelli, K.; Grigore, L.; Fracanzani, A.; Fargion, S.; Norata, G.D.; Catapano, A.L. Subclinical atherosclerosis is associated with Epicardial Fat Thickness and hepatic steatosis in the general population. *Nutr. Metab. Cardiovasc. Dis.* **2016**, *26*, 141–153. [CrossRef]

28. Alberti, K.G.M.M.; Eckel, R.H.; Grundy, S.M.; Zimmet, P.Z.; Cleeman, J.I.; Donato, K.A.; Fruchart, J.C.; James, W.P.T.; Loria, C.M.; Smith, S.C. Harmonizing the metabolic syndrome: A joint interim statement of the international diabetes federation task force on epidemiology and prevention; National heart, lung, and blood institute; American heart association; World heart federation; International atherosclerosis society; And international association for the study of obesity. *Circulation* **2009**, *120*, 1640–1645.

29. Stein, J.H.; Korcarz, C.E.; Hurst, R.T.; Lonn, E.; Kendall, C.B.; Mohler, E.R.; Najjar, S.S.; Rembold, C.M.; Post, W.S. Use of Carotid Ultrasound to Identify Subclinical Vascular Disease and Evaluate Cardiovascular Disease Risk: A Consensus Statement from the American Society of Echocardiography Carotid Intima-Media Thickness Task Force Endorsed by the Society for Vascular Medicine. *J. Am. Soc. Echocardiogr.* **2008**, *21*, 93–111. [CrossRef]

30. Grant, E.G.; Benson, C.B.; Moneta, G.L.; Alexandrov, A.V.; Baker, J.D.; Bluth, E.I.; Carroll, B.A.; Eliasziw, M.; Gocke, J.; Hertzberg, B.S.; et al. Carotid Artery Stenosis: Gray-Scale and Doppler US Diagnosis—Society of Radiologists in Ultrasound Consensus Conference. *Radiology* **2003**, *229*, 340–346. [CrossRef]

31. Redaelli, L.; Garlaschelli, K.; Grigore, L.; Norata, G.D.; Catapano, A.L. Association between the Adherence to AHA Step 1 Nutrition Criteria and the Cardiometabolic Outcome in the General Population a Two Year Follow-Up Study. *Food Nutr. Sci.* **2012**, *3*, 274–280. [CrossRef]

32. Atlante Fotografico delle Porzioni degli Alimenti. Available online: https://www.scottibassani.it/atlante-fotografico-delle-porzioni-degli-alimenti/ (accessed on 18 January 2021).

33. Banca Dati di Composizione degli Alimenti per Studi Epidemiologici in Italia. Available online: http://www.bda-ieo.it/wordpress/en/ (accessed on 18 January 2021).

34. Consolandi, C.; Turroni, S.; Emmi, G.; Severgnini, M.; Fiori, J.; Peano, C.; Biagi, E.; Grassi, A.; Rampelli, S.; Silvestri, E.; et al. Behçet's syndrome patients exhibit specific microbiome signature. *Autoimmun. Rev.* **2015**, *14*, 269–276. [CrossRef] [PubMed]

35. Klindworth, A.; Pruesse, E.; Schweer, T.; Peplies, J.; Quast, C.; Horn, M.; Glöckner, F.O. Evaluation of general 16S ribosomal RNA gene PCR primers for classical and next-generation sequencing-based diversity studies. *Nucleic Acids Res.* **2012**, *41*, e1. [CrossRef] [PubMed]

36. 16S Metagenomic Sequencing Library Preparation. Available online: https://support.illumina.com/documents/documentation/chemistry_documentation/16s/16s-metagenomic-library-prep-guide-15044223-b.pdf (accessed on 18 January 2021).

37. Masella, A.P.; Bartram, A.K.; Truszkowski, J.M.; Brown, D.G.; Neufeld, J.D. PANDAseq: Paired-end assembler for illumina sequences. *BMC Bioinform.* **2012**, *13*, 31. [CrossRef] [PubMed]

38. Caporaso, J.G.; Kuczynski, J.; Stombaugh, J.; Bittinger, K.; Bushman, F.D.; Costello, E.K.; Fierer, N.; Peña, A.G.; Goodrich, J.K.; Gordon, J.I.; et al. QIIME Allows Analysis of High-Throughput Community Sequencing data. *Nat. Methods* **2010**, *7*, 335–336. [CrossRef] [PubMed]

39. Available online: ftp://greengenes.microbio.me/greengenes_release/gg_13_8_otus (accessed on 19 January 2021).

40. Available online: https://cran.r-project.org/src/contrib/Archive/vegan/vegan_2.0--10.tar.gz (accessed on 19 January 2021).

41. The Human Microbiome Project Consortium. A framework for human microbiome research. *Nat. Cell Biol.* **2012**, *486*, 215–221. [CrossRef]

42. Available online: https://www.hmpdacc.org/hmp/resources/ (accessed on 19 January 2021).

43. Franzosa, E.A.; McIver, L.J.; Rahnavard, A.; Thompson, L.R.; Schirmer, M.; Weingart, G.; Lipson, K.S.; Knight, R.; Caporaso, J.G.; Segata, N.; et al. Species-level functional profiling of metagenomes and metatranscriptomes. *Nat. Methods* **2018**, *15*, 962–968. [CrossRef]

44. Caspi, R.; Billington, R.; Keseler, I.M.; Kothari, A.; Krummenacker, M.; Midford, P.E.; Ong, W.K.; Paley, S.; Subhraveti, P.; Karp, P.D. The MetaCyc database of metabolic pathways and enzymes—A 2019 update. *Nucleic Acids Research* **2020**, *48*, D445–D453. [CrossRef]

45. Smilde, A.K.; Kiers, H.A.L.; Bijlsma, S.; Rubingh, C.M.; Van Erk, M.J. Matrix correlations for high-dimensional data: The modified RV-coefficient. *Bioinformatics* **2008**, *25*, 401–405. [CrossRef]

46. Available online: http://www.ncbi.nlm.nih.gov/bioproject/PRJNA615842 (accessed on 19 January 2021).

47. Josse, J.; Pagès, J.; Husson, F. Testing the significance of the RV coefficient. *Comput. Stat. Data Anal.* **2008**, *53*, 82–91. [CrossRef]

48. Carnevale, R.; Nocella, C.; Petrozza, V.; Cammisotto, V.; Pacini, L.; Sorrentino, V.; Martinelli, O.; Irace, L.; Sciarretta, S.; Frati, G.; et al. Localization of lipopolysaccharide from Escherichia Coli into human atherosclerotic plaque. *Sci. Rep.* **2018**, *8*, 1–8. [CrossRef] [PubMed]

49. Schnorr, S.L.; Candela, M.; Rampelli, S.; Centanni, M.; Consolandi, C.; Basaglia, G.; Turroni, S.; Biagi, E.; Peano, C.; Severgnini, M.; et al. Gut microbiome of the Hadza hunter-gatherers. *Nat. Commun.* **2014**, *5*, 3654. [CrossRef] [PubMed]

50. De Filippis, F.; Pellegrini, N.; Vannini, L.; Jeffery, I.B.; La Storia, A.; Laghi, L.; Serrazanetti, D.I.; Di Cagno, R.; Ferrocino, I.; Lazzi, C.; et al. High-level adherence to a Mediterranean diet beneficially impacts the gut microbiota and associated metabolome. *Gut* **2016**, *65*, 1812–1821. [CrossRef] [PubMed]

51. Brahe, L.K.; Le Chatelier, E.; Prifti, E.; Pons, N.; Kennedy, S.B.; Hansen, T.; Pedersen, O.; Astrup, A.; Ehrlich, S.; Larsen, L.H. Specific gut microbiota features and metabolic markers in postmenopausal women with obesity. *Nutr. Diabetes* **2015**, *5*, e159. [CrossRef]

52. Karlsson, F.H.; Tremaroli, V.; Nookaew, I.; Bergström, G.; Behre, C.J.; Fagerberg, B.; Nielsen, J.B.; Bäckhed, F. Gut metagenome in European women with normal, impaired and diabetic glucose control. *Nat. Cell Biol.* **2013**, *498*, 99–103. [CrossRef]

53. Qin, J.; Li, Y.; Cai, Z.; Li, S.; Zhu, J.; Zhang, F.; Liang, S.; Zhang, W.; Guan, Y.; Shen, D.; et al. A metagenome-wide association study of gut microbiota in type 2 diabetes. *Nature* **2012**, *490*, 55–60. [CrossRef]

54. Vieira-Silva, S.; Consortium, M.; Falony, G.; Belda, E.; Nielsen, T.; Aron-Wisnewsky, J.; Chakaroun, R.; Forslund, S.K.; Assmann, K.; Valles-Colomer, M.; et al. Statin therapy is associated with lower prevalence of gut microbiota dysbiosis. *Nat. Cell Biol.* **2020**, *581*, 310–315. [CrossRef]

55. Aron-Wisnewsky, J.; Gaborit, B.; Dutour, A.; Clement, K. Gut microbiota and non-alcoholic fatty liver disease: New insights. *Clin. Microbiol. Infect.* **2013**, *19*, 338–348. [CrossRef]

56. Schneeberger, M.; Everard, A.; Gómez-Valadés, A.G.; Matamoros, S.; Ramírez, S.; Delzenne, N.M.; Gomis, R.; Claret, M.; Cani, P.D. Akkermansia muciniphila inversely correlates with the onset of inflammation, altered adipose tissue metabolism and metabolic disorders during obesity in mice. *Sci. Rep.* **2015**, *5*, 16643. [CrossRef]

57. Dao, M.C.; Everard, A.; Aron-Wisnewsky, J.; Sokolovska, N.; Prifti, E.; Verger, E.O.; Kayser, B.D.; Levenez, F.; Chilloux, J.; Hoyles, L.; et al. Akkermansia muciniphila and improved metabolic health during a dietary intervention in obesity: Relationship with gut microbiome richness and ecology. *Gut* **2015**, *65*, 426–436. [CrossRef]

58. Le Chatelier, E.; Nielsen, T.; Qin, J.; Prifti, E.; Hildebrand, F.; Falony, G.; Almeida, M.; Arumugam, M.; Batto, J.-M.; Kennedy, S.; et al. Richness of human gut microbiome correlates with metabolic markers. *Nature* **2013**, *500*, 541–546. [CrossRef] [PubMed]

59. Tang, W.W.; Wang, Z.; Levison, B.S.; Koeth, R.A.; Britt, E.B.; Fu, X.; Wu, Y.; Hazen, S.L. Intestinal Microbial Metabolism of Phosphatidylcholine and Cardiovascular Risk. *New Engl. J. Med.* **2013**, *368*, 1575–1584. [CrossRef] [PubMed]

60. Koeth, R.A.; Wang, Z.; Levison, B.S.; Buffa, J.A.; Org, E.; Sheehy, B.T.; Britt, E.B.; Fu, X.; Wu, Y.; Li, L.; et al. Intestinal microbiota metabolism of l-carnitine, a nutrient in red meat, promotes atherosclerosis. *Nat. Med.* **2013**, *19*, 576–585. [CrossRef] [PubMed]

61. Brunham, L.R.; Kruit, J.K.; Iqbal, J.; Fievet, C.; Timmins, J.M.; Pape, T.D.; Coburn, B.A.; Bissada, N.; Staels, B.; Groen, A.K.; et al. Intestinal ABCA1 directly contributes to HDL biogenesis in vivo. *J. Clin. Investig.* **2006**, *116*, 1052–1062. [CrossRef]

62. Carnevale, R.; Sciarretta, S.; Valenti, V.; Di Nonno, F.; Calvieri, C.; Nocella, C.; Frati, G.; Forte, M.; D'Amati, G.; Pignataro, M.G.; et al. Low-grade endotoxaemia enhances artery thrombus growth via Toll-like receptor 4: Implication for myocardial infarction. *Eur. Hear. J.* **2020**, *41*, 3156–3165. [CrossRef]
63. Li, C.; Gao, M.; Zhang, W.; Chen, C.; Zhou, F.; Hu, Z.; Zeng, C. Zonulin Regulates Intestinal Permeability and Facilitates Enteric Bacteria Permeation in Coronary Artery Disease. *Sci. Rep.* **2016**, *6*, 29142. [CrossRef]
64. Van Der Veen, J.N.; Kennelly, J.P.; Wan, S.; Vance, J.E.; Vance, D.E.; Jacobs, R.L. The critical role of phosphatidylcholine and phosphatidylethanolamine metabolism in health and disease. *Biochim. Biophys. Acta (BBA) Biomembr.* **2017**, *1859*, 1558–1572. [CrossRef]
65. Astrup, A.; Magkos, F.; Bier, D.M.; Brenna, J.T.; De Oliveira Otto, M.C.; Hill, J.O.; King, J.C.; Mente, A.; Ordovas, J.M.; Volek, J.S.; et al. Saturated Fats and Health: A Reassessment and Proposal for Food-Based Recommendations: JACC State-of-the-Art Review. *J. Am. Coll. Cardiol.* **2020**, *76*, 844–857. [CrossRef]
66. Mörkl, S.; Lackner, S.; Meinitzer, A.; Mangge, H.; Lehofer, M.; Halwachs, B.; Gorkiewicz, G.; Kashofer, K.; Painold, A.; Holl, A.K.; et al. Gut microbiota, dietary intakes and intestinal permeability reflected by serum zonulin in women. *Eur. J. Nutr.* **2018**, *57*, 2985–2997. [CrossRef]
67. Sokol, H.; Pigneur, B.; Watterlot, L.; Lakhdari, O.; Bermúdez-Humarán, L.G.; Gratadoux, J.-J.; Blugeon, S.; Bridonneau, C.; Furet, J.-P.; Corthier, G.; et al. Faecalibacterium prausnitzii is an anti-inflammatory commensal bacterium identified by gut microbiota analysis of Crohn disease patients. *Proc. Natl. Acad. Sci. USA* **2008**, *105*, 16731–16736. [CrossRef]
68. Geirnaert, A.; Calatayud, M.; Grootaert, C.; Laukens, D.; Devriese, S.; Smagghe, G.; De Vos, M.; Boon, N.; Van De Wiele, T. Butyrate-producing bacteria supplemented in vitro to Crohn's disease patient microbiota increased butyrate production and enhanced intestinal epithelial barrier integrity. *Sci. Rep.* **2017**, *7*, 1–14. [CrossRef] [PubMed]
69. Verhoog, S.; Taneri, P.E.; Díaz, Z.M.R.; Marques-Vidal, P.; Troup, J.P.; Bally, L.; Franco, O.H.; Glisic, M.; Muka, T. Dietary factors and modulation of bacteria strains of akkermansia muciniphila and faecalibac-terium prausnitzii: A systematic review. *Nutrients* **2019**, *11*, 1565. [CrossRef] [PubMed]
70. Ma, N.; Tian, Y.; Wu, Y.; Ma, X. Contributions of the Interaction Between Dietary Protein and Gut Microbiota to Intestinal Health. *Curr. Protein Pept. Sci.* **2017**, *18*, 795–808. [CrossRef] [PubMed]
71. Tuteja, S.; Ferguson, J.F. Gut Microbiome and Response to Cardiovascular Drugs. *Circ. Genom. Precis. Med.* **2019**, *12*, 421–429. [CrossRef] [PubMed]
72. Walker, M.E.; Song, R.J.; Xu, X.; Gerszten, R.E.; Ngo, D.; Clish, C.B.; Corlin, L.; Ma, J.; Xanthakis, V.; Jacques, P.F.; et al. Proteomic and Metabolomic Correlates of Healthy Dietary Patterns: The Framingham Heart Study. *Nutrients* **2020**, *12*, 1476. [CrossRef]

Review

Molecular Immune-Inflammatory Connections between Dietary Fats and Atherosclerotic Cardiovascular Disease: Which Translation into Clinics?

Elisa Mattavelli [1,2], Alberico Luigi Catapano [1,3] and Andrea Baragetti [1,3,*]

[1] Department of Pharmacological and Biomolecular Sciences, University of Milan, 20133 Milan, Italy; elisa.mattavelli@unimi.it (E.M.); alberico.catapano@unimi.it (A.L.C.)

[2] S.I.S.A. Centre for the Study of Atherosclerosis, Bassini Hospital, Cinisello Balsamo, Cinisello Balsamo, 20092 Milan, Italy

[3] IRCCS Multimedica Hospital, Sesto San Giovanni, 20092 Milan, Italy

* Correspondence: andrea.baragetti@unimi.it; Tel.: +39-0250318401 or +39-0257998779

Abstract: Current guidelines recommend reducing the daily intake of dietary fats for the prevention of ischemic cardiovascular diseases (CVDs). Avoiding saturated fats while increasing the intake of mono- or polyunsaturated fatty acids has been for long time the cornerstone of dietary approaches in cardiovascular prevention, mainly due to the metabolic effects of these molecules. However, recently, this approach has been critically revised. The experimental evidence, in fact, supports the concept that the pro- or anti-inflammatory potential of different dietary fats contributes to atherogenic or anti-atherogenic cellular and molecular processes beyond (or in addition to) their metabolic effects. All these aspects are hardly translatable into clinics when trying to find connections between the pro-/anti-inflammatory potential of dietary lipids and their effects on CVD outcomes. Interventional trials, although providing stronger potential for causal inference, are typically small sample-sized, and they have short follow-up, noncompliance, and high attrition rates. Besides, observational studies are confounded by a number of variables and the quantification of dietary intakes is far from optimal. A better understanding of the anatomic and physiological barriers for the absorption and the players involved in the metabolism of dietary lipids (e.g., gut microbiota) might be an alternative strategy in the attempt to provide a first step towards a personalized dietary approach in CVD prevention.

Keywords: dietary lipids; immune-inflammation; cardiovascular disease; microbiota

Citation: Mattavelli, E.; Catapano, A.L.; Baragetti, A. Molecular Immune-Inflammatory Connections between Dietary Fats and Atherosclerotic Cardiovascular Disease: Which Translation into Clinics?. *Nutrients* **2021**, *13*, 3768. https://doi.org/10.3390/nu13113768

Academic Editor: Arrigo Cicero

Received: 25 September 2021
Accepted: 22 October 2021
Published: 25 October 2021

Publisher's Note: MDPI stays neutral with regard to jurisdictional claims in published maps and institutional affiliations.

1. Introduction

The favorable transition from *hominids* to *homo sapiens* during evolution [1] prompted changes in the physiological functions and immune competences (including survival to pathogens and infections) to adapt to the intake of high-energy containing foods, principally dietary fats [2]. Today, the access to a variety of highly-caloric fatty foods is hardly balanced by energy consumption. As a consequence, the dominant genetic pathways evolved to favor the intake of calorie-rich diets and the storage of energy as fats in the adipose tissue are in some circumstances redundant, especially in affluent societies, giving rise to obesity, diabetes and cardiovascular disease (CVD)-comorbidities.

Currently, the Western lifestyle, including dietary habits, is believed to contribute a chronic state of low-grade inflammation [3] that eventually prompts the development of atherosclerosis, the etiopathological factor of ischemic CVDs. Of note is that the connection between the Western dietary lifestyle and onset of CVDs has been demonstrated to impact morbidity and mortality worldwide [4].

Dietary interventions are considered the first approach in preventing atherosclerotic CVDs. All guidelines, while recommending reduction of fat consumption, also advise avoidance of dietary trans-fats, reducing the intake of saturated fats, and preferring mono or poly-unsaturated long-chain fats [5]. Achieving these goals remains a challenge for physicians and patients. Furthermore, the level of evidence for these recommendations is backed-up by single randomized clinical trials [6] and mostly relies upon large non-randomized observational studies [7], which suffer from confounding [8] and difficulties in quantitatively measuring dietary intake [9].

In addition, other factors linked to dietary consumption which are less likely to be captured in epidemiological studies have emerged as being associated with the risk of atherosclerotic CVD, and include the type of food availability, personal knowledge of the impact of diet on health, and socio-economic status [10]. As an example, the availability of processed foods is associated with an increased risk of CVDs [11]; this is the case of fat-rich processed meats, whose consumption increases the risk of CVD as compared to fat-rich unprocessed meats [12], fatty fish, and poultry [13], whose consumption is not considered a CVD risk modifier.

Furthermore, in modern societies, we are continuously exposed to postprandial lipemia (PPL) [14], a condition that appears to be causally related to the risk of coronary artery disease [15], myocardial infarction, ischemic heart disease, and ischemic stroke [16]. The mechanisms by which an exaggerated PPL links to CVDs include the fostering of endothelial dysfunction [17–19], arterial inflammation, and a pro-atherogenic activation of myeloid cells [18]. In addition, the magnitude of PPL in response to high-fat- based meals in humans appears to be significantly affected by the taxonomic composition of intestinal microbiota [20], which also cross-talk with hematopoietic niches [21], ensuring the activity of the innate immune check-points in the intestine and lymphatics. Once absorbed in the intestine, the majority of dietary fats cross over a complex surveillance system (including the cells patrolling at the interface between enterocytes and lacteals like the mesenteric lymph node (MLN), as opposed to the carbohydrates, sugars, dietary amino acids, protein-rich foods and other dietary components that are believed to exert other pathophysiological mechanisms that do not engage these immune checkpoints [22–25]. For example, an elevated intestinal absorption of sugars promotes systemic inflammation by directly disrupting the intestinal barrier [26]. Observational studies have provided contrasting results, with some studies being in favor of t pro-inflammatory effect of dietary fats and others unable to show any effect. In this review, we aim at critically revising the available evidence and providing a platform to reconcile these findings.

2. Dietary Fats, Inflammation and Atherosclerosis

In the human body, dietary fats face a complex metabolic journey involving a number of cellular checkpoints (Figure 1).

Dietary fats (triglycerides, phospholipids and cholesterol) are digested in the upper part of the small intestine by the activity of multiple lipases and then absorbed by the enterocytes [27,28]. In addition, gut resident bacteria can contribute to circulating fats principally by producing short-chain fatty acids (SCFAs) from fibers/complex carbohydrates. For example, *Faecalibacterium prausnitzii* ferments fibers present in the food matrices of fatty foods (e.g., avocados, tree-nuts and peanuts, where over a third of carbohydrates are fibers) [29] and is the major producer of butyrate. Butyrate is known to regulate hematopoietic activity [21] and to control myeloid pro-inflammatory skewing [30], exerting anti-atherogenic properties [31]. Vice versa, *Ruminococcus bromii*, which is reduced in subjects with atherosclerosis [32], metabolizes complex carbohydrates (that are present in low-fat foods, including pinto beans, whole grains, and nuts with considerable proportion of fats) into propionate. This SCFA promotes insulin sensitivity and reduces the atherosclerotic burden in mice [33]. Beside the production of SCFAs, other gut microbial species express enzymatic systems that metabolize dietary lipids into inflammatory molecules. Among them, trimethylamine (TMA) lyase, an enzyme that converts dietary phosphatidylcholine and choline into TMA, is peculiarly expressed by *Eggerthella lenta* and *Eggerthella timonensis* [34,35]. Finally, gut resident Gram-negative commensals (e.g., *Escherichia coli*,

Salmonella minnesota, Salmonella typhimurium) synthetize lipid-containing molecules, such as lipopolysaccharide (LPS), which promote apoptotic signaling and trigger systemic immune-metabolic derangement and inflammation [36–38].

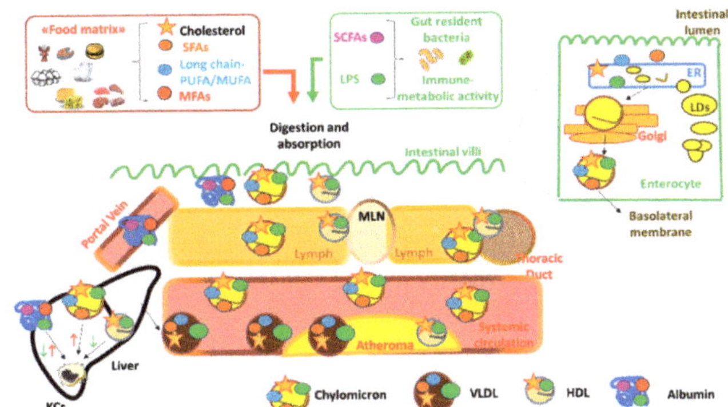

Figure 1. The routes for the absorption and the principal immune-inflammatory engagements of dietary lipids in intestinal villi, in the lacteals, MLN up to the liver. MLN = mesenteric lymph node; LDs: lipid droplets; SFAs: saturated fats; MUFA: mono-unsaturated fats; PUFA: poly-unsaturated fats; SCFAs: short chain fatty acids; MFA = medium chain fatty acids; LPS: lipopolysaccharide; ER: endoplasmic reticulum; LDs = lipid droplets; CM: chylomicrons; VLDL = very low density lipoproteins; HDL = high density lipoproteins; DC = dendritic cells; KCs = Kuppfer cells. Upward red arrows indicate activation of a cell or a pathway; downward green arrows indicate inhibition or regulation of a cell or a pathway. Both upward red and downward green arrows for SCFAs and MFAs carried by albumin in the liver indicate contrasting evidence depending on the type of dietary fat.

Once within the enterocytes, SCFAs directly reach the portal system and the liver where they are readily metabolized, while the majority of absorbed dietary fats are released into the lymphatic tree in large lipoproteins (chylomicrons and very-low density lipoproteins, VLDL) [39] (Figure 1). Fatty acids deriving from the hydrolysis of dietary triglycerides and phospholipids in the intestinal lumen are chaperoned to the intracellular endoplasmic reticulum (ER); there, diacylglycerol O-acyltransferase 1 (DGAT1) promotes their re-incorporation in triglycerides which are then transferred by the microsomial triglyceride transfer protein (MTTP) to nascent apolipoprotein B. In this way, chylomicrons are released by the enterocytes in their basolateral membrane. A small fraction of absorbed cholesterol is esterified by acyl CoA-transferase (ACAT) and packaged into chylomicrons. In the Golgi, other apolipoproteins, including apoCIII, apoCII, apoAV, and apoAIV, are added to chylomicrons, which then enter the bloodstream via the thoracic duct and will be eventually taken up by liver (Figure 1).

In addition to chylomicrons, the intestine also produces a small fraction of high density lipoproteins (HDL), through the activities of ATP binding cassette transporter A-1 (ABCA1) and phospholipid transfer protein (PLTP), which transfer cholesterol and phospholipids to apolipoprotein A-I [40]. It has been proposed that a fraction of HDL produced by the intestine moves to the liver through the portal system and antagonizes the binding of LPS to toll-like receptor 4 (TLR4) on the membrane of Kuppfer cells, liver-resident macrophages involved in the defense against gut-derived exogenous molecules [41], thus preventing the recruitment of pro-inflammatory myeloid cells [42] (Figure 1). Although further investigations are required, these findings are in line with the known anti-inflammatory function of HDL [43].

To date, the contribution of each single dietary fat in these multiple systems remains unclear. Moreover, a high quantity of data has been produced linking each single dietary fat to key molecular mechanisms and atherosclerosis, but principally providing scattered evidence from experimental models with poor translation to clinics. These aspects will be reviewed below.

2.1. Short Chain Fatty Acids

SCFAs (butyrate, acetate, and propionate) originate mostly from the fermentation of fibers and complex carbohydrates (that are abundant in vegetables, fruits, legumes, and whole grains), a process that is triggered by some gut resident bacteria [44,45]. SCFAs participate in multiple processes and their interaction with different receptor systems that are expressed in immune cells [46,47]. Accumulating experimental data have been used to analyze the immune-inflammatory cellular downstream effects of SCFAs using either purchased-SCFAs bound to bound to bovine serum albumin (BSA) or SCFAs from direct fermentation of commensals. In fact, SCFAs, by binding G protein-coupled free fatty acids receptors (FFARs; in particular isoform 2) [48], are significantly expressed in immune cells [49] and promote mechanisms engaged in the resolution of vascular inflammation. By interacting with these receptors, in one study, butyrate-BSA inhibited reactive oxygen species production in neutrophils, decreased the production of inflammatory cytokines (including monocyte chemoattractant protein-1 (MCP-1) and interleukin n-6 (IL-6)) [50]) and down-regulates the pro-inflammatory switching of macrophages [50]. Similarly, in an independent study, butyrate directly produced by gut commensals promotes the anti-inflammatory activity of regulatory T cells (Tregs) [51] (Figure 2).

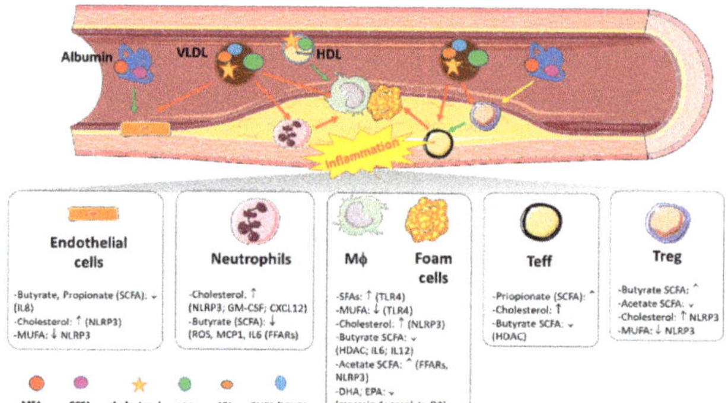

Figure 2. Principal immune-inflammatory pathways elicited by dietary fats in atherosclerosis. VLDL = very low density lipoproteins; HDL = high density lipoproteins; LPS = lipopolysaccharide; Mφ = macrophages; Teff = effector T cells; Treg = regulatory T cells; SFAs = saturated fatty acids; SCFAs = short chain fatty acids; MUFA = mono-unsaturated fatty acids; PUFA = poly-unsaturated fatty acids; DHA = docosahexaenoic acid; EPA = eicosapentaenoic acid; NLRP3 = NOD-like receptor protein 3; GM-SCF = granulocyte-coly stimulating factor; CXCL12 = C-X-C motif chemokine ligand 12; ROS = reactive oxygen species; MCP-1 = monocyte chemoattractant protein-1; FFARs = free fatty acids receptors; IL- = interleukin; HDAC = histone deacetylases. ↑: indicates an activated cell or pathway; ↓: indicates an inhibited cell or pathway.

By interacting with FFAR2, acetate-BSA promotes the phagocytic activity of macrophages, inhibits LPS-induced secretion of inflammatory cytokines by mononuclear cells [52], and triggers the production of oxygen free radicals at the sites of inflammation. Furthermore, mice fed a fiber-enriched diet with acetic acid in water ad libitum were protected against *C. difficile* infection due to interaction of gut commensal-derived acetate with FFAR2 that promotes the switching of the NOD-like receptor protein 3 (NLRP3)-inflammasome [53] (a central pathway in immune-metabolic derangements [54] and atherosclerosis [55]) (Figure 2).

In addition to FFARs, SCFAs also target histone deacetylases (HDACs), a group of deacetylating enzymes that regulate gene expression by removing acetyl groups from both histone and non-histone protein complexes in different genomic regions. Interestingly, butyrate and propionate reduce the production of pro-inflammatory IL-8 in activated endothelial cells by inhibiting HDAC activity [56] (Figure 2). With the same mechanism, butyrate (administered in mice in drinking water ad libitum) down-regulates LPS-induced secretion of pro-inflammatory mediators (such as nitric oxide, IL-6, and IL-12) by intestinal macrophages [57] (Figure 2).

Other in vivo observations in germ-free apoE-/- mice fed a fiber-rich diet and then inoculated with *Roseburia intestinalis* (a butyrate producer) show an increase in intestinal gluconeogenesis, reduction of LPS in the blood and aortic atherosclerotic lesions, decreased expression of inflammatory cytokines, and macrophage accumulation [58,59] (Figure 2). Similarly, apoE-/- mice fed a purified diet low in fibers and supplemented with propionic acid in the drinking water showed moderate cardiac hypertrophy, reduced aortic atherosclerosis and fewer effector lymphocytes (Teff) in the atheroma as compared to apoE-/- mice not receiving propionate supplementation [33]. By contrast to butyrate and propionate, intra-gastric infusion of acetate promotes glucose intolerance [60] and favors the polarization of naïve CD4 + T cells towards Teff, while inhibiting the immune suppressive activity of Treg [61] (Figure 2).

2.2. Medium and Long Chain Fatty Acids

Fatty acids can be defined as medium-chain, with 6–12 carbons, and long-chain, with up to 22 carbons. Dietary medium-chain fatty acids are majorly present, in the form of triglycerides, in coconut oil, palm kernel oil, and in dairy products; long-chain fatty acids are more abundant in vegetable oils, tree nuts, peanuts, and fish (principally salmon, tuna, mackerel, and sardines). Fatty acids can be saturated (SFA), monounsaturated (MUFA), or polyunsaturated (PUFA). Among PUFAs, those with a first double bond on the third carbon are referred to as n-3, whereas those with a first double bond on the sixth carbon are called n-6.

Both the degree of saturation and the chain length influence the effect of a specific fatty acid in the immune-inflammatory pathways. In adipocytes, for example, saturated lauric acid, myristic acid, and palmitic acid (12, 14, and 16 carbons, respectively) activate inflammatory genes [62], whereas stearic acid (a SFA with 18 carbons) does not [63]. The scenario, however, is complex, since contrasting data from the literature impede clearly discriminating the pro- or anti-inflammatory properties of medium-chain fatty acids.

SFAs stimulate the inflammatory activation of macrophages by a process that involves TLR4, a pattern recognition receptor that plays a key role in the innate patrolling of bacterial pathogens, including LPS. In fact, the activation of TLR4 by SFA induces an over-activation of IL-6 and TNF-α inflammatory genes through a nuclear factor κB (NFκB)–dependent mechanism [64–66]. These effects are reduced by docosahexaenoic acid (DHA, an n-3 PUFA with 22 carbons), which inhibits NFkB and down-regulates two isoforms of the TLR family, TLR2 and TLR6, by impeding their dimerization [67]. Similarly, in vitro, DHA and eicosapentaenoic acid (EPA, an n-3 PUFA with 20 carbons) inhibit the LPS-induced gene expression of cyclooxygenase-2 (COX-2), which is instead increased by treatment with lauric acid [68]. The effect of SFAs on TLR4 signaling is not due to a direct interaction [69], but it requires bacterial LPS binding to CD14, a complex that promotes the endocytosis and lysosomal degradation of TLR4 [70]. Palmitic acid prolongs the activation of TLR4 in macrophages pre-stimulated with LPS, enhancing the production of pro-inflammatory cytokines (MCP-1 and TNF-α) and apoptotic signals that are mediated by the ER [71]. Vice versa, macrophages pre-stimulated with palmitic acid and then treated with palmitoleic acid (MUFA) showed reduced pro-inflammatory activation [72].

The TLR4-mediated engagement of NFkB is tightly linked with the activation of NLRP3 in macrophages [73]. Palmitic acid fosters the activation of NLRP3, by inhibiting adenosine monophosphate-activated protein kinase (AMPK) [71], whereas unsaturated fats inhibit the activation of the inflammasome [74], reverting the apoptotic signals triggered by ER stress in macrophages [75]. In addition, maresin-1 (a DHA-derived metabolite produced by 12-lipoxygenase) and resolvin D2 (produced by 15-lipoxygenase) down-regulate the NLRP3 pathway with subsequent inhibition of caspase-1 and reduction of IL-1β secretion [76] (Figure 2). By this mechanism, maresin-1 and resolvin D2 induce an anti-inflammatory phenotype in macrophages from apoE-/- mice, an effect that results in the stabilization of the atherosclerotic lesion [77]. As compared to the n-3 series, n-6 PUFAs clearly demonstrated pro-inflammatory effects. Arachidonic acid (ARA) (a 20 carbon, n6 PUFA that can be found only in animal-derived foods) promotes oxidative metabolism [78] and stimulates the release of IL-6 and TNF-α by macrophages through the production of two downstream products of ARA, the leukotriene B4 (LTB4) and the prostaglandin E2 (PGE$_2$ [79]). Also, further in vitro experiments postulated that ARA is able to activate the Janus–Kinase pathway in macrophages, inducing cell cycle arrest [80]. In contrast to this pro-inflammatory vision, ARA might also support the efficiency of neutrophils in sites of inflammation [81] (Figure 2). In fact neutrophils, once activated at the site of inflammation through the granulocyte-colony stimulating factor (GM-CSF), increase the uptake of the ARA contained in triglycerides by fatty acid transport protein 2 (FATP2) and convert ARA to the PGE$_2$ that promotes the suppression of CD8$^+$ cytotoxic T cells [82].

All these cellular pathways have been described in vitro, but the understanding of the mechanisms by which dietary fats are absorbed might allow us to better define how they can exert such effects in vivo. The intestinal absorption of medium- and long-chain fatty acids depends on the length of their aliphatic tails. After hydrolysis of their triglyceride precursors, medium-chain caprylic acid (8 carbons) and capric acid (10 carbons) are absorbed by enterocytes and bind to albumin, being then directly transported to the hepatic portal system [83] (Figure 1). Vice versa, long-chain fatty acids are packaged into chylomicrons in the enterocytes, secreted into the lacteals, enter the intestinal villi, and pass through the MLN before reaching circulation through the thoracic duct (Figure 1). Within this system, long-chain fatty acids in chylomicrons undergo the surveillance of C-X3-C motif chemokine receptor 1 (CX3CR1)-expressing macrophages residing in the villi, the CD103/CD11b expressing DCs (representing up to 80% of the total population of gut resident DCs) [84,85], and the innate lymphoid cell family member 2 (ILC2) [86] (a subset that uses dietary fatty acids as fuel for cellular fatty acid oxidation and energy production to produce IL-5, IL-9, IL-13 against *Trichuris muris* helminth infection [87]). (Figure 3).

Chylomicrons engage the endothelial cells of lymphatic vessels to produce chemokines involved in the activation of effector T-helper 17 cells [88,89]; then, once in the MLN, chylomicrons induce macrophages to secrete pro-inflammatory cytokines and to switch into pro-inflammatory foam cells [90] (Figure 3). Furthermore, the inflammatory cascade involving the metabolic conversion of PUFAs to arachidonic acid (AA) produces different mediators promoting the activity of ILC2 [91–93]. Notwithstanding whether these effects are actually mediated by (and which type of) dietary fatty acids remains a matter of debate today. Mineral oils show pro-inflammatory effects in peritoneal macrophages, triggering caspase-1 and NLRP3 activation, whereas some vegetal oils induce foam cell formation and cell death via caspase-3 cleavage-dependent mechanisms [94]. However, macrophages not only take up fatty acids deriving from the hydrolysis of triglycerides contained in chylomicrons and VLDL [95], thus inducing pro-atherogenic effects in endothelial cells [17] and monocytes [96], but LPS produced by gut commensals [97]. It is therefore plausible that macrophages might be activated by this bacteria-derived component, as suggested by the observation that gut-derived LPS and TLR4 co-localize with CD68, a marker of macrophages, in carotid atheromas [38].

2.3. The Contribution of Dietary Cholesterol as Compared to Serum Cholesterol

The intestine is the source of exogenous cholesterol, representing up to 30% of the total cholesterol pool in the body [98,99]. Cholesterol in the intestine principally derives from the enterohepatic circulation (bile) and, secondly, from dietary sources (principally from animal and dairy food products). In the intestinal lumen, esterified cholesterol is hydrolyzed by the pancreatic cholesteryl ester hydrolase, producing free cholesterol [98]. Free cholesterol is then emulsified, along with other lipids and vitamins, into micelles and absorbed by the enterocytes through multiple transport systems such as Niemann-Pick C1-like 1 (NPC1L1) and ABCG5/ABCG8. After absorption, free cholesterol is re-esterified by ACAT and packaged into chylomicrons [98]. In the circulation, chylomicrons undergo the activity of lipases, which hydrolyze their lipids and reduce their diameter, generating chylomicron remnants which are removed from the circulation by the liver [98]. There, cholesterol is principally rewired to the intestine via the enterohepatic circulation, while a modest fraction is secreted in the systemic circulation, packaged into VLDL. VLDL are significantly produced during PPL and, following the activity of lipases, they become low density lipoproteins (LDL), smaller in diameter, and with a predominant cholesterol content. Despite the average content of Western-style meals being estimated between 20 and 40 g of total fats/meal and three-four meals/day being typically consumed [100,101], given the actual content of cholesterol in the majority of foods (by 4 to 700 mg per quantity of food consumed [102,103]), it appears clear that the dietary source of cholesterol is minor as compared to that re-cycled through this complex metabolic system. Historical data actually demonstrated that there is a poor effect of the physiological consumption of cholesterol by diet and changes in serum cholesterol [104]. In line with this, both the American Heart Association Guidelines of 2013 [105] and the more recent indications of the European Society of Cardiology (ESC) and European Atherosclerosis Society (EAS [5]) toned down the magnitude of the effect and the level of evidence of reducing dietary cholesterol intake for CVD prevention.

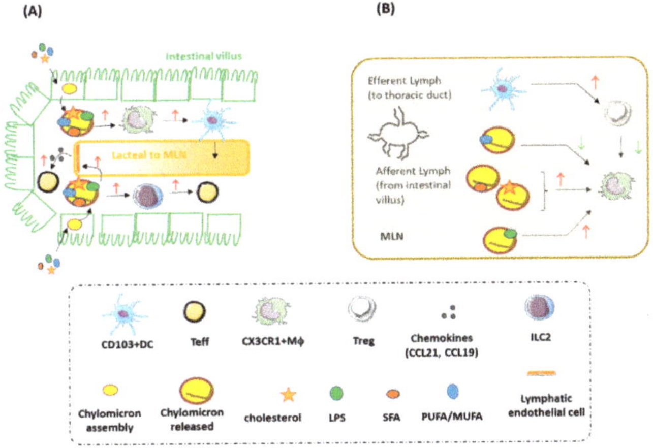

Figure 3. Summary of the immune-inflammatory pathways targeted by dietary fats in the intestinal villus (**A**) and in the MLN (**B**). LPS = lipopolysaccharide; CX3CR1 = C-X3-C motif chemokine receptor 1; Mϕ = macrophages; Teff = effector T cells; Treg = regulatory T cells; CCL- = C-C motif ligand-; ILC2 = innate lymphoid cell sub-type 2; SFA = saturated fatty acids; MUFA = mono-unsaturated fatty acids; PUFA = poly-unsaturated fatty acids; ↑: indicates an activated cell or pathway; ↓: indicates an inhibited cell or pathway.

Notwithstanding that, by virtue of their diameter and because they are significantly smaller than chylomicrons, VLDLs (30–70 nm) can enter the intima, and it is plausible that dietary cholesterol accumulating over time, can exert pro-inflammatory effects in the vasculature (Figure 2). When excessively produced during PPL, VLDL can foster atherogenic effects in endothelial cells by enhancing the expression of chemokine receptors and inducing apoptotic signals [17]. PPL extends the effects mediated by VLDL and furthermore promotes the accumulation of cholesterol in LDL. LDL are smaller than VLDL (20–30 nm in diameter) and their atherogenic potential is even higher [106]. Cholesterol can be oxidized into different types of oxysterols by a number of cardiovascular risk determinants as well as by factors related to the industrial processing of foods [3,11].

Oxysterols contribute to the formation of modified LDL (namely oxidized LDL, oxLDL), which are taken up by macrophages in the atheroma. Within cells, the crystallization of excess cholesterol occurs, thus further increasing its atherogenic potential and the ability to evoke the inflammatory activation of Teff [107–109] and the induction of NLRP3 [110,111] (Figure 2). Acute exposure of macrophages to oxLDL prolongs these mechanisms by inducing epigenetic priming of a complex set of inflammatory players [112]. Also, NLRP3 undergoes this epigenetic long-lasting activation, a process that has been described to favor an inflammatory phenotype of the myeloid hematopoietic immune compartment [113].

A small amount of robust experimental evidence suggests the intra-cellular effect of cholesterol in promoting the commitment of hematopoietic stem cells towards the expansion and pro-inflammatory activation of the myeloid compartment [114]. In fact, the apoE-/- mouse model of hypercholesterolemia actually displays aberrant hematopoietic commitment, a phenomenon that is mainly attributed to the remodeling of membrane lipid rafts and the over-expression of the GM-CSF receptors CD131 and CXCR4 (physiological regulators of HSC egress from medullary niches) [115] (Figure 2).

3. Data Linking Intake of Dietary Fats, Markers of Inflammation, and Risk of CVD

The evidence about the molecular and cellular mechanisms by which dietary fats participate in atherogenesis built up the rationale to unveil the connections between different dietary fats, the markers of systemic inflammation, and the risk of CVDs. Despite this aspect having been extensively discussed, data from both epidemiological studies and interventional clinical trials are, however, scarce and heterogeneous. Here, we will provide a separate discussion for results from epidemiological studies and interventional clinical trials.

3.1. Epidemiological Studies

An overall summary of the results from epidemiological studies on the relationship between the pro- or anti-inflammatory effects of dietary fats and CVD risk is reported in Table 1.

Table 1. Summary of the data from epidemiological studies on the association between dietary intake of fats, circulating markers of systemic inflammation and risk of CVD. ↑: Indicates data showing a positive association between the dietary fat intake and the outcome (either inflammatory markers or CVD risk factors); ↓ indicates data showing a positive association between the dietary fat intake and the outcome (either inflammatory markers or CVD risk factors). ↔ indicates that there are missing or contrasting data regarding association between the dietary fat intake and the outcome (either inflammatory markers or CVD risk factors).

	EPIDEMIOLOGICAL STUDIES	
Dietary fats	**Prevalent Effects on Inflammatory Markers**	**Effects on CVD Risk/Risk Factors**
Trans fats	↑ [116] [117] [118] ↓ ↔ [119,120] [121] [122]	↑ [116,119–121] [117] [118,123] ↓ ↔ [122]
Saturated fats	↔ [119,120] [121] [124,125]	↑ [119] ↔ [120] [121] [123,124]
Monounsaturated fats	↓ [119] [126,127]	↓ [119] [126,127] ↔ [121] [128] [123]
Polyunsaturated fats	↔ [119] [121] [126] [129]	↓ [119] [121] [123] [126] [130,131] [129]
n-3 and derivates	↓ [129] [132–135] ↔ [136]	↓ [132,137] [134] ↔ [136]
n-6 and derivates	↔ [119]	↓ [119]
Cholesterol	↑ [125] ↔ [138–141]	↑ [138] ↔ [139–142]

The association between the intake of SFAs, markers of inflammation, and CVD has been assessed in recent epidemiological studies. In two independent cohorts (the Nurses' Health Study and the Health Professionals Follow-up Study (HPFS)), increasing dietary intake of saturated and trans-fats was significantly associated with a higher risk of CVDs over a long-term follow-up (1980–2012), with dietary assessment evaluated every four years [119]. Another analysis from these two studies reported that a higher consumption of saturated dietary fats was not associated with higher circulating levels of IL-6 and C-reactive protein (CRP) both in women and men with a history of CHD, whereas this association was observed in subjects without previous CHD [143]. A lack of association was also found among older adults at elevated CV risk from the Health ABC Study [138], where the relation between elevated risk of heart failure and CHD and higher levels of inflammatory markers was independent from the intake of dietary fats [144,145]. In the Scottish Lothian Birth Cohort 1936 study, a higher consumption of saturated fats correlated with elevated CRP (but not with higher fibrinogen or IL-6) in older subjects; however, this association was significant in subjects with hypercholesterolemia, but not in subjects with previous CVDs and ischemic stroke [146]. In another study, in men with acceleration of aortic pulse wave velocity (aPWV) no relation was found between the dietary intake of saturated fats and CRP plasma levels [120]. In the SUN cohort, a higher intake of saturated fats, which contributed to higher inflammatory potential of diet (expressed by the dietary inflammatory index or DII) predicted higher occurrence of CVDs over an eight-year follow-up [147]. It has to be acknowledged that the lack of information about quantitative changes in circulating markers during follow-up does not allow to link the intake of dietary fats with systemic inflammation. The large Prospective Urban Rural Epidemiology (PURE) study, assessing the predictive role of dietary intake of fats on the occurrence of CV events over eight-year follow-up [148], found that lower intake of SFAs was associated with lower risk of ischemic stroke, whereas it did not associate with the risk of myocardial infarction. Also in this study, the lack of information on inflammatory markers does not allow to draw any conclusions.

The increased consumption of total trans-fats has been historically associated with an increased risk of CHD [119,121]. However, recent evidence attributed opposing effects to the two main classes of trans-fats. In fact, higher plasma phospholipids content of ruminant-trans fatty acids (rTFA), which are naturally found in dairy foods and meats, was associated with reduced CVD risk factors and with higher adiponectin levels (an anti-inflammatory adipokine). Conversely, a higher content of industrial-trans fatty acid (iTFA) (such as trans elaidic fatty acid) in plasma phospholipids was associated with higher CVD risk factors [116], although conflicting results have been reported. In the Akershus Cardiac Examination 1950 study, reduced intake of iTFA was associated with increased plasma levels of CV risk factors (triglycerides, fasting glucose, blood pressure and CRP) [117]. These discordant observations might be in part explained by the matrix effect, as multiple components, preservatives, and industrial processes in food might affect the link between inflammation and CVD as compared to a single dietary fatty acid [11]. This hypothesis might be also applied to previously discussed SFAs from non-processed dairy products whose consumption, despite not being associated with CVD risk [11], correlates with the reduction in inflammatory marker levels, such as CRP, IL-6, and TNF-α [149].

Several epidemiological studies have provided discordant data also for the association between dietary intake of unsaturated fats and CVD. The Nurse Health Study showed that a higher intake of both MUFAs and PUFAs was associated with reduced CV mortality; the Health Professional Follow-up study confirmed this association only for PUFAs [119]. These two large studies did not evaluate whether these results might be correlated to significant changes in markers of low-grade inflammation. In the Caerphilly Prospective study, the increased dietary intake of PUFAs was associated with reduced CRP and fibrinogen levels and with a higher aPWV. Vice versa, the intake of MUFAs, which are associated with fibrinogen levels, did not predict increase in aPWV [120]. The observation regarding the intake of PUFAs was also confirmed in the Whitehall study of London civil servants [150], while no information for the intake of MUFAs is available in this study.

The conflicting results reported for MUFAs might be attributed to the type of food source, as only MUFAs from animal sources (such as beef, pork, and processed meat, which also contributes to saturated fats intake), but not those from plant sources are associated to 16% higher risk of CVD mortality [128]. In addition, consumption of olive oil and nuts, which are the main plant food sources of MUFAs, is associated with both lower occurrence of nonfatal CVD events and lower inflammatory markers (IL-6) levels [126,127].

Among PUFAs, whether the potential anti-inflammatory effect is to be attributed to the n-3 or n-6 series is an actual matter of study. The adherence to the Mediterranean diet-based model of the PREDIMED study promoting the consumption of n-3 enriched based foods, was associated with reduced DII [130]. This finding was also reported in the PRE-diabetes and type 2 DIAbetes (SPREDIA-2) study, in which a higher consumption of n-3 PUFAs associated with reduced CRP levels [151]. In contrast, in patients in secondary prevention from the Western Norway B Vitamin Intervention Trial, a higher consumption of n-3 from either fatty fish or fish oil did not protect from recurring coronary events, nor it was associated with a significant reduction in CRP levels [136]. Vice versa, in men in primary prevention from the Health Professionals Follow-up Study HPFS, n-3 PUFAs from both seafood and plant sources predicted lower incidence of coronary events independently of the concomitant dietary intake of n-6 fatty acids, despite the effect on inflammatory markers not being investigated [137]. Also, in healthy women of the Nurses' Health Study I cohort, the intake of α-linolenic acid was inversely related to plasma concentrations of CRP and E-selectin, and EPA and DHA were inversely related to other markers of vascular endothelial dysfunction (ICAM-1 and VCAM-1) [132]. In the HPFS and the Nurses' Health Study II, a higher intake of EPA and DHA was associated with reduced plasma levels of both soluble isoforms of the TNF receptor, despite non-significant trends towards the reduction of CRP levels [133]. Of note, these associations were significant even in subjects with an elevated dietary intake of the n-6 series, suggesting that these fatty acids might not contrast the plausible anti-inflammatory effect of the n-3 series [133]. By contrast, in over three thousand community-dwelling Japanese individuals from the Hisayama Study, the decrement in the EPA/AA ratio significantly increased both the risk of occurring CVD and the serum levels of CRP [134]. In addition, low fish consumption, low EPA and DHA intake and a low intake of AA were also associated with higher levels of inflammatory markers in subjects with coronary artery disease [135]. Nuts represent an alternative food source of PUFAs, and accumulating evidence supports their anti-inflammatory and cardioprotective effects. High nut consumption (>120 g/week), particularly MUFA- and PUFA-enriched tree nuts and ground nuts, is associated with reduced CV risk factors, improved postprandial lipemia and significant reduction in CVD mortality [131]. Furthermore, the reduced risk of CVD mortality attributable to nut consumption (>2 times/week), was linked, although in a modest proportion (17.8%), to lower hs-CRP levels [129].

Dietary cholesterol present in multiple processed and non-processed food patterns, including eggs, butter, beef, cheese and shrimp, deserves a separate discussion. The established pro-inflammatory effects of cholesterol have been hardly confirmed in epidemiological studies, which report contrasting data on the relationships between dietary consumption of cholesterol, markers of inflammation and CVD risk [152]. In fact, the association between dietary cholesterol from eggs and markers of systemic low-grade inflammation is not clear, with studies in both healthy subjects and patients with type 2 diabetes reporting contrasting findings [139,140]. Bechthold et al. showed no correlation between the highest (75 g) and lowest intake of eggs (0 g) and the risk of CHD or stroke. In a dose-response sub-analysis, increased increments of egg intake (50 g) were predictive of higher risk of heart failure, but not CHD or stroke [153]. This observation was replicated in the Kuopio Ischaemic Heart Disease Risk Factors Study, where cholesterol intake from eggs was not associated with the risk of ischemic stroke [141]. Similarly, in the Nurses' Health Study and in the HPFS, the intake of dietary cholesterol (consumed as one egg daily) was not associated with increased risk of coronary artery disease in healthy men and women [142]. These data might be principally explained by the limited contribution of dietary cholesterol as compared to the endogenous fraction, which has been for long time the principal target of all the most effective pharmacological options. Furthermore, it cannot be ruled out that again the food matrix effect might reconcile the inflammatory effect of specific atherogenic components of eggs. This is the case of the immune-inflammatory and atherogenic trimethylamine N-oxide (TMAO) [32,154], which is significantly absorbed following the consumption of a fixed quantity of eggs, and its plasmatic concentration seems to depend on the abundance and activity of specific gut resident bacteria [155].

3.2. Interventional Clinical Trials

Table 2 reports a summary of the results from clinical trials, testing the impact of dietary fat consumption on markers of inflammation and CVD risk factors.

Table 2. Summary of data from interventional studies about the association between dietary intake of lipids, circulating markers of systemic inflammation and risk of cardiovascular diseases. ↑: Indicates data showing a positive association between the dietary intervention and the outcome (either inflammatory markers or CVD risk factors); ↓ indicates data showing a positive association between the dietary intervention and the outcome (either inflammatory markers or CVD risk factors). ↔ indicates that there are missing or contrasting data regarding association between the dietary intervention and the outcome (either inflammatory markers or CVD risk factors).

INTERVENTIONAL TRIALS		
Dietary Lipids	**Prevalent Effects on Inflammatory Markers**	**Effects on CVD Risk/Risk Factors**
Trans fats	↑ [156] ↔ [157]	↑ [155] [156]
Saturated fats	↑ [158] [159] [160] ↔ [156] [161,162] [163]	↑ [156] [158,161] [159,160,163] ↔ [162]
Monounsaturated fats	↓ [156] [158] ↔ [161,162] [163] [164] [165]	↓ [156] [158,161,162] [163] [164]
Polyunsaturated fats	↓ [159] ↔ [161] [163]	↓ [161] [159,163] [166]
Ω-3 and derivates	↓ [167–171] [165] ↔ [160] [172] [173,174] [175,176]	↓ [160] [168–171,173] ↔ [167,172] [165,174–176]
Ω-6 and derivates	↔ [160] [167]	↔ [160] [167]
Cholesterol	↓ [177] ↑ [178] ↔ [179]	↓ [180] ↑ [179]

Lower levels of adhesion molecules and CRP were observed when 8% of energy from trans-fats was replaced with the same amount of energy from MUFAs [156]. Vice versa, the substitution of SFAs (stearic acid) with trans-fats did not exert effects on inflammatory marker levels (CRP and IL-6) [157]. This study, however, did not discuss whether substitution with either rTFA or iTFA would have exerted different effects.

Also, replacing 8% of energy from digestible carbohydrates with the same amount of energy from either medium-chain or long-chain SFAs (stearic acid) did not reduce CRP and IL-6 levels over a five-week period [156,162]. In an independent cross-over randomized trial, consumption of SFAs from butter or cheese did not increase hs-CRP levels as compared to an isocaloric carbohydrate-enriched dietary regimen, despite inducing an increase in LDL-C [161]. A meta-analysis of 16 trials assessing the effects of the consumption of coconut oil (containing elevated amount of SFAs) or other fats found that coconut oil increased cholesterol, but had no effects on markers of systemic inflammation [181]. Similarly, CRP did not increase in cross-over-based trials comparing a high-cholesterol high-fat diet with a low-cholesterol high-fat diet, both in healthy subjects [178] and in patients with type 2 diabetes [179].

Trials comparing the substitution of SFAs with MUFAs reported a beneficial effect on LDL-C levels, but no changes in CRP and other inflammatory markers (including IL-6) [156,162,163]; other trials could not confirm this effect on LDL-C levels [164,182]. An increased intake of PUFAs was associated with lower risk of coronary artery disease [166], despite the potential pro- or anti-inflammatory effect not being assessed.

Both LDL-C-lowering effects and a trend towards CRP reduction were observed in studies evaluating consumption of n-6 PUFAs in place of SFAs [161,163]. Similarly, the replacement of SFAs with n-3 or n-6 PUFAs exerted significant reductions in CRP levels and lowered total cholesterol, LDL-C and triglycerides [160]; of note, these reductions were more robust following the substitution of SFAs with n-3 rather than n-6 PUFAs. In dyslipidaemic patients, n-6 PUFA supplementation lowered total cholesterol concentration from baseline without any effects on inflammatory markers [167].

N-3 supplementation trials failed to provide conclusive evidence on their effects on inflammation markers and CVD risk factors [172]. For example, in dyslipidaemic patients, the significant reduction in CRP and IL-6 observed with n-3 supplementation was not paralleled by a robust improvement in plasma lipids [167]. The REDUCE-IT trial demonstrated that the intervention with 4 g/day of a particular formulation of EPA provided a robust reduction in CRP levels and a significant reduction in the risk of atherosclerotic cardiovascular events in patients at elevated CV risk despite receiving the maximally tolerated statin therapy [168]. Besides these pharmacological findings, either nutraceutical or dietary supplementation with unsaturated fats and n-3 series exerted contrasting results on markers of inflammation and CVD risk. Recent data from the VITAL Research Group indicate that supplementation with 1 g/day of marine n-3 fats did not reduce the incidence of major cardiovascular events. A modest reduction in the risk of myocardial infarction was reported in subjects of the same study reporting dietary fish consumption less than 1.5 servings/week [183], but the lack of data on markers of inflammation does not allow the researchers to conclude whether this finding might be proportional to from the results of REDUCE–IT.

In parallel to these data, a beneficial effect of the dietary consumption of the n-3 series on markers of cellular inflammation has been proposed For example, increased n-3 consumption from nuts lowered blood pressure, plasma cholesterol levels, reduced markers of DNA oxidative damage in peripheral leukocytes, and improved endothelial function in subjects with metabolic syndrome [169–171,173]. Vice versa, in obese subjects, dietary n-3 supplementation at different dosages (although lower than that used in REDUCE-IT) had opposing effects on inflammatory markers and CV risk factors. At lower doses, when consumed as supplemented milk, there was no effect on inflammation, but a significant increase in HDL-C was observed [174]. At higher doses, obtained with n-3 rich flaxseed flour, reductions in CRP and serum amyloid A were reported in obese patients without any effects on body weight and other cardio-metabolic markers [165]. Furthermore, n-3 PUFA supplementation favorably influenced arterial stiffness and hs-CRP compared to corn oil supplementation in small sized trials, although not to a statistically significant degree [176]. Similarly, CRP was significantly reduced in statin-naïve subjects with dyslipidemia receiving n-3 fatty acids [176]. The consumption of n-3 PUFA-enriched foods (fish oil or fruit juice or fish pate), providing approximately 1 g EPA + DHA, reduced interferon-γ (IFN-γ) levels without any change in lipid profile [175].

The effects of n-3 PUFA are also influenced by the type of PUFA supplemented: a head-to-head comparison of the effects of EPA and DHA showed higher beneficial effects on inflammatory marker levels and CVD risk factors in the DHA-treated group, even though an increase in LDL-C was observed [184].

4. Conclusions

Despite the strong evidence at the cellular and molecular level, the relationship between dietary fats, immune-inflammation and atherosclerotic CVD is not well established at the clinical level [185]. Methodological and technical difficulties are, however, to be acknowledged. In fact, interventional trials, although designed with the attempt to provide strong causal inference, are oftentimes sample-sized, have short follow-up, have high attrition rates and have to cope with the difficulty of the standardization of the measurement of biomarkers. In parallel, observational studies are overly of misinterpreted dietary intakes, and the lack of sufficient follow-up oftentimes affect the robustness of the outcomes. Despite these difficulties, the principal take-home message from this complex background is that the paradigms the evolved during the last decades addressing pro-inflammatory effect to dietary SFAs, cholesterol and trans-fats while an anti–inflammatory potential to dietary MUFAs, PUFAs or n-3 probably need to be re-challenged with respect to the international recommendations for CVD prevention. As a matter of fact, recent large, multi-center observations questioned this archetype [10,186], supporting that additional factors, related to the individual environmental exposure, can be more likely effective. Among them, the matrix effect (the sum of the content in micro-, macronutrients and additives in food patterns) principally determines different associations between dietary fats and CVD [10,11]. Also, the metabolic postprandial response to foods, in particular to highly-caloric dietary fats, was recently demonstrated to vary largely among individuals because of strong interference from the gut [187,188]. Finally, even more recently, individual signatures of gut microbiota were connected to systemic markers of immune-inflammation in determining individual profiles of postprandial response to foods [20]. Altogether, these important observations suggest that the connection between dietary fats and CVD should be studied taking into account both the immune-inflammatory potential of these dietary sources and the individual predisposition for their metabolism. Under this perspective, a more personalized approach to diet could be pursued.

Author Contributions: Conceptualization, A.B.; validation, A.B.; investigation, A.B. and E.M. resources, A.B.; writing—original draft preparation, A.B. and E.M.; writing—review and editing, A.L.C. and A.B.; visualization, A.L.C.; supervision, A.L.C.; funding acquisition, A.L.C. and A.B. All authors have equally contributed to the development of the manuscript. All authors have read and agreed to the published version of the manuscript.

Funding: The work of A Baragetti is supported by "Cibo, Microbiota, Salute" by "Vini di Batasiolo S.p.A" AL_RIC19ABARA_01; by "Post-Doctoral Fellowship 2020" by "Fondazione Umberto Veronesi" 2020-3318; by a research award (2021) from "the Peanut Institute". The work of AL Catapano is supported by Ministry of Health-Ricerca Corrente-IRCCS MultiMedica, PRIN 2017H5F943, ERANET ER-2017-2364981, and Fondazione SISA. The work was supported by Regione Lombardia "PROGRAMMA OPERATIVO REGIONALE 2014–2020" OBIETTIVO "INVESTIMENTI IN FAVORE DELLA CRESCITA E DELL'OCCUPAZIONE" (226149- "Studio e messa punto di nuovi prodotti pro- e prebiotici per la prevenzione ed il trattamento di patologie infiammatorie quali la sindrome del colon irritabile e la dermatite atopica-SCIDA"). The work was supported by funding from the Russian Ministry of Science and Higher Education (Agreement No. 075-15-2020-901).

Institutional Review Board Statement: The work does not involve observational or experimental procedures nor in animals neither in humans.

Informed Consent Statement: The work does not involve observational or experimental procedures nor in animals neither in humans.

Data Availability Statement: The manuscript does not show data or analyses of data.

Conflicts of Interest: AL Catapano reports grants from Amgen, Sanofi, Regeneron and personal fees from Merck, Sanofi, Regeneron, AstraZeneca, Amgen, Novartis, outside the submitted work.

References

1. Aiello, L.C.; Wheeler, P. The Expensive-Tissue Hypothesis: The Brain and the Digestive System in Human and Primate Evolution. *Curr. Anthropol.* **1995**, *36*, 199–221. [CrossRef]
2. Holmes, E.; Li, J.V.; Marchesi, J.R.; Nicholson, J.K. Gut Microbiota Composition and Activity in Relation to Host Metabolic Phenotype and Disease Risk. *Cell Metab.* **2012**, *16*, 559–564. [CrossRef] [PubMed]
3. Christ, A.; Lauterbach, M.; Latz, E. Western Diet and the Immune System: An Inflammatory Connection. *Immunity* **2019**, *51*, 794–811. [CrossRef] [PubMed]
4. Meier, T.; Gräfe, K.; Senn, F.; Sur, P.; Stangl, G.I.; Dawczynski, C.; März, W.; Kleber, M.E.; Lorkowski, S. Cardiovascular mortality attributable to dietary risk factors in 51 countries in the WHO European Region from 1990 to 2016: A systematic analysis of the Global Burden of Disease Study. *Eur. J. Epidemiol.* **2019**, *34*, 37–55. [CrossRef]
5. Mach, F.; Baigent, C.; Catapano, A.L.; Koskinas, K.C.; Casula, M.; Badimon, L.; Chapman, M.J.; De Backer, G.G.; Delgado, V.; Ference, B.A.; et al. 2019 ESC/EAS Guidelines for the management of dyslipidaemias: Lipid modification to reduce cardiovascular risk. *Eur. Heart J.* **2020**, *41*, 111–188. [CrossRef] [PubMed]
6. Estruch, R.; Ros, E.; Salas-Salvadó, J.; Covas, M.-I.; Corella, D.; Arós, F.; Gómez-Gracia, E.; Ruiz-Gutiérrez, V.; Fiol, M.; Lapetra, J.; et al. Primary Prevention of Cardiovascular Disease with a Mediterranean Diet Supplemented with Extra-Virgin Olive Oil or Nuts. *N. Engl. J. Med.* **2018**, *378*, e34. [CrossRef] [PubMed]
7. Satija, A.; Yu, E.; Willett, W.C.; Hu, F.B. Understanding Nutritional Epidemiology and Its Role in Policy. *Adv. Nutr.* **2015**, *6*, 5–18. [CrossRef]
8. Maki, K.C.; Slavin, J.L.; Rains, T.M.; Kris-Etherton, P.M. Limitations of Observational Evidence: Implications for Evidence-Based Dietary Recommendations. *Adv. Nutr.* **2014**, *5*, 7–15. [CrossRef] [PubMed]
9. Hu, F.B.; Satija, A.; Rimm, E.B.; Spiegelman, D.; Sampson, L.; Rosner, B.; Camargo, C.A.; Stampfer, M.; Willett, W.C. Diet Assessment Methods in the Nurses' Health Studies and Contribution to Evidence-Based Nutritional Policies and Guidelines. *Am. J. Public Health* **2016**, *106*, 1567–1572. [CrossRef] [PubMed]
10. Yusuf, S.; Joseph, P.; Rangarajan, S.; Islam, S.; Mente, A.; Hystad, P.; Brauer, M.; Kutty, V.R.; Gupta, R.; Wielgosz, A.; et al. Modifiable risk factors, cardiovascular disease, and mortality in 155 722 individuals from 21 high-income, middle-income, and low-income countries (PURE): A prospective cohort study. *Lancet* **2020**, *395*, 795–808. [CrossRef]
11. Astrup, A.; Magkos, F.; Bier, D.M.; Brenna, J.T.; de Oliveira Otto, M.C.; Hill, J.O.; King, J.C.; Mente, A.; Ordovas, J.M.; Volek, J.S.; et al. Saturated Fats and Health: A Reassessment and Proposal for Food-Based Recommendations: JACC State-of-the-Art Review. *J. Am. Coll. Cardiol.* **2020**, *76*, 844–857. [CrossRef]
12. O'Connor, L.E.; Kim, J.E.; Campbell, W.W. Total red meat intake of ≥0.5 servings/d does not negatively influence cardiovascular disease risk factors: A systemically searched meta-analysis of randomized controlled trials. *Am. J. Clin. Nutr.* **2017**, *105*, 57–69. [CrossRef] [PubMed]
13. Zhong, V.W.; Van Horn, L.; Greenland, P.; Carnethon, M.R.; Ning, H.; Wilkins, J.T.; Lloyd-Jones, D.M.; Allen, N.B. Associations of Processed Meat, Unprocessed Red Meat, Poultry, or Fish Intake With Incident Cardiovascular Disease and All-Cause Mortality. *JAMA Intern. Med.* **2020**, *180*, 503. [CrossRef] [PubMed]
14. O'Keefe, J.H.; Torres-Acosta, N.; O'Keefe, E.L.; Saeed, I.M.; Lavie, C.J.; Smith, S.E.; Ros, E. A Pesco-Mediterranean Diet With Intermittent Fasting. *J. Am. Coll. Cardiol.* **2020**, *76*, 1484–1493. [CrossRef] [PubMed]
15. Patsch, J.R.; Miesenböck, G.; Hopferwieser, T.; Mühlberger, V.; Knapp, E.; Dunn, J.K.; Gotto, A.M.; Patsch, W. Relation of triglyceride metabolism and coronary artery disease. Studies in the postprandial state. *Arterioscler. Thromb. J. Vasc. Biol.* **1992**, *12*, 1336–1345. [CrossRef] [PubMed]
16. Nordestgaard, B.G.; Benn, M.; Schnohr, P.; Tybjærg-Hansen, A. Nonfasting triglycerides and risk of myocardial infarction, ischemic heart disease, and death in men and women. *J. Am. Med. Assoc.* **2007**, *298*, 299–308. [CrossRef] [PubMed]
17. Norata, G.D.; Grigore, L.; Raselli, S.; Redaelli, L.; Hamsten, A.; Maggi, F.; Eriksson, P.; Catapano, A.L. Post-prandial endothelial dysfunction in hypertriglyceridemic subjects: Molecular mechanisms and gene expression studies. *Atherosclerosis* **2007**, *193*, 321–327. [CrossRef] [PubMed]
18. Bernelot Moens, S.J.; Verweij, S.L.; Schnitzler, J.G.; Stiekema, L.C.A.; Bos, M.; Langsted, A.; Kuijk, C.; Bekkering, S.; Voermans, C.; Verberne, H.J.; et al. Remnant Cholesterol Elicits Arterial Wall Inflammation and a Multilevel Cellular Immune Response in Humans. *Arterioscler. Thromb. Vasc. Biol.* **2017**, *37*, 969–975. [CrossRef]
19. Norata, G.D.; Grigore, L.; Raselli, S.; Seccomandi, P.M.; Hamsten, A.; Maggi, F.M.; Eriksson, P.; Catapano, A.L. Triglyceride-rich lipoproteins from hypertriglyceridemic subjects induce a pro-inflammatory response in the endothelium: Molecular mechanisms and gene expression studies. *J. Mol. Cell. Cardiol.* **2006**, *40*, 484–494. [CrossRef]
20. Berry, S.E.; Valdes, A.M.; Drew, D.A.; Asnicar, F.; Mazidi, M.; Wolf, J.; Capdevila, J.; Hadjigeorgiou, G.; Davies, R.; Al Khatib, H.; et al. Human postprandial responses to food and potential for precision nutrition. *Nat. Med.* **2020**, *26*, 964–973. [CrossRef]
21. Zhang, D.; Chen, G.; Manwani, D.; Mortha, A.; Xu, C.; Faith, J.J.; Burk, R.D.; Kunisaki, Y.; Jang, J.-E.; Scheiermann, C.; et al. Neutrophil ageing is regulated by the microbiome. *Nature* **2015**, *525*, 528–532. [CrossRef] [PubMed]
22. Uehara, K.; Takahashi, T.; Fujii, H.; Shimizu, H.; Omori, E.; Matsumi, M.; Yokoyama, M.; Morita, K.; Akagi, R.; Sassa, S. The lower intestinal tract-specific induction of heme oxygenase-1 by glutamine protects against endotoxemic intestinal injury. *Crit. Care Med.* **2005**, *33*, 381–390. [CrossRef] [PubMed]

23. Ren, W.-K.; Yin, J.; Zhu, X.-P.; Liu, G.; Li, N.-Z.; Peng, Y.-Y.; Yin, Y.-Y. Glutamine on Intestinal Inflammation: A Mechanistic Perspective. *Eur. J. Inflamm.* **2013**, *11*, 315–326. [CrossRef]
24. Ren, W.; Yin, J.; Wu, M.; Liu, G.; Yang, G.; Xion, Y.; Su, D.; Wu, L.; Li, T.; Chen, S.; et al. Serum Amino Acids Profile and the Beneficial Effects of L-Arginine or L-Glutamine Supplementation in Dextran Sulfate Sodium Colitis. *PLoS ONE* **2014**, *9*, e88335. [CrossRef]
25. Wang, W.; Wu, Z.; Lin, G.; Hu, S.; Wang, B.; Dai, Z.; Wu, G. Glycine Stimulates Protein Synthesis and Inhibits Oxidative Stress in Pig Small Intestinal Epithelial Cells. *J. Nutr.* **2014**, *144*, 1540–1548. [CrossRef]
26. Thaiss, C.A.; Levy, M.; Grosheva, I.; Zheng, D.; Soffer, E.; Blacher, E.; Braverman, S.; Tengeler, A.C.; Barak, O.; Elazar, M.; et al. Hyperglycemia drives intestinal barrier dysfunction and risk for enteric infection. *Science* **2018**, *359*, 1376–1383. [CrossRef]
27. Wang, T.Y.; Liu, M.; Portincasa, P.; Wang, D.Q.-H. New insights into the molecular mechanism of intestinal fatty acid absorption. *Eur. J. Clin. Investig.* **2013**, *43*, 1203–1233. [CrossRef]
28. Beilstein, F.; Carrière, V.; Leturque, A.; Demignot, S. Characteristics and functions of lipid droplets and associated proteins in enterocytes. *Exp. Cell Res.* **2016**, *340*, 172–179. [CrossRef]
29. Foster-Powell, K.; Holt, S.H.; Brand-Miller, J.C. International table of glycemic index and glycemic load values: 2002. *Am. J. Clin. Nutr.* **2002**, *76*, 5–56. [CrossRef] [PubMed]
30. Cleophas, M.C.P.; Ratter, J.M.; Bekkering, S.; Quintin, J.; Schraa, K.; Stroes, E.S.; Netea, M.G.; Joosten, L.A.B. Effects of oral butyrate supplementation on inflammatory potential of circulating peripheral blood mononuclear cells in healthy and obese males. *Sci. Rep.* **2019**, *9*, 775. [CrossRef] [PubMed]
31. Du, Y.; Li, X.; Su, C.; Xi, M.; Zhang, X.; Jiang, Z.; Wang, L.; Hong, B. Butyrate protects against high-fat diet-induced atherosclerosis via up-regulating ABCA1 expression in apolipoprotein E-deficiency mice. *Br. J. Pharmacol.* **2020**, *177*, 1754–1772. [CrossRef] [PubMed]
32. Baragetti, A.; Severgnini, M.; Olmastroni, E.; Dioguardi, C.C.; Mattavelli, E.; Angius, A.; Rotta, L.; Cibella, J.; Caredda, G.; Consolandi, C.; et al. Gut microbiota functional dysbiosis relates to individual diet in subclinical carotid atherosclerosis. *Nutrients* **2021**, *13*, 304. [CrossRef] [PubMed]
33. Bartolomaeus, H.; Balogh, A.; Yakoub, M.; Homann, S.; Markó, L.; Höges, S.; Tsvetkov, D.; Krannich, A.; Wundersitz, S.; Avery, E.G.; et al. Short-Chain Fatty Acid Propionate Protects from Hypertensive Cardiovascular Damage. *Circulation* **2019**, *139*, 1407–1421. [CrossRef] [PubMed]
34. Wright, A.T. Gut commensals make choline too. *Nat. Microbiol.* **2019**, *4*, 4–5. [CrossRef]
35. Koeth, R.A.; Lam-Galvez, B.R.; Kirsop, J.; Wang, Z.; Levison, B.S.; Gu, X.; Copeland, M.F.; Bartlett, D.; Cody, D.B.; Dai, H.J.; et al. L-Carnitine in omnivorous diets induces an atherogenic gut microbial pathway in humans. *J. Clin. Investig.* **2019**, *129*, 373–387. [CrossRef]
36. Caesar, R.; Reigstad, C.S.; Bäckhed, H.K.; Reinhardt, C.; Ketonen, M.; Lundén, G.Ö.; Cani, P.D.; Bäckhed, F. Gut-derived lipopolysaccharide augments adipose macrophage accumulation but is not essential for impaired glucose or insulin tolerance in mice. *Gut* **2012**, *61*, 1701–1707. [CrossRef]
37. d'Hennezel, E.; Abubucker, S.; Murphy, L.O.; Cullen, T.W. Total Lipopolysaccharide from the Human Gut Microbiome Silences Toll-Like Receptor Signaling. *mSystems* **2017**, *2*, e00046-17. [CrossRef] [PubMed]
38. Carnevale, R.; Sciarretta, S.; Valenti, V.; di Nonno, F.; Calvieri, C.; Nocella, C.; Frati, G.; Forte, M.; d'Amati, G.; Pignataro, M.G.; et al. Low-grade endotoxaemia enhances artery thrombus growth via toll-like receptor 4: Implication for myocardial infarction. *Eur. Heart J.* **2020**, *41*, 3156–3165. [CrossRef]
39. Schönfeld, P.; Wojtczak, L. Short- and medium-chain fatty acids in energy metabolism: The cellular perspective. *J. Lipid Res.* **2016**, *57*, 943–954. [CrossRef]
40. Randolph, G.J.; Miller, N.E. Lymphatic transport of high-density lipoproteins and chylomicrons. *J. Clin. Investig.* **2014**, *124*, 929–935. [CrossRef]
41. Bonnardel, J.; T'Jonck, W.; Gaublomme, D.; Browaeys, R.; Scott, C.L.; Martens, L.; Vanneste, B.; De Prijck, S.; Nedospasov, S.A.; Kremer, A.; et al. Stellate Cells, Hepatocytes, and Endothelial Cells Imprint the Kupffer Cell Identity on Monocytes Colonizing the Liver Macrophage Niche. *Immunity* **2019**, *51*, 638–654. [CrossRef] [PubMed]
42. Han, Y.-H.; Onufer, E.J.; Huang, L.-H.; Sprung, R.W.; Davidson, W.S.; Czepielewski, R.S.; Wohltmann, M.; Sorci-Thomas, M.G.; Warner, B.W.; Randolph, G.J. Enterically derived high-density lipoprotein restrains liver injury through the portal vein. *Science* **2021**, *373*, eabe6729. [CrossRef]
43. Norata, G.D.; Catapano, A.L. Molecular mechanisms responsible for the antiinflammatory and protective effect of HDL on the endothelium. *Vasc. Health Risk Manag.* **2005**, *1*, 119–129. [CrossRef] [PubMed]
44. Chakraborti, C.K. New-found link between microbiota and obesity. *World J. Gastrointest. Pathophysiol.* **2015**, *6*, 110. [CrossRef] [PubMed]
45. Den Besten, G.; Van Eunen, K.; Groen, A.K.; Venema, K.; Reijngoud, D.J.; Bakker, B.M. The role of short-chain fatty acids in the interplay between diet, gut microbiota, and host energy metabolism. *J. Lipid Res.* **2013**, *54*, 2325–2340. [CrossRef]
46. Garrett, W.S. Immune recognition of microbial metabolites. *Nat. Rev. Immunol.* **2020**, *20*, 91–92. [CrossRef]
47. Minihane, A.M.; Vinoy, S.; Russell, W.R.; Baka, A.; Roche, H.M.; Tuohy, K.M.; Teeling, J.L.; Blaak, E.E.; Fenech, M.; Vauzour, D.; et al. Low-grade inflammation, diet composition and health: Current research evidence and its translation. *Br. J. Nutr.* **2015**, *114*, 999–1012. [CrossRef]

48. Björkman, L.; Mårtensson, J.; Winther, M.; Gabl, M.; Holdfeldt, A.; Uhrbom, M.; Bylund, J.; Højgaard Hansen, A.; Pandey, S.K.; Ulven, T.; et al. The Neutrophil Response Induced by an Agonist for Free Fatty Acid Receptor 2 (GPR43) Is Primed by Tumor Necrosis Factor Alpha and by Receptor Uncoupling from the Cytoskeleton but Attenuated by Tissue Recruitment. *Mol. Cell. Biol.* **2016**, *36*, 2583–2595. [CrossRef]
49. Nilsson, N.E.; Kotarsky, K.; Owman, C.; Olde, B. Identification of a free fatty acid receptor, FFA2R, expressed on leukocytes and activated by short-chain fatty acids. *Biochem. Biophys. Res. Commun.* **2003**, *303*, 1047–1052. [CrossRef]
50. Ohira, H.; Fujioka, Y.; Katagiri, C.; Mamoto, R.; Aoyama-Ishikawa, M.; Amako, K.; Izumi, Y.; Nishiumi, S.; Yoshida, M.; Usami, M.; et al. Butyrate attenuates inflammation and lipolysis generated by the interaction of adipocytes and macrophages. *J. Atheroscler. Thromb.* **2013**, *20*, 425–442. [CrossRef] [PubMed]
51. Arpaia, N.; Campbell, C.; Fan, X.; Dikiy, S.; Liu, H.; Cross, J.R.; Pfeffer, K.; Coffer, P.J.; Rudensky, A.Y.; Donald, B.; et al. Metabolites produced by commensal bacteria promote peripheral regulatory T cell generation. *Nature* **2013**, *504*, 451–455. [CrossRef] [PubMed]
52. Masui, R.; Sasaki, M.; Funaki, Y.; Ogasawara, N.; Mizuno, M.; Iida, A.; Izawa, S.; Kondo, Y.; Ito, Y.; Tamura, Y.; et al. G protein-coupled receptor 43 moderates gut inflammation through cytokine regulation from mononuclear cells. *Inflamm. Bowel Dis.* **2013**, *19*, 2848–2856. [CrossRef]
53. Fachi, J.L.; Sécca, C.; Rodrigues, P.B.; de Mato, F.C.P.; Di Luccia, B.; de Souza Felipe, J.; Pral, L.P.; Rungue, M.; de Melo Rocha, V.; Sato, F.T.; et al. Acetate coordinates neutrophil and ILC3 responses against C. difficile through FFAR2. *J. Exp. Med.* **2020**, *217*, e20190489. [CrossRef] [PubMed]
54. Strowig, T.; Henao-Mejia, J.; Elinav, E.; Flavell, R. Inflammasomes in health and disease. *Nature* **2012**, *481*, 278–286. [CrossRef]
55. Baragetti, A.; Catapano, A.L.; Magni, P. Multifactorial activation of nlrp3 inflammasome: Relevance for a precision approach to atherosclerotic cardiovascular risk and disease. *Int. J. Mol. Sci.* **2020**, *21*, 4459. [CrossRef]
56. Li, M.; van Esch, B.C.A.M.; Henricks, P.A.J.; Folkerts, G.; Garssen, J. The anti-inflammatory effects of short chain fatty acids on lipopolysaccharide- or tumor necrosis factor α-stimulated endothelial cells via activation of GPR41/43 and inhibition of HDACs. *Front. Pharmacol.* **2018**, *9*, 533. [CrossRef] [PubMed]
57. Chang, P.V.; Hao, L.; Offermanns, S.; Medzhitov, R. The microbial metabolite butyrate regulates intestinal macrophage function via histone deacetylase inhibition. *Proc. Natl. Acad. Sci. USA* **2014**, *111*, 2247–2252. [CrossRef] [PubMed]
58. Kasahara, K.; Krautkramer, K.A.; Org, E.; Romano, K.A.; Kerby, R.L.; Vivas, E.I.; Mehrabian, M.; Denu, J.M.; Bäckhed, F.; Lusis, A.J.; et al. Interactions between Roseburia intestinalis and diet modulate atherogenesis in a murine model. *Nat. Microbiol.* **2018**, *3*, 1461–1471. [CrossRef]
59. Bultman, S.J. Bacterial butyrate prevents atherosclerosis. *Nat. Microbiol.* **2018**, *3*, 1332–1333. [CrossRef]
60. Perry, R.J.; Peng, L.; Barry, N.A.; Cline, G.W.; Zhang, D.; Cardone, R.L.; Petersen, K.F.; Kibbey, R.G.; Goodman, A.L.; Shulman, G.I. Acetate mediates a microbiome–brain–β-cell axis to promote metabolic syndrome. *Nature* **2016**, *534*, 213–217. [CrossRef] [PubMed]
61. Park, J.; Kim, M.; Kang, S.G.; Jannasch, A.H.; Cooper, B.; Patterson, J.; Kim, C.H. Short-chain fatty acids induce both effector and regulatory T cells by suppression of histone deacetylases and regulation of the mTOR–S6K pathway. *Mucosal Immunol.* **2015**, *8*, 80–93. [CrossRef] [PubMed]
62. Han, C.Y.; Kargi, A.Y.; Omer, M.; Chan, C.K.; Wabitsch, M.; O'Brien, K.D.; Wight, T.N.; Chait, A. Differential effect of saturated and unsaturated free fatty acids on the generation of monocyte adhesion and chemotactic factors by adipocytes: Dissociation of adipocyte hypertrophy from inflammation. *Diabetes* **2010**, *59*, 386–396. [CrossRef]
63. Suganami, T.; Nishida, J.; Ogawa, Y. A paracrine loop between adipocytes and macrophages aggravates inflammatory changes: Role of free fatty acids and tumor necrosis factor α. *Arterioscler. Thromb. Vasc. Biol.* **2005**, *25*, 2062–2068. [CrossRef]
64. Suganami, T.; Tanimoto-Koyama, K.; Nishida, J.; Itoh, M.; Yuan, X.; Mizuarai, S.; Kotani, H.; Yamaoka, S.; Miyake, K.; Aoe, S.; et al. Role of the Toll-like receptor 4/NF-κB pathway in saturated fatty acid-induced inflammatory changes in the interaction between adipocytes and macrophages. *Arterioscler. Thromb. Vasc. Biol.* **2007**, *27*, 84–91. [CrossRef] [PubMed]
65. Song, M.J.; Kim, K.H.; Yoon, J.M.; Kim, J.B. Activation of Toll-like receptor 4 is associated with insulin resistance in adipocytes. *Biochem. Biophys. Res. Commun.* **2006**, *346*, 739–745. [CrossRef] [PubMed]
66. Ajuwon, K.M.; Spurlock, M.E. Palmitate Activates the NF-κB Transcription Factor and Induces IL-6 and TNFα Expression in 3T3-L1 Adipocytes. *J. Nutr.* **2005**, *135*, 1841–1846. [CrossRef]
67. Lee, J.Y.; Zhao, L.; Youn, H.S.; Weatherill, A.R.; Tapping, R.; Feng, L.; Lee, W.H.; Fitzgerald, K.A.; Hwang, D.H. Saturated Fatty Acid Activates but Polyunsaturated Fatty Acid Inhibits Toll-like Receptor 2 Dimerized with Toll-like Receptor 6 or 1. *J. Biol. Chem.* **2004**, *279*, 16971–16979. [CrossRef]
68. Lee, J.Y.; Plakidas, A.; Lee, W.H.; Heikkinen, A.; Chanmugam, P.; Bray, G.; Hwang, D.H. Differential modulation of Toll-like receptors by fatty acids: Preferential inhibition by n-3 polyunsaturated fatty acids. *J. Lipid Res.* **2003**, *44*, 479–486. [CrossRef]
69. Schaeffler, A.; Gross, P.; Buettner, R.; Bollheimer, C.; Buechler, C.; Neumeier, M.; Kopp, A.; Schoelmerich, J.; Falk, W. Fatty acid-induced induction of Toll-like receptor-4/nuclear factor-κB pathway in adipocytes links nutritional signalling with innate immunity. *Immunology* **2009**, *126*, 233–245. [CrossRef]
70. Zanoni, I.; Ostuni, R.; Marker, L.R.; Barresi, S.; Barbalat, R.; Barton, G.M.; Granucci, F.; Kagan, J.C. CD14 controls the LPS-induced endocytosis of Tool-like Receptor 4. *Cell* **2011**, *147*, 868–880. [CrossRef]
71. Wen, H.; Gris, D.; Lei, Y.; Jha, S.; Zhang, L.; Huang, M.T.H.; Brickey, W.J.; Ting, J.P.-Y. Fatty acid-induced NLRP3-ASC inflammasome activation interferes with insulin signaling. *Nat. Immunol.* **2011**, *12*, 408–415. [CrossRef]

72. Talbot, N.A.; Wheeler-Jones, C.P.; Cleasby, M.E. Palmitoleic acid prevents palmitic acid-induced macrophage activation and consequent p38 MAPK-mediated skeletal muscle insulin resistance. *Mol. Cell. Endocrinol.* **2014**, *393*, 129–142. [CrossRef]

73. Ciesielska, A.; Matyjek, M.; Kwiatkowska, K. TLR4 and CD14 trafficking and its influence on LPS-induced pro-inflammatory signaling. *Cell. Mol. Life Sci.* **2021**, *78*, 1233–1261. [CrossRef]

74. Finucane, O.M.; Lyons, C.L.; Murphy, A.M.; Reynolds, C.M.; Klinger, R.; Healy, N.P.; Cooke, A.A.; Coll, R.C.; Mcallan, L.; Nilaweera, K.N.; et al. Monounsaturated fatty acid-enriched high-fat diets impede adipose NLRP3 inflammasome-mediated IL-1β secretion and insulin resistance despite obesity. *Diabetes* **2015**, *64*, 2116–2128. [CrossRef] [PubMed]

75. Ishiyama, J.; Taguchi, R.; Akasaka, Y.; Shibata, S.; Ito, M.; Nagasawa, M.; Murakami, K. Unsaturated FAs prevent palmitate-induced LOX-1 induction via inhibition of ER stress in macrophages. *J. Lipid Res.* **2011**, *52*, 299–307. [CrossRef]

76. Zheng, S.; Ma, M.; Li, Z.; Hao, Y.; Li, H.; Fu, P.; Jin, S. Posttreatment of Maresin1 Inhibits NLRP3 inflammasome activation via promotion of NLRP3 ubiquitination. *FASEB J.* **2020**, *34*, 11944–11956. [CrossRef]

77. Viola, J.R.; Lemnitzer, P.; Jansen, Y.; Csaba, G.; Winter, C.; Neideck, C.; Silvestre-Roig, C.; Dittmar, G.; Döring, Y.; Drechsler, M.; et al. Resolving Lipid Mediators Maresin 1 and Resolvin D2 Prevent Atheroprogression in Mice. *Circ. Res.* **2016**, *119*, 1030–1038. [CrossRef] [PubMed]

78. Adam, A.-C.; Lie, K.K.; Moren, M.; Skjærven, K.H. High dietary arachidonic acid levels induce changes in complex lipids and immune-related eicosanoids and increase levels of oxidised metabolites in zebrafish (Danio rerio). *Br. J. Nutr.* **2017**, *117*, 1075–1085. [CrossRef]

79. Holladay, C.S.; Wright, R.M.; Spangelo, B.L. Arachidonic acid stimulates interleukin-6 release from rat peritoneal macrophages in vitro: Evidence for a prostacyclin-dependent mechanism. *Prostaglandins Leukot. Essent. Fat. Acids* **1993**, *49*, 915–922. [CrossRef]

80. Shen, Z.; Ma, Y.; Ji, Z.; Hao, Y.; Yan, X.; Zhong, Y.; Tang, X.; Ren, W. Arachidonic acid induces macrophage cell cycle arrest through the JNK signaling pathway. *Lipids Health Dis.* **2018**, *17*, 26. [CrossRef] [PubMed]

81. Wellenstein, M.D.; de Visser, K.E. Fatty Acids Corrupt Neutrophils in Cancer. *Cancer Cell* **2019**, *35*, 827–829. [CrossRef] [PubMed]

82. Veglia, F.; Tyurin, V.A.; Blasi, M.; De Leo, A.; Kossenkov, A.V.; Donthireddy, L.; To, T.K.J.; Schug, Z.; Basu, S.; Wang, F.; et al. Fatty acid transport protein 2 reprograms neutrophils in cancer. *Nature* **2019**, *569*, 73–78. [CrossRef]

83. Guillot, E.; Vaugelade, P.; Lemarchali, P.; Re Rat, A. Intestinal absorption and liver uptake of medium-chain fatty acids in non-anaesthetized pigs. *Br. J. Nutr.* **1993**, *69*, 431–442. [CrossRef]

84. Cerovic, V.; Bain, C.C.; Mowat, A.M.; Milling, S.W.F. Intestinal macrophages and dendritic cells: What's the difference? *Trends Immunol.* **2014**, *35*, 270–277. [CrossRef]

85. Bogunovic, M.; Ginhoux, F.; Helft, J.; Shang, L.; Greter, M.; Liu, K.; Jakubzick, C.; Ingersoll, M.A.; Leboeuf, M.; Stanley, E.R.; et al. Origin of the Lamina Propria Dendritic Cell Network. *Immunity* **2009**, *31*, 513–525. [CrossRef]

86. Bernier-Latmani, J.; Petrova, T.V. Intestinal lymphatic vasculature: Structure, mechanisms and functions. *Nat. Rev. Gastroenterol. Hepatol.* **2017**, *14*, 510–526. [CrossRef] [PubMed]

87. Wilhelm, C.; Harrison, O.J.; Schmitt, V.; Pelletier, M.; Spencer, S.P.; Urban, J.F.; Ploch, M.; Ramalingam, T.R.; Siegel, R.M.; Belkaid, Y. Critical role of fatty acid metabolism in ILC2-mediated barrier protection during malnutrition and helminth infection. *J. Exp. Med.* **2016**, *213*, 1409–1418. [CrossRef] [PubMed]

88. Bsat, M.; Chapuy, L.; Rubio, M.; Wassef, R.; Richard, C.; Schwenter, F.; Loungnarath, R.; Soucy, G.; Mehta, H.; Sarfati, M. Differential Pathogenic Th17 Profile in Mesenteric Lymph Nodes of Crohn's Disease and Ulcerative Colitis Patients. *Front. Immunol.* **2019**, *10*, 1177. [CrossRef]

89. Farnsworth, R.H.; Karnezis, T.; Maciburko, S.J.; Mueller, S.N.; Stacker, S.A. The Interplay Between Lymphatic Vessels and Chemokines. *Front. Immunol.* **2019**, *10*, 518. [CrossRef]

90. Ghoshal, S.; Witta, J.; Zhong, J.; de Villiers, W.; Eckhardt, E. Chylomicrons promote intestinal absorption of lipopolysaccharides. *J. Lipid Res.* **2009**, *50*, 90–97. [CrossRef]

91. Mjösberg, J.M.; Trifari, S.; Crellin, N.K.; Peters, C.P.; van Drunen, C.M.; Piet, B.; Fokkens, W.J.; Cupedo, T.; Spits, H. Human IL-25- and IL-33-responsive type 2 innate lymphoid cells are defined by expression of CRTH2 and CD161. *Nat. Immunol.* **2011**, *12*, 1055–1062. [CrossRef]

92. Xue, L.; Salimi, M.; Panse, I.; Mjösberg, J.M.; McKenzie, A.N.J.; Spits, H.; Klenerman, P.; Ogg, G. Prostaglandin D2 activates group 2 innate lymphoid cells through chemoattractant receptor-homologous molecule expressed on TH2 cells. *J. Allergy Clin. Immunol.* **2014**, *133*, 1184–1194.e7. [CrossRef]

93. Pelly, V.S.; Kannan, Y.; Coomes, S.M.; Entwistle, L.J.; Rückerl, D.; Seddon, B.; MacDonald, A.S.; McKenzie, A.; Wilson, M.S. IL-4-producing ILC2s are required for the differentiation of TH2 cells following Heligmosomoides polygyrus infection. *Mucosal Immunol.* **2016**, *9*, 1407–1417. [CrossRef]

94. Alsina-Sanchis, E.; Mülfarth, R.; Moll, I.; Mogler, C.; Rodriguez-Vita, J.; Fischer, A. Intraperitoneal oil application causes local inflammation with depletion of resident peritoneal macrophages. *Mol. Cancer Res.* **2021**, *19*, 288–300. [CrossRef]

95. Pirillo, A.; Norata, G.D.; Catapano, A.L. Postprandial lipemia as a cardiometabolic risk factor. *Curr. Med. Res. Opin.* **2014**, *30*, 1489–1503. [CrossRef]

96. Gower, R.M.; Wu, H.; Foster, G.A.; Devaraj, S.; Jialal, I.; Ballantyne, C.M.; Knowlton, A.A.; Simon, S.I. CD11c/CD18 Expression Is Upregulated on Blood Monocytes During Hypertriglyceridemia and Enhances Adhesion to Vascular Cell Adhesion Molecule-1. *Arterioscler. Thromb. Vasc. Biol.* **2011**, *31*, 160–166. [CrossRef]

97. Lipopolysaccharide (LPS)-Binding Protein Mediates LPS Detoxification by Chylomicrons. Available online: https://www.jimmunol.org/content/170/3/1399 (accessed on 2 August 2021). [CrossRef]
98. Shepherd, J. The role of the exogenous pathway in hypercholesterolaemia. *Eur. Heart J. Suppl.* **2001**, *3*, E2–E5. [CrossRef]
99. Kapourchali, F.R.; Surendiran, G.; Goulet, A.; Moghadasian, M.H. The Role of Dietary Cholesterol in Lipoprotein Metabolism and Related Metabolic Abnormalities: A Mini-review. *Crit. Rev. Food Sci. Nutr.* **2016**, *56*, 2408–2415. [CrossRef]
100. Lopez-Miranda, J.; Williams, C.; Lairon, D. Dietary, physiological, genetic and pathological influences on postprandial lipid metabolism. *Br. J. Nutr.* **2007**, *98*, 458–473. [CrossRef]
101. Sharrett, A.R.; Heiss, G.; Chambless, L.E.; Boerwinkle, E.; Coady, S.A.; Folsom, A.R.; Patsch, W. Metabolic and lifestyle determinants of postprandial lipemia differ from those of fasting triglycerides the Atherosclerosis Risk in Communities (ARIC) study. *Arterioscler. Thromb. Vasc. Biol.* **2001**, *21*, 275–281. [CrossRef]
102. U.S. Department of Agriculture. FoodData Central. Available online: https://fdc.nal.usda.gov/ (accessed on 2 August 2021).
103. European Institute of Oncology Food Composition Database for Epidemiological Studies in Italy. Available online: http://www.bda-ieo.it/wordpress/en/ (accessed on 2 August 2021).
104. Berger, S.; Raman, G.; Vishwanathan, R.; Jacques, P.F.; Johnson, E.J. Dietary cholesterol and cardiovascular disease: A systematic review and meta-analysis. *Am. J. Clin. Nutr.* **2015**, *102*, 276–294. [CrossRef]
105. Eckel, R.H.; Jakicic, J.M.; Ard, J.D.; de Jesus, J.M.; Miller, N.H.; Hubbard, V.S.; Lee, I.-M.; Lichtenstein, A.H.; Loria, C.M.; Millen, B.E.; et al. 2013 AHA/ACC Guideline on Lifestyle Management to Reduce Cardiovascular Risk. *Circulation* **2014**, *129*, S76–S99. [CrossRef]
106. Borén, J.; Chapman, M.J.; Krauss, R.M.; Packard, C.J.; Bentzon, J.F.; Binder, C.J.; Daemen, M.J.; Demer, L.L.; Hegele, R.A.; Nicholls, S.J.; et al. Low-density lipoproteins cause atherosclerotic cardiovascular disease: Pathophysiological, genetic, and therapeutic insights: A consensus statement from the European Atherosclerosis Society Consensus Panel. *Eur. Heart J.* **2020**, *41*, 2313–2330. [CrossRef]
107. Sheedy, F.J.; Grebe, A.; Rayner, K.J.; Kalantari, P.; Ramkhelawon, B.; Carpenter, S.B.; Becker, C.E.; Ediriweera, H.N.; Mullick, A.E.; Golenbock, D.T.; et al. CD36 coordinates NLRP3 inflammasome activation by facilitating intracellular nucleation of soluble ligands into particulate ligands in sterile inflammation. *Nat. Immunol.* **2013**, *14*, 812–820. [CrossRef]
108. Yvan-Charvet, L.; Pagler, T.; Gautier, E.L.; Avagyan, S.; Siry, R.L.; Han, S.; Welch, C.L.; Wang, N.; Randolph, G.J.; Snoeck, H.W.; et al. ATP-binding cassette transporters and HDL suppress hematopoietic stem cell proliferation. *Science* **2010**, *328*, 1689–1693. [CrossRef]
109. Tall, A.R.; Yvan-Charvet, L. Cholesterol, inflammation and innate immunity. *Nat. Rev. Immunol.* **2015**, *15*, 104–116. [CrossRef]
110. Duewell, P.; Kono, H.; Rayner, K.J.; Sirois, C.M.; Vladimer, G.; Bauernfeind, F.G.; Abela, G.S.; Franchi, L.; Núez, G.; Schnurr, M.; et al. NLRP3 inflammasomes are required for atherogenesis and activated by cholesterol crystals. *Nature* **2010**, *464*, 1357–1361. [CrossRef]
111. Rajamäki, K.; Lappalainen, J.; Öörni, K.; Välimäki, E.; Matikainen, S.; Kovanen, P.T.; Kari, E.K. Cholesterol crystals activate the NLRP3 inflammasome in human macrophages: A novel link between cholesterol metabolism and inflammation. *PLoS ONE* **2010**, *5*, e11765. [CrossRef]
112. Bekkering, S.; Quintin, J.; Joosten, L.A.B.; van der Meer, J.W.M.; Netea, M.G.; Riksen, N.P. Oxidized Low-Density Lipoprotein Induces Long-Term Proinflammatory Cytokine Production and Foam Cell Formation via Epigenetic Reprogramming of Monocytes. *Arterioscler. Thromb. Vasc. Biol.* **2014**, *34*, 1731–1738. [CrossRef]
113. Christ, A.; Günther, P.; Lauterbach, M.A.R.; Duewell, P.; Biswas, D.; Pelka, K.; Scholz, C.J.; Oosting, M.; Haendler, K.; Baßler, K.; et al. Western Diet Triggers NLRP3-Dependent Innate Immune Reprogramming. *Cell* **2018**, *172*, 162–175. [CrossRef]
114. Baragetti, A.; Bonacina, F.; Catapano, A.L.; Norata, G.D. Effect of Lipids and Lipoproteins on Hematopoietic Cell Metabolism and Commitment in Atherosclerosis. *Immunometabolism* **2021**, *3*, e210014. [CrossRef]
115. Murphy, A.J.; Akhtari, M.; Tolani, S.; Pagler, T.; Bijl, N.; Kuo, C.-L.; Wang, M.; Sanson, M.; Abramowicz, S.; Welch, C.; et al. ApoE regulates hematopoietic stem cell proliferation, monocytosis, and monocyte accumulation in atherosclerotic lesions in mice. *J. Clin. Investig.* **2011**, *121*, 4138–4149. [CrossRef]
116. Da Silva, M.S.; Julien, P.; Pérusse, L.; Vohl, M.C.; Rudkowska, I. Natural Rumen-Derived trans Fatty Acids Are Associated with Metabolic Markers of Cardiac Health. *Lipids* **2015**, *50*, 873–882. [CrossRef]
117. Chandra, A.; Lyngbakken, M.N.; Eide, I.A.; Røsjø, H.; Vigen, T.; Ihle-Hansen, H.; Orstad, E.B.; Rønning, O.M.; Berge, T.; Schmidt, E.B.; et al. Plasma trans fatty acid levels, cardiovascular risk factors and lifestyle: Results from the Akershus cardiac examination 1950 study. *Nutrients* **2020**, *12*, 1419. [CrossRef]
118. Lopez-Garcia, E.; Schulze, M.B.; Manson, J.A.E.; Meigs, J.B.; Albert, C.M.; Rifai, N.; Willett, W.C.; Hu, F.B. Consumption of trans fatty acids is related to plasma biomarkers of inflammation and endothelial dysfunction. *J. Nutr.* **2005**, *135*, 562–566. [CrossRef]
119. Wang, D.D.; Li, Y.; Chiuve, S.E.; Stampfer, M.J.; Manson, J.A.E.; Rimm, E.B.; Willett, W.C.; Hu, F.B. Association of specific dietary fats with total and cause-specific mortality. *JAMA Intern. Med.* **2016**, *176*, 1134–1145. [CrossRef]
120. Livingstone, K.M.; Givens, D.I.; Cockcroft, J.R.; Pickering, J.E.; Lovegrove, J.A. Is fatty acid intake a predictor of arterial stiffness and blood pressure in men? Evidence from the Caerphilly Prospective Study. *Nutr. Metab. Cardiovasc. Dis.* **2013**, *23*, 1079–1085. [CrossRef]

121. Li, Y.; Hruby, A.; Bernstein, A.M.; Ley, S.H.; Wang, D.D.; Chiuve, S.E.; Sampson, L.; Rexrode, K.M.; Rimm, E.B.; Willett, W.C.; et al. Saturated Fats Compared with Unsaturated Fats and Sources of Carbohydrates in Relation to Risk of Coronary Heart Disease A Prospective Cohort Study. *J. Am. Coll. Cardiol.* **2015**, *66*, 1538–1548. [CrossRef]
122. Nielsen, B.M.; Nielsen, M.M.; Jakobsen, M.U.; Nielsen, C.J.; Holst, C.; Larsen, T.M.; Bendsen, N.T.; Bysted, A.; Leth, T.; Hougaard, D.M.; et al. A cross-sectional study on trans-fatty acids and risk markers of CHD among middle-aged men representing a broad range of BMI. *Br. J. Nutr.* **2011**, *106*, 1245–1252. [CrossRef]
123. Oh, K.; Hu, F.B.; Manson, J.E.; Stampfer, M.J.; Willett, W.C. Dietary fat intake and risk of coronary heart disease in women: 20 Years of follow-up of the nurses' health study. *Am. J. Epidemiol.* **2005**, *161*, 672–679. [CrossRef]
124. Sohn, C.; Kim, J.; Bae, W. The framingham risk score, diet, and inflammatory markers in Korean men with metabolic syndrome. *Nutr. Res. Pract.* **2012**, *6*, 246–253. [CrossRef]
125. Khayyatzadeh, S.S.; Kazemi-Bajestani, S.M.R.; Bagherniya, M.; Mehramiz, M.; Tayefi, M.; Ebrahimi, M.; Ferns, G.A.; Safarian, M.; Ghayour-Mobarhan, M. Serum high C reactive protein concentrations are related to the intake of dietary macronutrients and fiber: Findings from a large representative Persian population sample. *Clin. Biochem.* **2017**, *50*, 750–755. [CrossRef] [PubMed]
126. Salas-Salvadó, J.; Garcia-Arellano, A.; Estruch, R.; Marquez-Sandoval, F.; Corella, D.; Fiol, M.; Gómez-Gracia, E.; Viñoles, E.; Arós, F.; Herrera, C.; et al. Components of the mediterranean-type food pattern and serum inflammatory markers among patients at high risk for cardiovascular disease. *Eur. J. Clin. Nutr.* **2008**, *62*, 651–659. [CrossRef] [PubMed]
127. Guasch-Ferré, M.; Liu, G.; Li, Y.; Sampson, L.; Manson, J.A.E.; Salas-Salvadó, J.; Martínez-González, M.A.; Stampfer, M.J.; Willett, W.C.; Sun, Q.; et al. Olive Oil Consumption and Cardiovascular Risk in U.S. Adults. *J. Am. Coll. Cardiol.* **2020**, *75*, 1729–1739. [CrossRef] [PubMed]
128. Guasch-Ferré, M.; Zong, G.; Willett, W.C.; Zock, P.L.; Wanders, A.J.; Hu, F.B.; Sun, Q. Associations of Monounsaturated Fatty Acids From Plant and Animal Sources With Total and Cause-Specific Mortality in Two US Prospective Cohort Studies. *Circ. Res.* **2019**, *124*, 1266–1275. [CrossRef]
129. Imran, T.F.; Kim, E.; Buring, J.E.; Lee, I.M.; Gaziano, J.M.; Djousse, L. Nut consumption, risk of cardiovascular mortality, and potential mediating mechanisms: The Women's Health Study. *J. Clin. Lipidol.* **2021**, *15*, 266–274. [CrossRef]
130. Garcia-Arellano, A.; Ramallal, R.; Ruiz-Canela, M.; Salas-Salvadó, J.; Corella, D.; Shivappa, N.; Schröder, H.; Hébert, J.R.; Ros, E.; Gómez-Garcia, E.; et al. Dietary inflammatory index and incidence of cardiovascular disease in the PREDIMED study. *Nutrients* **2015**, *7*, 4124–4138. [CrossRef]
131. De Souza, R.J.; Dehghan, M.; Mente, A.; Bangdiwala, S.I.; Ahmed, S.H.; Alhabib, K.F.; Altuntas, Y.; Basiak-Rasała, A.; Dagenais, G.R.; Diaz, R.; et al. Association of nut intake with risk factors, cardiovascular disease, and mortality in 16 countries from 5 continents: Analysis from the Prospective Urban and Rural Epidemiology (PURE) study. *Am. J. Clin. Nutr.* **2020**, *112*, 208–219. [CrossRef]
132. Lopez-Garcia, E.; Schulze, M.B.; Manson, J.A.E.; Meigs, J.B.; Albert, C.M.; Rifai, N.; Willett, W.C.; Hu, F.B. Consumption of (n-3) fatty acids is related to plasma biomarkers of inflammation and endothelial activation in women. *J. Nutr.* **2004**, *134*, 1806–1811. [CrossRef]
133. Pischon, T.; Hankinson, S.E.; Hotamisligil, G.S.; Rifai, N.; Willett, W.C.; Rimm, E.B. Habitual dietary intake of n-3 and n-6 fatty acids in relation to inflammatory markers among US men and women. *Circulation* **2003**, *108*, 155–160. [CrossRef]
134. Ninomiya, T.; Nagata, M.; Hata, J.; Hirakawa, Y.; Ozawa, M.; Yoshida, D.; Ohara, T.; Kishimoto, H.; Mukai, N.; Fukuhara, M.; et al. Association between ratio of serum eicosapentaenoic acid to arachidonic acid and risk of cardiovascular disease: The Hisayama Study. *Atherosclerosis* **2013**, *231*, 261–267. [CrossRef]
135. Madsen, T.; Skou, H.A.; Hansen, V.E.; Fog, L.; Christensen, J.H.; Toft, E.; Schmidt, E.B. C-Reactive Protein, Dietary n-3 Fatty Acid, and the Extent of Coronary Artery Disease. *Am. J. Cardiol.* **2001**, *88*, 1139–1142. [CrossRef]
136. Manger, M.S.; Strand, E.; Ebbing, M.; Seifert, R.; Refsum, H.; Nordrehaug, J.E.; Nilsen, D.W.; Drevon, C.A.; Tell, G.S.; Bleie, Ø.; et al. Dietary intake of n-3 long-chain polyunsaturated fatty acids and coronary events in Norwegian patients with coronary artery disease. *Am. J. Clin. Nutr.* **2010**, *92*, 244–251. [CrossRef] [PubMed]
137. Mozaffarian, D.; Ascherio, A.; Hu, F.B.; Stampfer, M.J.; Willett, W.C.; Siscovick, D.S.; Rimm, E.B. Interplay between different polyunsaturated fatty acids and risk of coronary heart disease in men. *Circulation* **2005**, *111*, 157–164. [CrossRef]
138. Bergendal, E. Dietary Fat and Cholesterol and Risk of Cardiovascular Disease in Older Adults: The Health ABC Study. *Bone* **2008**, *23*, 1–7. [CrossRef]
139. Bagherniya, M.; Khayyatzadeh, S.S.; Bakavoli, A.R.H.; Ferns, G.A.; Ebrahimi, M.; Safarian, M.; Nematy, M.; Ghayour-Mobarhan, M. Serum high-sensitive C-reactive protein is associated with dietary intakes in diabetic patients with and without hypertension: A cross-sectional study. *Ann. Clin. Biochem.* **2018**, *55*, 422–429. [CrossRef]
140. Virtanen, J.K.; Mursu, J.; Tuomainen, T.P.; Virtanen, H.E.K.; Voutilainen, S. Egg consumption and risk of incident type 2 diabetes in men: The kuopio ischaemic heart disease risk factor study. *Am. J. Clin. Nutr.* **2015**, *101*, 1088–1096. [CrossRef]
141. Abdollahi, A.M.; Virtanen, H.E.K.; Voutilainen, S.; Kurl, S.; Tuomainen, T.P.; Salonen, J.T.; Virtanen, J.K. Egg consumption, cholesterol intake, and risk of incident stroke in men: The Kuopio Ischaemic Heart Disease Risk Factor Study. *Am. J. Clin. Nutr.* **2019**, *110*, 169–176. [CrossRef]
142. Hu, F.B.; Stampfer, M.J.; Rimm, E.B.; Manson, J.A.E.; Ascherio, A.; Colditz, G.A.; Rosner, B.A.; Spiegelman, D.; Speizer, F.E.; Sacks, F.M.; et al. A prospective study of egg consumption and risk of cardiovascular disease in men and women. *J. Am. Med. Assoc.* **1999**, *281*, 1387–1394. [CrossRef]

143. Pai, J.K.; Pischon, T.; Ma, J.; Manson, J.E.; Hankinson, S.E.; Joshipura, K.; Curhan, G.C.; Rifai, N.; Cannuscio, C.C.; Stampfer, M.J.; et al. Inflammatory Markers and the Risk of Coronary Heart Disease in Men and Women. *N. Engl. J. Med.* **2004**, *351*, 2599–2610. [CrossRef] [PubMed]

144. Kalogeropoulos, A.; Georgiopoulou, V.; Psaty, B.M.; Rodondi, N.; Smith, A.L.; Harrison, D.G.; Liu, Y.; Hoffmann, U.; Bauer, D.C.; Newman, A.B.; et al. Inflammatory Markers and Incident Heart Failure Risk in Older Adults. The Health ABC (Health, Aging, and Body Composition) Study. *J. Am. Coll. Cardiol.* **2010**, *55*, 2129–2137. [CrossRef] [PubMed]

145. Rodondi, N.; Marques-Vidal, P.; Butler, J.; Sutton-Tyrrell, K.; Cornuz, J.; Satterfield, S.; Harris, T.; Bauer, D.C.; Ferrucci, L.; Vittinghoff, E.; et al. Markers of atherosclerosis and inflammation for prediction of coronary heart disease in older adults. *Am. J. Epidemiol.* **2010**, *171*, 540–549. [CrossRef]

146. Corley, J.; Shivappa, N.; Hébert, J.R.; Starr, J.M.; Deary, I.J. Associations between Dietary Inflammatory Index Scores and Inflammatory Biomarkers among Older Adults in the Lothian Birth Cohort 1936 Study. *J. Nutr. Health Aging* **2019**, *23*, 628–636. [CrossRef] [PubMed]

147. Ramallal, R.; Toledo, E.; Martínez-González, M.A.; Hernández-Hernández, A.; García-Arellano, A.; Shivappa, N.; Hébert, J.R.; Ruiz-Canela, M. Dietary inflammatory index and incidence of cardiovascular disease in the SUN cohort. *PLoS ONE* **2015**, *10*, e0135221. [CrossRef]

148. Dehghan, M.; Mente, A.; Zhang, X.; Swaminathan, S.; Li, W.; Mohan, V.; Iqbal, R.; Kumar, R.; Wentzel-Viljoen, E.; Rosengren, A.; et al. Associations of fats and carbohydrate intake with cardiovascular disease and mortality in 18 countries from five continents (PURE): A prospective cohort study. *Lancet* **2017**, *390*, 2050–2062. [CrossRef]

149. Kouvari, M.; Panagiotakos, D.B.; Chrysohoou, C.; Georgousopoulou, E.N.; Yannakoulia, M.; Tousoulis, D.; Pitsavos, C.; Skoumas, Y.; Katinioti, N.; Papadimitriou, L.; et al. Dairy products, surrogate markers, and cardiovascular disease; a sex-specific analysis from the ATTICA prospective study. *Nutr. Metab. Cardiovasc. Dis.* **2020**, *30*, 2194–2206. [CrossRef] [PubMed]

150. Clarke, R.; Shipley, M.; Armitage, J.; Collins, R.; Harris, W. Plasma phospholipid fatty acids and CHD in older men: Whitehall study of London civil servants. *Br. J. Nutr.* **2009**, *102*, 279–284. [CrossRef]

151. Lahoz, C.; Castillo, E.; Mostaza, J.M.; de Dios, O.; Salinero-Fort, M.A.; González-Alegre, T.; García-Iglesias, F.; Estirado, E.; Laguna, F.; Sanchez, V.; et al. Relationship of the adherence to a mediterranean diet and its main components with CRP levels in the Spanish population. *Nutrients* **2018**, *10*, 379. [CrossRef]

152. Soliman, G.A. Dietary cholesterol and the lack of evidence in cardiovascular disease. *Nutrients* **2018**, *10*, 780. [CrossRef]

153. Bechthold, A.; Boeing, H.; Schwedhelm, C.; Hoffmann, G.; Knüppel, S.; Iqbal, K.; De Henauw, S.; Michels, N.; Devleesschauwer, B.; Schlesinger, S.; et al. Food groups and risk of coronary heart disease, stroke and heart failure: A systematic review and dose-response meta-analysis of prospective studies. *Crit. Rev. Food Sci. Nutr.* **2019**, *59*, 1071–1090. [CrossRef]

154. Tang, W.H.W.; Wang, Z.; Levison, B.S.; Koeth, R.A.; Britt, E.B.; Fu, X.; Wu, Y.; Hazen, S.L. Intestinal Microbial Metabolism of Phosphatidylcholine and Cardiovascular Risk. *N. Engl. J. Med.* **2013**, *368*, 1575–1584. [CrossRef]

155. Wang, Z.; Klipfell, E.; Bennett, B.J.; Koeth, R.; Levison, B.S.; Dugar, B.; Feldstein, A.E.; Britt, E.B.; Fu, X.; Chung, Y.M.; et al. Gut flora metabolism of phosphatidylcholine promotes cardiovascular disease. *Nature* **2011**, *472*, 57–65. [CrossRef]

156. Baer, D.J.; Judd, J.T.; Clevidence, B.A.; Tracy, R.P. Dietary fatty acids affect plasma markers of inflammation in healthy men fed controlled diets: A randomized crossover study. *Am. J. Clin. Nutr.* **2004**, *79*, 969–973. [CrossRef] [PubMed]

157. Gebauer, S.K.; Dionisi, F.; Krauss, R.M.; Baer, D.J. Vaccenic acid and trans fatty acid isomers from partially hydrogenated oil both adversely affect LDL cholesterol: A double-blind, randomized. *Am. J. Clin. Nutr.* **2015**, *102*, 1339–1346. [CrossRef]

158. Kien, C.L.; Bunn, J.Y.; Fukagawa, N.K.; Anathy, V.; Matthews, D.E.; Crain, K.I.; Ebenstein, D.B.; Tarleton, E.K.; Pratley, R.E.; Poynter, M.E. Lipidomic evidence that lowering the typical dietary palmitate to oleate ratio in humans decreases the leukocyte production of proinflammatory cytokines and muscle expression of redox-sensitive genes. *J. Nutr. Biochem.* **2015**, *26*, 1599–1606. [CrossRef] [PubMed]

159. Nicholls, S.J.; Lundman, P.; Harmer, J.A.; Ons, B.S.C.H.; Cutri, B.; Ons, B.M.E.D.S.C.H.; Griffiths, K.A.; Rye, K.; Barter, P.J.; Celermajer, D.S. Consumption of Saturated Fat Impairs the Anti-Inflammatory Properties of High-Density Lipoproteins and Endothelial Function. *J. Am. Coll. Cardiol.* **2006**, *48*, 715–720. [CrossRef]

160. Zhao, G.; Etherton, T.D.; Martin, K.R.; West, S.G.; Gillies, P.J.; Kris-Etherton, P.M. Dietary α-Linolenic Acid Reduces Inflammatory and Lipid Cardiovascular Risk Factors in Hypercholesterolemic Men and Women. *J. Nutr.* **2004**, *134*, 2991–2997. [CrossRef] [PubMed]

161. Brassard, D.; Tessier-Grenier, M.; Allaire, J.; Rajendiran, E.; She, Y.; Ramprasath, V.; Gigleux, I.; Talbot, D.; Levy, E.; Tremblay, A.; et al. Comparison of the impact of SFAs from cheese and butter on cardiometabolic risk factors: A randomized controlled trial. *Am. J. Clin. Nutr.* **2017**, *105*, 800–809. [CrossRef] [PubMed]

162. Teng, K.; Faun, L.; Ratna, S.; Nesaretnam, K.; Sanders, T.A.B. Effects of exchanging carbohydrate or monounsaturated fat with saturated fat on inflammatory and thrombogenic responses in subjects with abdominal obesity: A randomized controlled trial. *Clin. Nutr.* **2017**, *36*, 1250–1258. [CrossRef]

163. Vafeiadou, K.; Weech, M.; Altowaijri, H.; Todd, S.; Yaqoob, P.; Jackson, K.G.; Lovegrove, J.A. Replacement of saturated with unsaturated fats had no impact on vascular function but beneficial effects on lipid biomarkers, E-selectin, and blood pressure: Results from the randomized, controlled Dietary Intervention and VAScular function (DIVAS). *Am. J. Clin. Nutr.* **2015**, *102*, 40–48. [CrossRef]

164. Voon, P.T.; Ng, T.K.W.; Lee, V.K.M.; Nesaretnam, K. Diets high in palmitic acid (16:0), lauric and myristic acids (12:0 + 14:0), or oleic acid (18:1) do not alter postprandial or fasting plasma homocysteine and inflammatory markers in healthy Malaysian adults. *Am. J. Clin. Nutr.* **2012**, *95*, 780. [CrossRef]

165. Faintuch, J.; Horie, L.M.; Barbeiro, H.V.; Barbeiro, D.F.; Soriano, F.G.; Ishida, R.K.; Cecconello, I. Systemic inflammation in morbidly obese subjects: Response to oral supplementation with alpha-linolenic acid. *Obes. Surg.* **2007**, *17*, 341–347. [CrossRef] [PubMed]

166. Abdelhamid, A.S.; Brown, T.J.; Brainard, J.S.; Biswas, P.; Thorpe, G.C.; Moore, H.J.; Deane, K.H.O.; Summerbell, C.D.; Worthington, H.V.; Song, F.; et al. Polyunsaturated fatty acids for the primary and secondary prevention of cardiovascular disease. *Cochrane Database Syst. Rev.* **2020**. [CrossRef]

167. Rallidis, L.S.; Paschos, G.; Liakos, G.K.; Velissaridou, A.H.; Anastasiadis, G.; Zampelas, A. Dietary α-linolenic acid decreases C-reactive protein, serum amyloid A and interleukin-6 in dyslipidaemic patients. *Atherosclerosis* **2003**, *167*, 237–242. [CrossRef]

168. Bhatt, D.L.; Steg, P.G.; Miller, M.; Brinton, E.A.; Jacobson, T.A.; Ketchum, S.B.; Doyle, R.T.; Juliano, R.A.; Jiao, L.; Granowitz, C.; et al. Cardiovascular Risk Reduction with Icosapent Ethyl for Hypertriglyceridemia. *N. Engl. J. Med.* **2019**, *380*, 11–22. [CrossRef] [PubMed]

169. Mitjavila, M.T.; Fandos, M.; Salas-Salvadó, J.; Covas, M.-I.; Borrego, S.; Estruch, R.; Lamuela-Raventós, R.; Corella, D.; Martínez-Gonzalez, M.Á.; Sánchez, J.M.; et al. The Mediterranean diet improves the systemic lipid and DNA oxidative damage in metabolic syndrome individuals. A randomized, controlled, trial. *Clin. Nutr.* **2013**, *32*, 172–178. [CrossRef]

170. Esposito, K.; Marfella, R.; Ciotola, M.; Di Palo, C.; Giugliano, F.; Giugliano, G.; D'Armiento, M.; D'Andrea, F.; Giugliano, D. Effect of a Mediterranean-Style Diet on Endothelial Dysfunction and Markers of Vascular Inflammation in the Metabolic Syndrome. *JAMA* **2004**, *292*, 1440. [CrossRef]

171. Davis, C.R.; Hodgson, J.M.; Woodman, R.; Bryan, J.; Wilson, C.; Murphy, K.J. A Mediterranean diet lowers blood pressure and improves endothelial function: Results from the MedLey randomized intervention trial. *Am. J. Clin. Nutr.* **2017**, *105*, 1305–1313. [CrossRef] [PubMed]

172. Weinberg, R.L.; Brook, R.D.; Rubenfire, M.; Eagle, K.A. Cardiovascular Impact of Nutritional Supplementation With Omega-3 Fatty Acids: JACC Focus Seminar. *J. Am. Coll. Cardiol.* **2021**, *77*, 593–608. [CrossRef] [PubMed]

173. Storniolo, C.E.; Casillas, R.; Bulló, M.; Castañer, O.; Ros, E.; Sáez, G.T.; Toledo, E.; Estruch, R.; Ruiz-Gutiérrez, V.; Fitó, M.; et al. A Mediterranean diet supplemented with extra virgin olive oil or nuts improves endothelial markers involved in blood pressure control in hypertensive women. *Eur. J. Nutr.* **2017**, *56*, 89–97. [CrossRef]

174. Lambert, C.; Cubedo, J.; Padró, T.; Sánchez-Hernández, J.; Antonijoan, R.; Perez, A.; Badimon, L. Phytosterols and Omega 3 Supplementation Exert Novel Regulatory Effects on Metabolic and Inflammatory Pathways: A Proteomic Study. *Nutrients* **2017**, *9*, 599. [CrossRef]

175. Kirkhus, B.; Lamglait, A.; Eilertsen, K.; Falch, E.; Haider, T.; Vik, H.; Hoem, N.; Hagve, T.; Basu, S.; Olsen, E.; et al. Effects of similar intakes of marine n-3 fatty acids from enriched food products and fish oil on cardiovascular risk markers in healthy human subjects. *Br. J. Nutr.* **2012**, *107*, 1339–1349. [CrossRef]

176. Krantz, M.J.; Havranek, E.P.; Pereira, R.I.; Beaty, B.; Mehler, P.S.; Long, C.S. Effects of omega-3 fatty acids on arterial stiffness in patients with hypertension: A randomized pilot study. *J. Negat. Results Biomed.* **2017**, *14*, 12–17. [CrossRef] [PubMed]

177. Blesso, C.N.; Andersen, C.J.; Barona, J.; Volk, B.; Volek, J.S.; Fernandez, M.L. Effects of carbohydrate restriction and dietary cholesterol provided by eggs on clinical risk factors in metabolic syndrome. *J. Clin. Lipidol.* **2013**, *7*, 463–471. [CrossRef]

178. Morgantini, C.; Trifirò, S.; Tricò, D.; Meriwether, D.; Baldi, S.; Mengozzi, A.; Reddy, S.T.; Natali, A. A short-term increase in dietary cholesterol and fat intake affects high-density lipoprotein composition in healthy subjects. *Nutr. Metab. Cardiovasc. Dis.* **2018**, *28*, 575–581. [CrossRef] [PubMed]

179. Tricò, D.; Trifirò, S.; Mengozzi, A.; Morgantini, C.; Baldi, S.; Mari, A.; Natali, A. Reducing cholesterol and fat intake improves glucose tolerance by enhancing B cell function in nondiabetic subjects. *J. Clin. Endocrinol. Metab.* **2018**, *103*, 622–631. [CrossRef]

180. Blesso, C.N.; Andersen, C.J.; Barona, J.; Volek, J.S.; Luz, M. Whole egg consumption improves lipoprotein profiles and insulin sensitivity to a greater extent than yolk-free egg substitute in individuals with metabolic syndrome. *Metabolism* **2013**, *62*, 400–410. [CrossRef]

181. Neelakantan, N.; Seah, J.Y.H.; van Dam, R.M. The Effect of Coconut Oil Consumption on Cardiovascular Risk Factors. *Circulation* **2020**, *141*, 803–814. [CrossRef]

182. Desroches, S.; Archer, W.R.; Paradis, M.; Dériaz, O.; Couture, P.; Bergeron, J.; Bergeron, N.; Lamarche, B. Baseline Plasma C-Reactive Protein Concentrations Influence Lipid and Lipoprotein Responses to Low-Fat and High Monounsaturated Fatty Acid Diets in Healthy Men. *J. Nutr.* **2006**, *136*, 1005–1011. [CrossRef]

183. Manson, J.E.; Cook, N.R.; Lee, I.; Christen, W.; Bassuk, S.S.; Mora, S.; Gibson, H.; Albert, C.M.; Gordon, D.; Copeland, T.; et al. Marine Omega-3 Fatty Acids and Prevention of Vascular Disease and Cancer. *N. Engl. J. Med.* **2019**, *380*, 23–32. [CrossRef]

184. Allaire, J.; Couture, P.; Leclerc, M.; Charest, A.; Marin, J.; Marie-claude, L.; Talbot, D.; Tchernof, A. A randomized, crossover, head-to-head comparison of eicosapentaenoic acid and docosahexaenoic acid supplementation to reduce inflammation markers in men and women: The Comparing EPA to DHA (ComparED). *Am. J. Clin. Nutr.* **2016**, *3*, 280–287. [CrossRef]

185. Swann, J.R.; Rajilic-Stojanovic, M.; Salonen, A.; Sakwinska, O.; Gill, C.; Meynier, A.; Fança-Berthon, P.; Schelkle, B.; Segata, N.; Shortt, C.; et al. Considerations for the design and conduct of human gut microbiota intervention studies relating to foods. *Eur. J. Nutr.* **2020**, *59*, 3347–3368. [CrossRef] [PubMed]

186. Estruch, R.; Ros, E.; Salas-Salvadó, J.; Covas, M.-I.; Corella, D.; Arós, F.; Gómez-Gracia, E.; Ruiz-Gutiérrez, V.; Fiol, M.; Lapetra, J.; et al. Primary Prevention of Cardiovascular Disease with a Mediterranean Diet. *N. Engl. J. Med.* **2013**, *368*, 1279–1290. [CrossRef]
187. Zeevi, D.; Korem, T.; Zmora, N.; Israeli, D.; Rothschild, D.; Weinberger, A.; Ben-Yacov, O.; Lador, D.; Avnit-Sagi, T.; Lotan-Pompan, M.; et al. Personalized Nutrition by Prediction of Glycemic Responses. *Cell* **2015**, *163*, 1079–1094. [CrossRef]
188. Asnicar, F.; Berry, S.E.; Valdes, A.M.; Nguyen, L.H.; Piccinno, G.; Drew, D.A.; Leeming, E.; Gibson, R.; Le Roy, C.; Khatib, H.A.; et al. Microbiome connections with host metabolism and habitual diet from 1,098 deeply phenotyped individuals. *Nat. Med.* **2021**, *27*, 321–332. [CrossRef] [PubMed]

Review

Nutraceuticals for the Control of Dyslipidaemias in Clinical Practice

Peter E. Penson [1,2] **and Maciej Banach** [3,4,*]

1   School of Pharmacy & Biomolecular Sciences, Liverpool John Moores University, Byrom Street, Liverpool L3 3AF, UK; P.Penson@ljmu.ac.uk
2   Liverpool Centre for Cardiovascular Science, William Henry Duncan Building, 6 West Derby Street, Liverpool L7 8TX, UK
3   Department of Preventive Cardiology and Lipidology, Medical University of Lodz (MUL), Rzgowska 281/289, 93-338 Lodz, Poland
4   Cardiovascular Research Centre, University of Zielona Gora, 65-046 Zielona Gora, Poland
*   Correspondence: maciejbanach77@gmail.com; Tel./Fax: +48-42-639-37-71

Abstract: Dyslipidaemias result in the deposition of cholesterol and lipids in the walls of blood vessels, chronic inflammation and the formation of atherosclerotic plaques, which impede blood flow and (when they rupture) result in acute ischaemic episodes. Whilst recent years have seen enormous success in the reduction of cardiovascular risk using conventional pharmaceuticals, there is increasing interest amongst patients and practitioners in the use of nutraceuticals to combat dyslipidaemias and inflammation in cardiovascular disease. Nutraceutical is a portmanteau term: 'ceutical' indicate pharmaceutical-grade preparations, and 'nutra' indicates that the products contain nutrients from food. Until relatively recently, little high-quality evidence relating to the safety and efficacy of nutraceuticals has been available to prescribers and policymakers. However, as a result of recent randomised-controlled trials, cohort studies and meta-analyses, this situation is changing, and nutraceuticals are now recommended in several mainstream guidelines relating to dyslipidaemias and atherosclerosis. This article will summarise recent clinical-practice guidance relating to the use of nutraceuticals in this context and the evidence which underlies them. Particular attention is given to position papers and recommendations from the International Lipid Expert Panel (ILEP), which has produced several practical and helpful recommendations in this field.

Keywords: nutraceuticals; dyslipidaemia; atherosclerosis

Citation: Penson, P.E.; Banach, M. Nutraceuticals for the Control of Dyslipidaemias in Clinical Practice. *Nutrients* **2021**, *13*, 2957. https://doi.org/10.3390/nu13092957

Academic Editors: Paolo Magni, Andrea Baragetti and Andrea Poli

Received: 4 August 2021
Accepted: 22 August 2021
Published: 25 August 2021

Publisher's Note: MDPI stays neutral with regard to jurisdictional claims in published maps and institutional affiliations.

1. Introduction

Cardiovascular diseases (CVD) are responsible for an estimated 17.9 million deaths each year and represent the largest overall cause of mortality worldwide [1,2]. Although CVD are a heterogenous range of pathologies, many share a common atherosclerotic pathophysiology. More than four-fifths of these deaths occur as a result of myocardial infarction and stroke [1,2]. Worryingly, one-third of these deaths occur prematurely in people under 70 years of age [1]. However, the picture is not entirely bleak. It has been estimated that over the thirty years to 2016, life expectancy increased by eight years, and the majority of this increase (at least six years) was attributable to cardiology [3]. However, despite the great success of conventional approaches to therapeutics, an enormous disease burden remains. Moreover, in many highly developed countries we may observe a plateau, or even a slight reduction in life expectancy in recent years, which may only be partly attributable to the coronavirus pandemic [4,5].

Nutraceuticals have a role to play in managing and reducing this disease burden [6–8]. Until recently, high-quality evidence relating to the safety and efficacy of nutraceuticals has not been available to prescribers and policymakers. However, as a result of recent randomised-controlled trials (RCTs), this situation is changing, and nutraceuticals are now recommended in several mainstream guidelines relating to atherosclerosis and preventive cardiology. This article will summarise recent clinical-practice guidance relating to the use of nutraceuticals in this context and the evidence which underlies them.

2. Dyslipidaemias, Inflammation, and Atherosclerosis

The associations between plasma lipoproteins, atherosclerosis and CVD have been topics of intense research since the observations by Gofman [9] and the results of the Framingham research group [10,11] identified associations between low-density lipoprotein cholesterol (LDL-C) and atherosclerotic events. Interventions to reduce LDL-C, including statins [12–15], ezetimibe [16] and inhibitors of proprotein convertase subtilisin-kexin type 9 [17] have demonstrated remarkable reductions in CV risk when used alone or in combination [18]. Increasing evidence is emerging for newer therapies, including inclisiran [19] and bempedoic acid [20], which can be used to further optimise LDL-C reduction and to enable the achievement of clinical goals, which are still achieved only in about 1/3 patients [21].

Furthermore, inflammation, which occurs when dyslipidaemia results in the deposition of LDL-C in the blood vessel wall is increasingly recognised as a therapeutic target [22,23] in the management of atherosclerosis. Additionally, recent research has broadened beyond LDL-C with a renewed interest in the cardiovascular risk conveyed by hypertriglyceridaemia [24–26] and hyper-lipoproteinaemia(a) [27]. A wide range of nutraceuticals have been shown to exert biological actions at targets relevant to these processes, and therefore have the potential to treat dyslipidaemias and ameliorate the severity of their consequences.

3. The Use of Nutraceuticals in Dyslipidaemias and Atherosclerosis

The term 'nutraceutical' was coined by DeFelice in 1989 as a portmanteau of "NUTRient" and "pharmACEUTICAL". DeFelice defined nutraceutical as "food, or parts of a food, that provide medical or health benefits, including the prevention and treatment of disease" [1,2] but which are *"produced from foods but sold in pills, powders, (potions) and other medicinal forms not generally associated with food"* [3]. Importantly, nutraceuticals are different from 'functional foods' which are *"similar in appearance to conventional foods . . . consumed as part of a usual diet"* [3]. Nutraceuticals should also be distinguished from other molecules, which come from nature but are not derived from foods. Synthetic molecules, which closely resemble natural molecules also cannot be considered as nutraceuticals. For example, aspirin (derived from willow bark and used in preventative cardiology for its antiplatelet action [28]) and colchicine (derived from the crocus, and which has seen a remarkable renaissance since it has been demonstrated to be effective at addressing inflammation in atherosclerosis [29]) cannot be considered nutraceuticals. Furthermore, although such definitions are somewhat arbitrary, nutraceuticals are generally micronutrients, thus despite extensive research about the health benefits of modifying the proportion of energy derived from different dietary sources [30,31], such macronutrients cannot be considered as nutraceuticals.

The development of nutraceuticals, therefore, attempts to use biologically active druglike molecules from nature to prevent or treat disease. However, instead of consuming such molecules in the diet (by eating foods known to contain them), nutraceuticals enable control of dose, quality, and composition of formulations, thus enabling the principles of Good Manufacturing Practice (GMP) to be applied, using the same approach as is used in the production of conventional pharmaceutical products.

The International Lipid Expert Panel (ILEP) position paper on lipid-lowering nutraceuticals in clinical practice for the first time provides a comprehensive and detailed overview of a very wide range of nutraceuticals with the potential to elicit favourable effects on plasma lipoproteins. These include inhibitors of intestinal cholesterol absorption (plant sterols and stanols, soluble fibres, chitosan and probiotics); inhibitors of liver cholesterol synthesis (red yeast rice, garlic, pantethine, bergamot and policosanols); compounds that promote LDL-excretion (berberine, green tea extracts, soy and lupin); and a range of other nutraceuticals including ω-3 fatty acids, spirulina and curcumin [32]. Helpfully, the position paper makes clear and detailed recommendations for each agent and classifies the recommendations according to the strength of recommendation and level of evidence supporting it. The experts from the panel that was founded in 2015 [33], aimed in this paper to present only these nutraceuticals that have real beneficial effect on LDL-C based on high quality data; they also discussed in detail their safety (nutrivigilance), and present recommendations that patients might benefit from the most and in what clinical settings [32].

A more recent position paper by the same group takes a very similar approach but focuses instead on the evidence supporting the anti-inflammatory actions of nutraceuticals [34], a topic of great importance during the current pandemic and in a situation where there is a lack of effective drugs to reduce inflammatory markers [29,35]. Taken together, these two documents provide a wealth of useful information relating to the management of plasma lipids and the associated risk of atherosclerotic disease in clinical practice.

It is beyond the scope of the present article to review this evidence again; however, the reader is directed to the ILEP papers (including a supplementary paper on the role of nutraceuticals in heart failure patients [36]). The evidence supporting the clinical use of three well-evaluated nutraceuticals (red-yeast rice, ω-3 fatty acids and phytosterols) is described in the sections below, and brief mechanistic details for these agents are provided in Figure 1.

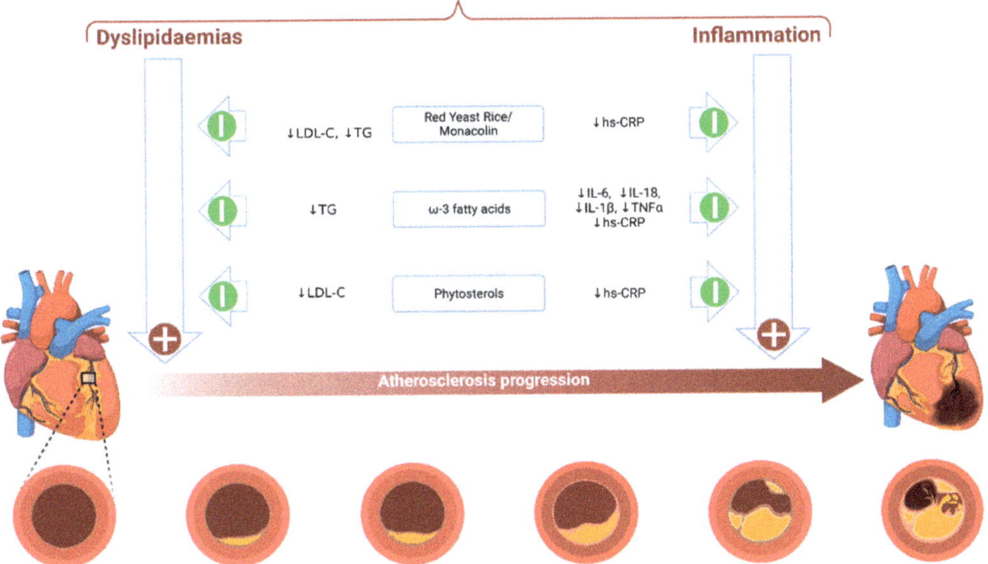

Figure 1. Summary of biological mechanisms involved in the anti-atherosclerotic actions of key evidence-based nutraceuticals. Image Adapted from "Atherosclerosis Progression", by BioRender.com (2021). Retrieved from https://app.biorender.com/biorender-templates (Accessed date: 23 July 2021).

3.1. Red Yeast Rice

Red yeast rice (RYR) is a traditional fermented ingredient of Chinese food which contains a bioactive molecule called monacolin K, a naturally occurring inhibitor of 3-hydroxy-3-methyl-glutaryl-coenzyme A reductase, the target of statins. As such, the consumption of red yeast rice and its nutraceutical preparations reduces plasma LDL-C [37]. Remarkably, an RCT that evaluated a nutraceutical extract of red yeast rice in 5000 individuals in China reported a 45% relative reduction in cardiovascular events [38].

If this effect size can be achieved in other populations, it suggests that RYR preparations might have an important contribution to make to CVD risk reduction.

In the past, concerns have been raised about the safety of red-yeast rice preparations. These generally relate to the batch-to-batch variability in monacolin content and the presence of toxic by-products of rice fermentation (citrinin) [39]. These issues can be resolved through the employment of the principles of GMP. Indeed, a recent nutrivigilance study of a high-quality nutraceutical preparation containing red-yeast rice and other lipid-lowering nutraceuticals found that only 0.037% of 2,287,449 consumers reported adverse events [40]. The recent draft recommendations to European Food Safety Authority (EFSA) suggest that RYR preparations should contain no more than 3 mg of monacolin K.

3.2. ω-3 Fatty Acids

Polyunsaturated ω-3 fatty acids (PUFAs), which are derived from oily fish, have been the subject of intense recent research. Eicosapentaenoic acid (EPA) and docosahexaenoic acid (DHA) reduce circulating concentrations of triglycerides [41] and therefore have been evaluated for the potential to reduce the CV risk associated with hypertriglyceridemia. Unusually for nutraceuticals, polyunsaturated ω-3 fatty acids have been evaluated in numerous cardiovascular outcomes trials (CVOTs), however, the results of the trials have not been consistent, resulting in considerable discussion within the scientific and medical community.

The VITAL study [42] and ASCEND trial [43] randomised patients to receive placebo or a preparation of ω-3 fatty acids approved by the American Heart Association (AHA) (460 mg EPA, 380 mg DHA). The trials were similar, although a higher risk population was recruited in ASCEND. Neither trial showed a difference in composite CVOTs over a relatively long period of follow-up (7.4 years in ASCEND, or 5.3 years in VITAL). Recently, STRENGTH, a similar trial (however, with the high dose of mixed omega-3 fatty acids of 4 g/day), was halted based on an interim analysis, which suggested no likely benefit of the intervention a combination of EPA/DHA) [26,44].

Conversely, a high dose (4 g) preparation of pure EPA (prepared as an ethyl ester and known as 'icosapent ethyl') has demonstrated benefit on physiological parameters and cardiovascular outcomes [24,25]. The EVAPORATE trial demonstrated [45] that icosapent ethyl resulted in atherosclerotic plaque regression (measured using coronary computed tomography angiography). However, more importantly, the REDUCE-IT randomised, placebo-controlled trial evaluated icosapent ethyl in 8179 participants and demonstrated a 25% relative reduction in cardiovascular events over five years [24,25]. However, the most recent meta-analysis suggest (similarly to the one presented by authors [46]), that omega-3 FAs reduced CV mortality and improved CVOTs, however, with more prominent cardiovascular risk reduction with EPA monotherapy than with EPA+DHA [47].

There are some safety concerns for omega-3 acids in the available studies. In the REDUCE-IT trial, an increase in the rate of hospitalization for atrial fibrillation and peripheral edema in the icosapent ethyl vs. placebo group were observed [24,25].

3.3. Phytosterols

Phytosterols and phytostanols have been widely used in nutraceuticals and in functional foods (such as phytosterol-enriched yoghurts and spreads). Unfortunately, there is no randomised-controlled trial data evaluating the effects of these compounds on hard clinical outcomes [48]. As such, the evidence basis for these interventions relies on surrogate outcomes. Nevertheless, such evidence is convincing. A meta-analysis of 113 RCTs demonstrated a dose-dependent effect of plant sterols on LDL (by even 15%), with optimal daily doses above 2 g [49].

Also, in case of phytosterols some safety concerns need to be mentioned. They can lead to hypercholesterolemia and elevated CVD risk in some rare genetic abnormalities of the ABCG5/ABCG8 transporters [50].

4. General Recommendations for Nutraceuticals in International Guidelines

International guidelines vary in the extent to which they recommend nutraceuticals. The 2019 European Society of Cardiology/European Atherosclerosis Society (ESC/EAS) Guidelines for the management of dyslipidaemia give Class A recommendations to the three classes of nutraceuticals described above (red-yeast rice, ω-3 fatty acids and phytosterols) [48]. The American College of Cardiology/American Heart Association (ACC/AHA) guidelines for the primary prevention of cardiovascular disease [51] and the management of blood cholesterol [52] do not recommend any nutraceutical preparations. In the United Kingdom, the National Institute for Health and Care Excellence (NICE) guidelines do not make any recommendations favouring nutraceuticals [53] and specifically advise against the use of ω-3 fatty acid compounds [53]; however, it should be noted that these guidelines precede the publication of IMPROVE-IT and EVAPORATE, and therefore do not reflect the latest evidence relating to EPA. The guidelines are currently undergoing a process of update and revision.

More specifically, ILEP recommendations, which are focused directly on this issue, have suggested a range of specific situations, in which nutraceuticals may be considered in the management of dyslipidaemias and atherosclerosis. These are discussed briefly below (Section 5, Table 1), and the reader is encouraged to access the relevant ILEP position papers for practical advice on the control of dyslipidaemias in clinical practice.

Table 1. Summary of recommendations on the use of nutraceuticals in dyslipidaemias and atherosclerosis.

Organisation	Patient Population	Recommendation	Ref
ILEP	Statin intolerance	Nutraceuticals may be used in combination with other lipid-lowering drugs.	[50]
ILEP	Nocebo/drucebo	Nutraceuticals may be used in combination with other lipid-lowering drugs.	[54]
ILEP	Low CVD risk	Nutraceuticals may be appropriate to control lipids in patients ineligible for statins (and for those not willing using statins).	[55]
ILEP	High CVD risk	Nutraceuticals may be used in combination with other lipid-lowering drugs and may be useful to control residual risk.	[7]

Abbreviations, CVD, cardiovascular disease; ILEP, International Lipid Expert Panel.

5. Specific Situations in Which Nutraceuticals May Be Considered

5.1. Statin Intolerance

Statin intolerance occurs when adverse effects on statin therapy limit the ability of an individual to take statins at guideline-recommended doses [56,57]. Many adverse symptoms reported by patients receiving statin therapy are coincidental and not caused by statins. Therefore, careful diagnostic workup must be undertaken to exclude other causes of symptoms [58,59] (see also Section 5.2 'Nocebo/Drucebo effect', below). However, in the rare cases of complete statin intolerance (usually up to 5% where patients cannot tolerate any statin at any dose) or the more common situation of partial intolerance (where a patient can tolerate a statin, but not at a sufficiently high dose to reach their treatment targets), additional lipid-lowering therapies are necessary. Ideally, such patients may receive PCSK9 inhibitors; however, these are not available or reimbursable in all jurisdictions, and therefore, with other drugs still unavailable and not reimbursed (bempedoic acid and inclisiran), nutraceuticals may be considered in combination with other lipid-lowering drugs. Helpful guidelines, including classification of level and class of evidence, have been published by ILEP relating to the use of nutraceuticals in this context [50].

Nutraceuticals may also be of benefit in patients who are nonadherent to statin therapy (as statin intolerance is the most common reason of statin nonadherence), as well as those who are not willing to use statins, despite indications and the physician's attempts to convince them (statin deniers). Unfortunately, this group (often young people who feel healthy in general) can constitute even 5–7% of patients who should be on statin therapy [54,58,60].

5.2. Nocebo/Drucebo Effect

In relation to statin intolerance (see Section 5.1 'Statin intolerance' above), it is important to note that most reported muscle pain in statin therapy is attributable to the nocebo/drucebo effect and occurs as a result of the patient's expectation of adverse effects [61]. Whilst the ultimate aim of treatment should be to help patients to use life-saving statins [62], nutraceuticals may be useful in the control of plasma lipids in these patients. Forthcoming ILEP recommendations will address this issue in detail [54].

5.3. Patients Considered 'Low-Risk' by Conventional Risk Scores

In most jurisdictions, patient eligibility for statin therapy is decided based on their risk of cardiovascular disease, usually calculated over a 10-year period, based on their demographic characteristics and physical and biomarker measurements. Such an approach almost certainly underestimates the life-long risk of cardiovascular disease in younger individuals with dyslipidaemias but without other risk factors. Epidemiological and interventional trials have unambiguously demonstrated that CVD risk is a function of lifelong exposure to LDL-C, and this can be helpfully summarised as 'lower is better for longer'. ILEP recommendations have been produced to enable the optimal management of patients in this situation. Initially, LDL-C should be controlled by lifestyle interventions (including exercise, bodyweight reduction, smoking cessation and adherence to a diet low in saturated fat and high in plant protein and fibre); however, if adherence to this regimen is poor, or LDL-C is not sufficiently reduced, then nutraceuticals (especially in the form of polypills) may be considered [55].

5.4. Optimisation of Therapy in High-Risk Patients

ILEP has also produced advice about the use of nutraceuticals in patients at high risk of CVD. They summarise the potential uses of nutraceuticals as follows: (1) managing residual risk associated with lipids other than low-density lipoprotein cholesterol (for example, risk mediated by triglycerides may be ameliorated by EPA (see Section 3.2 'ω-3 fatty acids')); (2) managing non-lipid-mediated residual risk (such as inflammatory risk [34]); (3) optimising LDL-C treatment in statin intolerance (see Section 5.1 'Statin intolerance' and Section 5.2 Nocebo/Drucebo effect, above); (4) optimising LDL-C treatment when add-on therapies for statins are not available (e.g., with the limited reimbursement criteria); (5) as adjuncts to lifestyle for individuals at high lifetime risk of atherosclerotic cardiovascular disease (ASCVD) [7].

5.5. Patient-Initiated Nutraceutical Use

A wide variety of nutraceutical preparations are available for patients to purchase over the counter. Prescribers and physicians should always question patients about nutraceutical use as part of routine history taking. This is important to avoid potential adverse effects relating to nutraceutical use. Whilst most preparations are thought to be generally safe, long-term safety data is often not available (see information above). Harm may also result as a result of batch-to-batch variation in the content of the active ingredient in poor quality preparations or because of contamination of preparations with unwanted ingredients [39]. Both problems can be overcome by adherence to the principles of GMP; however, the relatively light regulatory framework for such products in some jurisdictions means that low-quality preparations may reach the market. When patients do choose to use nutraceuticals, it is important that this occurs as a supplement to evidence-based guideline-directed therapies and not as a replacement for them.

6. Conclusions

The development of nutraceuticals from the micronutrient components of food presents an opportunity to target dyslipidaemias and atherosclerosis through direct effects on plasma lipids, and through the modification of pathophysiological processes elicited through atherogenic lipoproteins. Until recently, limited evidence has been available to evaluate the efficacy of nutraceuticals on clinical outcomes and their safety at doses above those usually consumed in the diet. However, recent data from RCTs and nutrivigilance studies has changed this picture. The strongest evidence exists for plant sterols and stanols, eicosapentaenoic acid and red-yeast rice. As such, guidelines and position papers (especially those produced by the ILEP) have recommended roles for nutraceuticals. Whilst many practitioners prescribe or recommend nutraceuticals in their practice, the increasing availability of such preparations and their use by patients makes it imperative that all practitioners treating dyslipidaemias and atherosclerosis are aware of major nutraceuticals, their indications, and the evidence supporting their use. It should be stressed, however, that nutraceuticals should only be used to supplement, not to replace guideline-directed evidence-based therapeutics.

Author Contributions: Writing—original draft preparation, P.E.P.; writing—review and editing, M.B. All authors have read and agreed to the published version of the manuscript.

Funding: This research received no external funding.

Institutional Review Board Statement: Not applicable.

Informed Consent Statement: Not applicable.

Conflicts of Interest: P.E.P. owns four shares in AstraZeneca PLC and has received honoraria and/or travel reimbursement for events sponsored by AKCEA, Amgen, AMRYT, Link Medical, Mylan, Napp, and Sanofi, M.B.—speakers bureau: Amgen, Esperion, Herbapol, Kogen, KRKA, Novartis, Polpharma, Sanofi, Servier, Teva, Viatris, and Zentiva; consultant to Akcea, Amgen, Daichii Sankyo, Esperion, Freia Pharmaceuticals, Polfarmex, and Sanofi; grants from Amgen, Viatris, Sanofi, and Valeant.

References

1. World Health Organisation. Cardiovascular Diseases. 2021. Available online: https://www.who.int/health-topics/cardiovascular-diseases/ (accessed on 21 July 2021).
2. Roth, G.A.; Mensah, G.A.; Johnson, C.O.; Addolorato, G.; Ammirati, E.; Baddour, L.M.; Barengo, N.C.; Beaton, A.Z.; Benjamin, E.J.; Benziger, C.P.; et al. Global burden of cardiovascular diseases and risk factors, 1990–2019: Update from the gbd 2019 study. *J. Am. Coll. Cardiol.* **2020**, *76*, 2982–3021. [CrossRef]
3. Ferrari, R.; Luscher, T.F. Reincarnated medicines: Using out-dated drugs for novel indications. *Eur. Heart. J.* **2016**, *37*, 2571–2576. [CrossRef]
4. Lewek, J.; Jatczak-Pawlik, I.; Maciejewski, M.; Jankowski, P.; Banach, M. COVID-19 and cardiovascular complications-preliminary results of the Late-COVID study. *Arch. Med. Sci.* **2021**, *17*, 818–822. [CrossRef]
5. Banach, M.; Penson, P.E.; Fras, Z.; Vrablik, M.; Pella, D.; Reiner, Z.; Nabavi, S.M.; Sahebkar, A.; Kayikcioglu, M.; Daccord, M.; et al. Brief recommendations on the management of adult patients with familial hypercholesterolemia during the COVID-19 pandemic. *Pharm. Res.* **2020**, *158*, 104891. [CrossRef]
6. Penson, P.E.; Banach, M. Natural compounds as anti-atherogenic agents: Clinical evidence for improved cardiovascular outcomes. *Atherosclerosis* **2021**, *316*, 58–65. [CrossRef] [PubMed]
7. Penson, P.E.; Banach, M. The role of nutraceuticals in the optimization of lipid-lowering therapy in high-risk patients with dyslipidaemia. *Curr. Atheroscler Rep.* **2020**, *22*, 67. [CrossRef]
8. Sosnowska, B.; Penson, P.; Banach, M. The role of nutraceuticals in the prevention of cardiovascular disease. *Cardiovasc. Diagn. Ther.* **2017**, *7*, S21–S31. [CrossRef]
9. Gofman, J.W.; Young, W.; Tandy, R. Ischemic heart disease, atherosclerosis, and longevity. *Circulation* **1966**, *34*, 679–697. [CrossRef] [PubMed]
10. Levy, D.; Brink, S. *A Change of Heart: How the Framingham Heart Study Helped Unravel the Mysteries of Cardiovascular Disease*, 1st ed.; Distributed by Random House; Knopf: New York, NY, USA, 2005.
11. Wilson, P.W.; D'Agostino, R.B.; Levy, D.; Belanger, A.M.; Silbershatz, H.; Kannel, W.B. Prediction of coronary heart disease using risk factor categories. *Circulation* **1998**, *97*, 1837–1847. [CrossRef]
12. Collins, R.; Reith, C.; Emberson, J.; Armitage, J.; Baigent, C.; Blackwell, L.; Blumenthal, R.; Danesh, J.; Smith, G.D.; DeMets, D.; et al. Interpretation of the evidence for the efficacy and safety of statin therapy. *Lancet* **2016**, *388*, 2532–2561. [CrossRef]
13. Cybulska, B.; Cybulska, B.; Kłosiewicz-Latoszek, L.; Penson, P.E.; Nabavi, S.M.; Lavie, C.J.; Banach, M. International Lipid Expert. How much should LDL cholesterol be lowered in secondary prevention? Clinical efficacy and safety in the era of pcsk9 inhibitors. *Prog. Cardiovasc. Dis.* **2021**, *67*, 65–74. [CrossRef]
14. Banach, M.; Penson, P.E. Statins and ldl-c in secondary prevention-so much progress, so far to go. *JAMA Netw. Open* **2020**, *3*, e2025675. [CrossRef] [PubMed]

15. Ursoniu, S.; Mikhailidis, D.P.; Serban, M.C.; Penson, P.; Toth, P.P.; Ridker, P.M.; Ray, K.K.; Hovingh, G.K.; Kastelein, J.J.; Hernandez, A.V.; et al. The effect of statins on cardiovascular outcomes by smoking status: A systematic review and meta-analysis of randomized controlled trials. *Pharmacol. Res.* **2017**, *122*, 105–117. [CrossRef]

16. Cannon, C.P.; Blazing, M.A.; Giugliano, R.P.; McCagg, A.; White, J.A.; Theroux, P.; Darius, H.; Lewis, B.S.; Ophuis, T.O.; Jukema, J.W.; et al. Ezetimibe added to statin therapy after acute coronary syndromes. *N. Engl. J. Med.* **2015**, *372*, 2387–2397. [CrossRef]

17. Banach, M.; Penson, P.E. What have we learned about lipids and cardiovascular risk from pcsk9 inhibitor outcome trials: Odyssey and fourier? *Cardiovasc. Res.* **2019**, *115*, e26–e31. [CrossRef] [PubMed]

18. Banach, M.; Penson, P.E. Lipid-lowering therapies: Better together. *Atherosclerosis* **2021**, *320*, 86–88. [CrossRef]

19. Dyrbuś, K.; Gąsior, M.; Penson, P.; Ray, K.K.; Banach, M. Inclisiran-new hope in the management of lipid disorders? *J. Clin. Lipidol.* **2020**, *14*, 16–27. [CrossRef]

20. Penson, P.; McGowan, M.; Banach, M. Evaluating bempedoic acid for the treatment of hyperlipidaemia. *Expert Opin. Investig. Drugs* **2017**, *26*, 251–259. [CrossRef]

21. Ray, K.K.; Molemans, B.; Schoonen, W.M.; Giovas, P.; Bray, S.; Kiru, G.; Murphy, J.; Banach, M.; De Servi, S.; Gaita, D.; et al. Eu-wide cross-sectional observational study of lipid-modifying therapy use in secondary and primary care: The da vinci study. *Eur. J. Prev. Cardiol.* **2020**. [CrossRef]

22. Ridker, P.M.; Everett, B.M.; Thuren, T.; MacFadyen, J.G.; Chang, W.H.; Ballantyne, C.; Fonseca, F.; Nicolau, J.; Koenig, W.; Anker, S.D.; et al. Antiinflammatory therapy with canakinumab for atherosclerotic disease. *N. Engl. J. Med.* **2017**, *377*, 1119–1131. [CrossRef]

23. Penson, P.; Long, D.L.; Howard, G.; Toth, P.P.; Muntner, P.; Howard, V.J.; Safford, M.M.; Jones, S.R.; Martin, S.S.; Mazidi, M.; et al. Associations between very low concentrations of low density lipoprotein cholesterol, high sensitivity c-reactive protein, and health outcomes in the reasons for geographical and racial differences in stroke (regards) study. *Eur. Heart J.* **2018**, *39*, 3641–3653. [CrossRef] [PubMed]

24. Bhatt, D.L.; Steg, P.G.; Miller, M.; Brinton, E.A.; Jacobson, T.A.; Ketchum, S.B.; Doyle, R.T., Jr.; Juliano, R.A.; Jiao, L.; Granowitz, C.; et al. Cardiovascular risk reduction with icosapent ethyl for hypertriglyceridemia. *N. Engl. J. Med.* **2019**, *380*, 11–22. [CrossRef] [PubMed]

25. Bhatt, D.L.; Steg, P.G.; Miller, M.; Brinton, E.A.; Jacobson, T.A.; Ketchum, S.B.; Doyle, R.T.; Juliano, R.A.; Jiao, L.; Granowitz, C.; et al. Effects of icosapent ethyl on total ischemic events: From reduce-it. *J. Am. Coll. Cardiol.* **2019**, *73*, 2791–2802. [CrossRef] [PubMed]

26. Nicholls, S.J.; Lincoff, A.M.; Bash, D.; Ballantyne, C.M.; Barter, P.J.; Davidson, M.H.; Kastelein, J.J.P.; Koenig, W.; McGuire, D.K.; Mozaffarian, D.; et al. Assessment of omega-3 carboxylic acids in statin-treated patients with high levels of triglycerides and low levels of high-density lipoprotein cholesterol: Rationale and design of the strength trial. *Clin. Cardiol.* **2018**, *41*, 1281–1288. [CrossRef]

27. Cybulska, B.; Kłosiewicz-Latoszek, L.; Penson, P.E.; Banach, M. What do we know about the role of lipoprotein(a) in atherogenesis 57 years after its discovery? *Prog. Cardiovasc. Dis.* **2020**, *63*, 219–227. [CrossRef]

28. Raber, I.; McCarthy, C.P.; Vaduganathan, M.; Bhatt, D.L.; Wood, D.A.; Cleland, J.G.F.; Blumenthal, R.S.; McEvoy, J.W. The rise and fall of aspirin in the primary prevention of cardiovascular disease. *Lancet* **2019**, *393*, 2155–2167. [CrossRef]

29. Banach, M.; Penson, P.E. Colchicine and cardiovascular outcomes: A critical appraisal of recent studies. *Curr. Atheroscler. Rep.* **2021**, *23*, 32. [CrossRef]

30. Mazidi, M.; Mikhailidis, D.P.; Sattar, N.; Toth, P.P.; Judd, S.; Blaha, M.J.; Hernandez, A.V.; Penson, P.; Banach, M. Association of types of dietary fats and all-cause and cause-specific mortality: A prospective cohort study and meta-analysis of prospective studies with 1,164,029 participants. *Clin. Nutr.* **2020**, *39*, 3677–3686. [CrossRef]

31. Gjuladin-Hellon, T.; Davies, I.G.; Penson, P.; Baghbadorani, R.A. Effects of carbohydrate-restricted diets on low-density lipoprotein cholesterol levels in overweight and obese adults: A systematic review and meta-analysis. *Nutr. Rev.* **2019**, *77*, 161–180. [CrossRef]

32. Cicero, A.F.G.; Colletti, A.; Bajraktari, G.; Descamps, O.; Djuric, D.M.; Ezhov, M.; Fras, Z.; Katsiki, N.; Langlois, M.; Latkovskis, G.; et al. Lipid-lowering nutraceuticals in clinical practice: Position paper from an international lipid expert panel. *Nutr. Rev.* **2017**, *75*, 731–767. [CrossRef]

33. Banach, M. The international lipid expert panel (ilep)-the role of 'optimal' collaboration in the effective diagnosis and treatment of lipid disorders. *Eur. Heart J.* **2021**. [CrossRef] [PubMed]

34. Ruscica, M.; Penson, P.E.; Ferri, N.; Sirtori, C.R.; Pirro, M.; Mancini, G.J.; Sattar, N.; Toth, P.P.; Sahebkar, A.; Lavie, C.J.; et al. Impact of nutraceuticals on markers of systemic inflammation: Potential relevance to cardiovascular diseases-a position paper from the international lipid expert panel (ilep). *Prog. Cardiovasc. Dis.* **2021**, *67*, 40–52. [CrossRef]

35. Fiolet, A.T.; Opstal, T.S.; Mosterd, A.; Eikelboom, J.W.; Jolly, S.S.; Keech, A.C.; Kelly, P.; Tong, D.C.; Layland, J.; Nidorf, S.M.; et al. Efficacy and safety of colchicine in patients with coronary artery disease: A systematic review and meta-analysis of randomized controlled trials. *Br. J. Clin. Pharmacol.* **2021**, *42*, 2765–2775. [CrossRef]

36. Cicero, A.F.G.; Colletti, A.; Von Haehling, S.; Vinereanu, D.; Bielecka-Dabrowa, A.; Sahebkar, A.; Toth, P.P.; Reiner, Ž.; Wong, N.D.; Mikhailidis, D.P.; et al. Nutraceutical support in heart failure: A position paper of the international lipid expert panel (ilep). *Nutr. Res. Rev.* **2020**, *33*, 155–179. [CrossRef]

37. Li, Y.; Jiang, L.; Jia, Z.; Xin, W.; Yang, S.; Yang, Q.; Wang, L. A meta-analysis of red yeast rice: An effective and relatively safe alternative approach for dyslipidemia. *PLoS ONE* **2014**, *9*, e98611. [CrossRef]

38. Lu, Z.; Kou, W.; Du, B.; Wu, Y.; Zhao, S.; Brusco, O.A.; Morgan, J.M.; Capuzzi, D.M. Chinese Coronary Secondary Prevention Study and S. Li. Effect of xuezhikang, an extract from red yeast chinese rice, on coronary events in a chinese population with previous myocardial infarction. *Am. J. Cardiol.* **2008**, *101*, 1689–1693. [CrossRef]

39. De Backer, G.G. Food supplements with red yeast rice: More regulations are needed. *Eur. J. Prev. Cardiol.* **2017**, *24*, 1429–1430. [CrossRef] [PubMed]

40. Banach, M.; Katsiki, N.; Latkovskis, G.; Rizzo, M.; Pella, D.; Penson, P.E.; Reiner, Ž.; Cicero, A.F. Postmarketing nutrivigilance safety profile: A line of dietary food supplements containing red yeast rice for dyslipidemia. *Arch. Med. Sci.* **2021**, *17*, 856–863. [CrossRef] [PubMed]

41. Mason, R.P. New insights into mechanisms of action for omega-3 fatty acids in atherothrombotic cardiovascular disease. *Curr. Atheroscler. Rep.* **2019**, *21*, 2. [CrossRef] [PubMed]

42. Manson, J.E.; Cook, N.R.; Lee, I.-M.; Christen, W.; Bassuk, S.S.; Mora, S.; Gibson, H.; Albert, C.M.; Gordon, D.; Copeland, T.; et al. Marine n-3 fatty acids and prevention of cardiovascular disease and cancer. *N. Engl. J. Med.* **2019**, *380*, 23–32. [CrossRef] [PubMed]

43. ASCEND Study Collaborative Group; Bowman, L.; Mafham, M.; Wallendszus, K.; Stevens, W.; Buck, G.; Barton, J.; Murphy, K.; Aung, T.; Haynes, R.; et al. Effects of n-3 fatty acid supplements in diabetes mellitus. *N. Engl. J. Med.* **2018**, *379*, 1540–1550. [CrossRef]

44. Nicholls, S.J.; Lincoff, A.M.; Garcia, M.; Bash, D.; Ballantyne, C.M.; Barter, P.J.; Davidson, M.H.; Kastelein, J.J.P.; Koenig, W.; McGuire, D.K.; et al. Effect of high-dose omega-3 fatty acids vs corn oil on major adverse cardiovascular events in patients at high cardiovascular risk: The strength randomized clinical trial. *JAMA* **2020**, *324*, 2268–2280. [CrossRef] [PubMed]

45. Budoff, M.J.; Bhatt, D.L.; Kinninger, A.; Lakshmanan, S.; Muhlestein, J.B.; Le, V.T.; May, H.T.; Shaikh, K.; Shekar, C.; Roy, S.K.; et al. Effect of icosapent ethyl on progression of coronary atherosclerosis in patients with elevated triglycerides on statin therapy: Final results of the evaporate trial. *Eur. Heart J.* **2020**, *41*, 3925–3932. [CrossRef] [PubMed]

46. Mazidi, M.; Mikhailidis, D.P.; Banach, M. Omega-3 fatty acids and risk of cardiovascular disease: Systematic review and meta-analysis of randomized controlled trials with 127,447 individuals and a mendelian randomization study. *Circulation* **2019**, *140*, e965–e1011, (Poster No. 20948). [CrossRef]

47. Khan, S.U.; Lone, A.N.; Khan, M.S.; Virani, S.S.; Blumenthal, R.S.; Nasir, K.; Miller, M.; Michos, E.D.; Ballantyne, C.M.; Boden, W.E.; et al. Effect of omega-3 fatty acids on cardiovascular outcomes: A systematic review and meta-analysis. *EClinicalMedicine* **2021**. In Press. [CrossRef]

48. Mach, F.; Baigent, C.; Catapano, A.L.; Koskinas, K.C.; Casula, M.; Badimon, L.; Chapman, M.J.; De Backer, G.G.; Delgado, V.; Ference, B.A.; et al. 2019 ESC/EAS guidelines for the management of dyslipidaemias: Lipid modification to reduce cardiovascular risk. *Eur. Heart J.* **2020**, *41*, 111–188. [CrossRef] [PubMed]

49. Musa-Veloso, K.; Poon, T.H.; Elliot, J.A.; Chung, C. A comparison of the ldl-cholesterol lowering efficacy of plant stanols and plant sterols over a continuous dose range: Results of a meta-analysis of randomized, placebo-controlled trials. *Prostaglandins Leukot. Essent. Fat. Acids* **2011**, *85*, 9–28. [CrossRef]

50. Banach, M.; Patti, A.M.; Giglio, R.V.; Cicero, A.F.; Atanasov, A.; Bajraktari, G.; Bruckert, E.; Descamps, O.; Djuric, D.M.; Ezhov, M.; et al. The role of nutraceuticals in statin intolerant patients. *J. Am. Coll. Cardiol.* **2018**, *72*, 96–118. [CrossRef]

51. Arnett, D.K.; Blumenthal, R.S.; Albert, M.A.; Buroker, A.B.; Goldberger, Z.D.; Hahn, E.J.; Himmelfarb, C.D.; Khera, A.; Lloyd-Jones, D.; McEvoy, J.W.; et al. 2019 acc/aha guideline on the primary prevention of cardiovascular disease: A report of the american college of cardiology/american heart association task force on clinical practice guidelines. *Circulation* **2019**, *140*, e596–e646. [CrossRef] [PubMed]

52. Grundy, S.M.; Stone, N.J.; Bailey, A.L.; Beam, C.; Birtcher, K.K.; Blumenthal, R.S.; Braun, L.T.; De Ferranti, S.; Faiella-Tommasino, J.; Forman, D.E.; et al. 2018 aha/acc/aacvpr/aapa/abc/acpm/ada/ags/apha/aspc/nla/pcna guideline on the management of blood cholesterol: A report of the american college of cardiology/american heart association task force on clinical practice guidelines. *Circulation* **2019**, *73*, e285–e350. [CrossRef]

53. National Institute for Health and Care Excellence (NICE). Cardiovascular Disease: Risk Assessment and Reduction, Including Lipid Modification. CG81. 2014 updated 2016. Available online: https://www.nice.org.uk/guidance/cg181 (accessed on 22 July 2021).

54. Penson, P.E.; Banach, M. Nocebo/drucebo effect in statin-intolerant patients: An attempt at recommendations. *Eur. Heart. J.* **2021**, ehab358. [CrossRef]

55. Penson, P.E.; Pirro, M.; Banach, M. Ldl-c: Lower is better for longer-even at low risk. *BMC Med.* **2020**, *18*, 1–6. [CrossRef]

56. Mancini, G.J.; Baker, S.; Bergeron, J.; Fitchett, D.; Frohlich, J.; Genest, J.; Gupta, M.; Hegele, R.A.; Ng, D.; Pearson, G.J.; et al. Diagnosis, prevention, and management of statin adverse effects and intolerance: Canadian consensus working group update. *Can. J. Cardiol.* **2016**, *32*, S35–S65. [CrossRef]

57. Banach, M.; Rizzo, M.; Toth, P.P.; Farnier, M.; Davidson, M.H.; Al-Rasadi, K.; Aronow, W.S.; Athyros, V.; Djuric, D.M.; Ezhov, M.V.; et al. Statin intolerance-an attempt at a unified definition. Position paper from an international lipid expert panel. *Arch. Med. Sci.* **2015**, *11*, 1–23. [CrossRef] [PubMed]

58. Penson, P.; Toth, P.; Mikhailidis, D.; Ezhov, M.; Fras, Z.; Mitchenko, O.; Pella, D.; Sahebkar, A.; Rysz, J.; Reiner, Z.; et al. P705step by step diagnosis and management of statin intolerance: Position paper from an international lipid expert panel. *Eur. Heart J.* **2019**, *40*, ehz747-0310. [CrossRef]

59. Banach, M.; Mikhailidis, D.P. Statin intolerance: Some practical hints. *Cardiol. Clin.* **2018**, *36*, 225–231. [CrossRef] [PubMed]
60. Banach, M.; Stulc, T.; Dent, R.; Toth, P.P. Statin non-adherence and residual cardiovascular risk: There is need for substantial improvement. *Int. J. Cardiol.* **2016**, *225*, 184–196. [CrossRef] [PubMed]
61. Penson, P.E.; Mancini, G.B.J.; Toth, P.P.; Martin, S.S.; Watts, G.F.; Sahebkar, A.; Mikhailidis, D.P.; Banach, M.; Lipid, G.; Blood Pressure Meta-Analysis Collaboration; et al. Introducing the 'drucebo' effect in statin therapy: A systematic review of studies comparing reported rates of statin-associated muscle symptoms, under blinded and open-label conditions. *J. Cachex. Sarcopenia Muscle* **2018**, *9*, 1023–1033. [CrossRef]
62. Banach, M.; Penson, P.E. Drucebo effect-the challenge we should all definitely face! *Arch. Med. Sci.* **2021**, *17*, 542–543. [CrossRef]

 *nutrients*

Article

Interactions of Oxysterols with Atherosclerosis Biomarkers in Subjects with Moderate Hypercholesterolemia and Effects of a Nutraceutical Combination (*Bifidobacterium longum* BB536, Red Yeast Rice Extract) (Randomized, Double-Blind, Placebo-Controlled Study)

Stefania Cicolari [1], Chiara Pavanello [1,2], Elena Olmastroni [3], Marina Del Puppo [4], Marco Bertolotti [5], Giuliana Mombelli [6], Alberico L. Catapano [1,7], Laura Calabresi [1,2] and Paolo Magni [1,7,*]

[1] Dipartimento di Scienze Farmacologiche e Biomolecolari, Università degli Studi di Milano, 20133 Milan, Italy; stefania.cicolari@unimi.it (S.C.); chiara.pavanello@unimi.it (C.P.); alberico.catapano@unimi.it (A.L.C.); laura.calabresi@unimi.it (L.C.)

[2] Centro E. Grossi Paoletti, Dipartimento di Scienze Farmacologiche e Biomolecolari, Università degli Studi di Milano, 20133 Milan, Italy

[3] Servizio di Epidemiologia e Farmacologia Preventiva (SEFAP), Dipartimento di Scienze Farmacologiche e Biomolecolari, Università degli Studi di Milano, 20133 Milan, Italy; elena.olmastroni@unimi.it

[4] Dipartimento di Medicina e Chirurgia, Università degli Studi di Milano Bicocca, 20900 Monza, Italy; marina.delpuppo@unimib.it

[5] Dipartimento di Scienze Biomediche, Metaboliche e Neuroscienze, Università degli Studi di Modena e Reggio Emilia, 41126 Modena, Italy; marco.bertolotti@unimore.it

[6] Centro Dislipidemie, ASST Grande Ospedale Metropolitano Niguarda, 20162 Milan, Italy; giuliana.mombelli@ospedaleniguarda.it

[7] IRCCS MultiMedica, Sesto S. Giovanni, 20099 Milan, Italy

* Correspondence: paolo.magni@unimi.it; Tel.: +39-02-50318229

Citation: Cicolari, S.; Pavanello, C.; Olmastroni, E.; Puppo, M.D.; Bertolotti, M.; Mombelli, G.; Catapano, A.L; Calabresi, L.; Magni, P. Interactions of Oxysterols with Atherosclerosis Biomarkers in Subjects with Moderate Hypercholesterolemia and Effects of a Nutraceutical Combination (*Bifidobacterium longum* BB536, Red Yeast Rice Extract) (Randomized, Double-Blind, Placebo-Controlled Study). *Nutrients* **2021**, *13*, 427. https://doi.org/10.3390/nu13020427

Academic Editor: Roberto Cangemi
Received: 21 December 2020
Accepted: 26 January 2021
Published: 28 January 2021

Publisher's Note: MDPI stays neutral with regard to jurisdictional claims in published maps and institutional affiliations.

Abstract: Background: Oxysterol relationship with cardiovascular (CV) risk factors is poorly explored, especially in moderately hypercholesterolaemic subjects. Moreover, the impact of nutraceuticals controlling hypercholesterolaemia on plasma levels of 24-, 25- and 27-hydroxycholesterol (24-OHC, 25-OHC, 27-OHC) is unknown. Methods: Subjects (n = 33; 18–70 years) with moderate hypercholesterolaemia (low-density lipoprotein cholesterol (LDL-C:): 130–200 mg/dL), in primary CV prevention as well as low CV risk were studied cross-sectionally. Moreover, they were evaluated after treatment with a nutraceutical combination (*Bifidobacterium longum* BB536, red yeast rice extract (10 mg/dose monacolin K)), following a double-blind, randomized, placebo-controlled design. We evaluated 24-OHC, 25-OHC and 27-OHC levels by gas chromatography/mass spectrometry analysis. Results: 24-OHC and 25-OHC were significantly correlated, 24-OHC was correlated with apoB. 27-OHC and 27-OHC/total cholesterol (TC) were higher in men (median 209 ng/mL and 77 ng/mg, respectively) vs. women (median 168 ng/mL and 56 ng/mg, respectively); 27-OHC/TC was significantly correlated with abdominal circumference, visceral fat and, negatively, with high-density lipoprotein cholesterol (HDL-C). Triglycerides were significantly correlated with 24-OHC, 25-OHC and 27-OHC and with 24-OHC/TC and 25-OHC/TC. After intervention, 27-OHC levels were significantly reduced by 10.4% in the nutraceutical group Levels of 24-OHC, 24-OHC/TC, 25-OHC, 25-OHC/TC and 27-OHC/TC were unchanged. Conclusions: In this study, conducted in moderate hypercholesterolemic subjects, we observed novel relationships between 24-OHC, 25-OHC and 27-OHC and CV risk biomarkers. In addition, no adverse changes of OHC levels upon nutraceutical treatment were found.

Keywords: oxysterols; 24-OHC; 25-OHC; 27-OHC; cholesterol metabolism; probiotic; cardiovascular risk; hypercholesterolemia; monacolin K; nutraceutical

---

1. Introduction

Several lipid biomarkers may contribute to atherosclerosis-related cardiovascular diseases (ASCVD) and, among them, low-density lipoprotein (LDL) is a well-established causal factor [1]. Cholesterol metabolism includes de-esterification in lysosomes to generate free cholesterol which is used for several cellular processes [2] and could also undergo enzymatic oxidation in the mitochondria, leading to the formation of oxysterols [3]. These cholesterol metabolites, precursors of bile acids, are involved both in physiological mechanisms, interdependent with lipid and glucose metabolism, as well as in biological functions such as immune and cerebral homeostasis [4,5]. Additionally, oxysterols have been related to some pathological processes (e.g., atherosclerosis, type 2 diabetes mellitus, neurodegenerative disorders, cancer), for which they may represent potential innovative biomarkers [6]. We focused our attention on 24-hydroxycholesterol (24-OHC), 25-hydroxycholesterol (25-OHC) and 27-hydroxycholesterol (27-OHC), which are mainly synthesized by cytochrome P450 family 46 subfamily A member 1 (CYP46A1), cholesterol 25-hydroxylase (CH25H) and cytochrome P450 family 27 subfamily A member 1 (CYP27A1), respectively. Interestingly, these oxysterols have been shown to act as a link between cholesterol metabolism and different physiological systems [4,7].

Moderate hypercholesterolemia is frequently observed in subjects with medium/low 10 years CV risk, represents a significant population burden, particularly when combined with unhealthy lifestyle habits [8] and is often underdiagnosed and undertreated, therefore highly contributing significantly to ASCVD prevalence [9]. Few studies have evaluated the circulating levels of 24-, 25- and 27-OHC in subjects with moderate hypercholesterolemia, also after statin treatment [10,11].

Therapeutical strategies for this condition may include the use of low-efficacy/low-dose statins and/or nutraceutics [12–14]. This treatment may offer multi-faceted effects and significant advantages over no-treatment or inadequate adherence to drug treatment, sometimes due to adverse effects and other reasons [15–17]. To date, information is lacking about the impact of nutraceuticals, targeted to improve the atherogenic lipid profile, on the synthesis of oxysterols, downstream of inhibition of cholesterol biosynthesis and absorption. Based on these considerations, the main objective of the present study was to evaluate the interactions of oxysterols with cardiovascular biomarkers and subsequently to study the effects of a nutraceutical treatment (*Bifidobacterium longum* BB536, RYR extract, niacin, coenzyme Q10) on their circulating levels. This nutraceutical combination was previously found to be quite effective in reducing LDL cholesterol (LDL-C) and total cholesterol (TC) levels in moderately hypercholesterolemic subjects [18]. The study yielded novel data on these biochemical events, with potential health implications and supported the safety profile of this nutraceutical combination.

2. Materials and Methods

2.1. Study Design and Population

The study was conducted at the Centro Dislipidemie (ASST Grande Ospedale Metropolitano Niguarda, Milan, Italy) in the period 2015–2017, according to the guidelines of the Declaration of Helsinki. The study cohort included 33 subjects (16 males and 17 females) with moderate hypercholesterolemia, median age: 57 years (Q1 = 48 and Q3 = 63 years) and low total CVD risk at (0%: 8 subjects; 1%: 15; 2%: 5; 3%: 3; 4%: 1; 5%: 1), assessed by the SCORE Risk Charts for low risk countries (like Italy) [19]. Inclusion criteria were subjects in primary CV prevention, age: 18–70 years, non-smokers, LDL-C: 130–200 mg/dL. Exclusion criteria included: pregnancy, smoking (current or previous), diagnosis of diabetes mellitus, chronic liver disease, renal disease, or severe renal impairment treated with insulin or antidiabetic drugs; untreated, uncontrolled or severe arterial hypertension; obesity (body mass index (BMI) >30 kg/m^2); any pharmacological treatments known to interfere with the study treatment (including statins, ezetimibe, fibrates, thyroid hormones); and patients enrolled in another research study in the past 3 months. The study cohort included

5/32 subjects (15.6%) undergoing drug therapy for arterial hypertension, as reported in [18], together with average food intake according to sex.

The same cohort was also included in a 12-week intervention study (a randomized controlled trial (RCT) design with parallel-groups (NCT02689934)). Subjects were randomly assigned to receive either placebo (1 sachet/d; $n = 17$) or a nutraceutical combination (Lactoflorene Colesterolo®-1 sachet/d; granules for oral suspension; with taste/appearance identical to the placebo sachet) composed of 1 bn UFC *Bifidobacterium longum* BB536, RYR extract (10 mg monacolin K), 16 mg niacin, 20 mg coenzyme Q10; $n = 16$) or (Figure 1; CONSORT flow diagram). The randomization table was produced by computer-generated random numbers. The prevalence of subjects with drug-controlled hypertension was 18.8% in the placebo arm and 12.5% in the intervention arm. The study was approved by the Ethics Committee of ASST Grande Ospedale Metropolitano Niguarda. A written informed consent was obtained from all subjects.

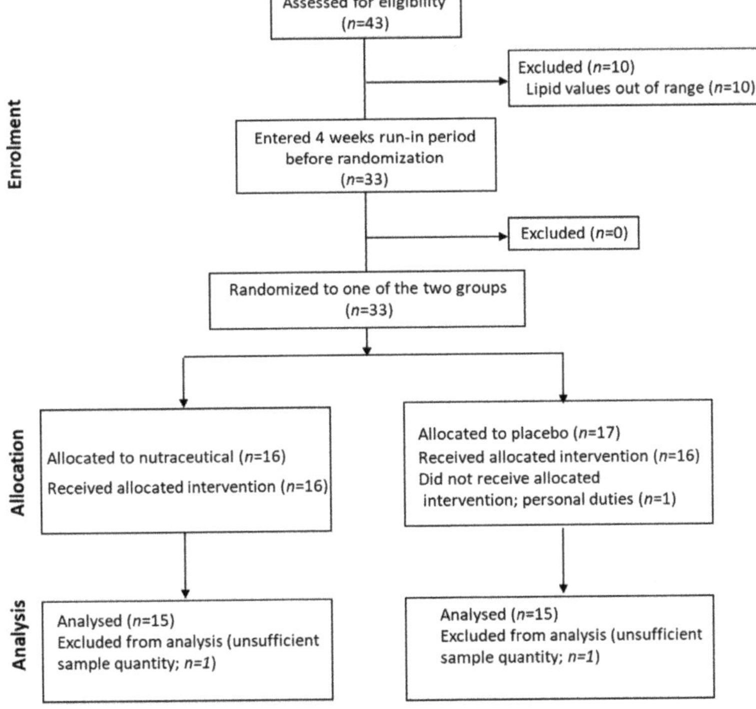

Figure 1. CONSORT statement flow diagram.

2.2. Clinical Procedures

Patients underwent a fasting venous blood sampling and a full clinical examination, with determination of height, body weight, waist circumference, heart rate, and systolic and diastolic blood pressure (SBP, DBP). Bioelectric impedance analysis (ViScan device-Tanita Inc., Tokyo, Japan) was used to estimate % abdominal fat mass (BIA (%)) and % visceral fat rating (VFR (%)), according to reported procedures [12]. Plasma samples were immediately separated by centrifugation, and aliquots were immediately stored at −20 °C. In the present analysis, based upon the study reported in [18], we evaluated basal and post-intervention circulating oxysterol levels. CV biomarkers (TC, non-HDL-C, triglycerides (TG), HDL-C, apolipoprotein (apo)AI, apoB, lipoprotein(a) (Lp(a)), proprotein convertase subtilisin/kexin type 9 (PCSK9)) from this study were used for correlation analysis [18]. Data retrieval, analysis, and the preparation of the manuscript were solely the responsibility of the authors.

2.3. Immunometric and Biochemical Assays

In all blood samples, TC, HDL-C, TG, apoAI, apoB, Lp(a), fasting plasma glucose (FPG), uric acid were measured according to a standard automated clinical procedure (Cobas system, Roche, Italy). LDL-C was calculated according to the Friedewald formula. Non-HDL-C was calculated as TC minus HDL-C. Enzyme-linked immunosorbent assay (ELISA) kits were used according to manufacturer's specifications to quantify fibroblast growth factor (FGF) 19, FGF21 and PCSK9 [20] (R&D System, Minneapolis, MN, USA). Oxidized LDL (oxLDL), and insulin were measured by ELISA kits (Mercodia, Sweden). The homeostasis model assessment of insulin resistance (HOMA-IR) index was calculated according to this equation: HOMA-IR = [fasting glucose (mg/dL) × insulin (mUI/L)/405].

2.4. Determination of Serum Levels of Oxysterols

Oxysterols were analyzed as previous described [21] using a Thermofinnigan GC-Q instrument supplied with an ion trap source. Oxysterols separation was obtained with an HP5 (Agilent, Lexington, MN, USA) capillary column 0.25 mm i.d., 0.25 μm film thickness, 30 m length, operating at 1 mL/min helium flow rate. Column temperature was programmed from 200 to 300 °C at 20 °C/min. Ions were recorded at m/z 353 for 19-hydroxycholesterol and m/z 462 for deuterated 27-OHC (internal standards), m/z 456 for 27- and 25-OHC and m/z 413 for 24-OHC. Endogenous hydroxysterol concentrations were calculated with a standard curve, prepared as described [21], and the peak area ratio (sterol/IS) found in the sample. Noteworthy, in the literature the same chemical compound is called both 26-OHC and 27-OHC [22]. This second term over time has become common and to prevent misunderstandings in this study the term "27-OHC" will be used [23].

2.5. Statistical Analysis

Sample size calculation. According to [18], a group sample size of 16 per arm achieves 80% power to detect a difference of 20 mg/mL in absolute changes (12 weeks-0 week) in LDL-C levels (mg/mL), between the null hypothesis that in both arms the means of change in LDL-C are 10 mg/mL and the alternative hypothesis that the mean of change in LDL-C in the treatment arms is −10 mg/mL [12]. The estimated group standard deviations were 25 mg/mL per arm, with a significance level of 5% using a two-sided two-sample t-test.

Results are shown as median and interquartile ranges (Q1 and Q3) for all parameters. Correlations between circulating oxysterols as well as oxysterols normalized for TC and several covariates were analyzed using a Pearson correlation coefficient. Oxysterol levels are expressed both as absolute values as well as oxysterol-to-TC ratios, in order to correct for differences in plasma TC concentration and for the evidence that sterols are transported by plasma lipoproteins, in line with the available literature [24].

Differences in median values between treatment and placebo arms at baseline were assessed by Wilcoxon-rank sum test. Distributions of percent changes in oxysterols and lipid levels from baseline (0 week) to the end of follow-up (12 weeks treatment) were compared between the placebo and the treatment groups with Wilcoxon rank sum tests. At the end of the follow-up timeframe, the difference in the median percent change observed in the treatment group minus the median percent change in the placebo group at that time was used to summarize the treatment effect. Similar comparisons of percent changes in lipid levels between arms used the same approach. All tests are 2-sided; p values 0.05 are considered statistically significant. Statistical analysis was conducted by using both SAS Software version 9.3 (SAS, Cary, NC, USA) and R Software version 3.6.2.

3. Results

3.1. Study Population

All subjects were in primary CV prevention and showed a moderate hypercholesterolemia. The data analysis was conducted on the 30/33 subjects who also completed the interventional study (Figure 1). The clinical and biochemical data suggest that this cohort showed normal body weight, BMI, waist circumference, BIA and VFR (Table 1). Median TC was 270 (246, 288) mg/dL and LDL-C was 179 (169, 195) mg/dL. TG, HDL-C, insulin sensitivity and blood pressure were in the reference range [25,26].

Table 1. Main clinical and biochemical characteristics of the whole study population and sorted in the two arms.

	Whole Cohort	Placebo Arm	Nutraceutical Combination Arm	Difference between Arms at Baseline (*p*-Value)
Sex (M/F)	16/14	8/7	8/7	-
Age (years)	57.5 (48, 64)	48 (41, 58)	63 (57, 65)	0.006
Weight (kg)	65 (62, 79)	65 (63, 80)	63 (60, 77)	0.57
BMI (kg/m^2)	23.84 (21.3, 27.7)	23.58 (20.75, 27.94)	23.9 (22.77, 27.63)	0.71
Abdominal Circumference (cm)	87.25 (83.5, 94)	87.5 (82, 94)	87 (84, 94)	0.75
Waist Circumference (cm)	94.5 (88, 99)	95 (87, 101.5)	94 (90, 99)	0.91
BIA (%)	30.9 (27.2, 38.5)	30.6 (25.5, 39.7)	32.3 (27.3, 38.5)	0.62
VFR (%)	10.5 (7.5, 11.5)	9.5 (6, 13.5)	10.5 (8.5, 11.5)	0.72
SBP (mmHg)	120 (120, 130)	125 (110, 130)	120 (120, 130)	0.53
DBP (mmHg)	80 (80, 80)	80 (80, 80)	80 (80, 80)	0.97
HR (bpm)	64 (60, 68)	65 (64, 68)	64 (60, 68)	0.28
TC (mg/mL)	270 (246, 288)	270 (255, 290)	270 (233, 288)	0.51
LDL-C (mg/mL)	179 (169, 195)	187 (172, 195)	176 (165, 196)	0.33
HDL-C (mg/mL)	56.5 (43, 77)	54 (47, 67)	65 (42, 83)	0.67
non-HDL-C (mg/mL)	209.5 (192, 230)	214 (196, 234)	207 (188, 216)	0.31
TG (mg/mL)	114.5 (95, 153)	112 (92, 130)	127 (95, 159)	0.57
apoAI (mg/dL)	114.5 (95, 132)	113 (95, 129)	125 (91, 141)	0.39
apoB (mg/dL)	146 (135, 155)	143 (133, 146)	155 (142, 158)	0.03
oxLDL (U/L)	76.6 (70, 85.2)	71.7 (67.3, 84.6)	76.8 (74.5, 123.8)	0.17
24-OHC (ng/mL)	89 (73, 109)	89 (73, 103)	90.9 (71.8, 110)	0.72
24-OHC/TC (ng/mg)	34 (27, 41)	33 (27, 39)	35 (26, 43)	0.71
25-OHC (ng/mL)	84.2 (60.5, 96)	86 (63, 106)	81 (57, 96)	0.55
25-OHC/TC (ng/mg)	29 (25, 38)	31 (25, 39)	28 (22, 38)	0.87
27-OHC (ng/mL)	183.5 (152, 211)	174 (115, 219)	190 (166.3, 211)	0.6
27-OHC/TC (ng/mg)	72 (51, 80)	72 (47, 74)	73 (51, 88)	0.3
Lp(a) (mg/dL)	6 (4, 11)	4 (2, 9)	7 (5, 13)	0.09
PCSK9 (ng/dL)	339.87 (283.17, 403.96)	340.04 (279.63, 402.52)	339.7 (283.17, 410.09)	0.89
FPG (mg/dL)	93.5 (89, 97)	95 (89, 97)	92 (89, 103)	0.98
Insulin (mUI/L)	3.38 (2.42, 5.04)	3.08 (2.49, 6.2)	3.38 (2.13, 5.04)	0.77
HOMA-IR	0.75 (0.56, 1.18)	0.72 (0.57, 1.44)	0.77 (0.47, 1.18)	0.81
FGF19 (pg/mL)	222.87 (173.05, 330.6)	232.4 (106.58, 333.98)	215.61 (176.79, 237.12)	0.94
FGF21 (pg/mL)	174.86 (118.32, 237.14)	157.08 (71.66, 226.57)	178.62 (158.16, 364.15)	0.25
Creatinine (mg/dL)	0.8 (0.7, 0.9)	0.8 (0.7, 1)	0.8 (0.7, 0.9)	0.76

Data are median (Q1, Q3). BMI: body mass index, BIA: bioelectrical impedance analysis/abdominal fat mass, VFR: visceral fat rating, SBP: systolic blood pressure, DBP: diastolic blood pressure, HR: heart rate, TC: total cholesterol, LDL-C: low-density lipoprotein. cholesterol, HDL-C: high-density lipoprotein cholesterol, TG: triglycerides, apoAI: apolipoprotein AI, apoB: apolipoprotein B, oxLDL: oxidized LDL, 24-OHC: 24-hydroxycholesterol, 25-OHC: 25-hydroxycholesterol, 27-OHC: 27-hydroxycholesterol, Lp(a): lipoprotein (a), PCSK9: proprotein convertase subtilisin/kexin type 9, FPG: fasting plasma glucose, HOMA-IR: Homeostatic Model Assessment of Insulin Resistance, FGF: fibroblast growth factor.

3.2. Analysis of Oxysterols in the Study Population

The 24-OHC values (89 (73, 109) ng/mL) of the study cohort are reported in Table 1. They did not differ according to sex (males: 89.4 (73.5, 110.5) ng/mL; females: 89 (73, 109) ng/mL) (Figure 2A) and did not correlate with age (Table 2). The 24-OHC/TC ratio was 34 (27, 41) ng/mg when considering the entire cohort (Table 1), without sex difference (males 37 (27, 43) ng/mg; females 31 (27, 38)) (Figure 2B) and no correlation with age (Table 2). Intriguingly, 24-OHC was positively correlated with TG ($p = 0.004$) and apoB ($p = 0.012$) and, after normalizing the values for TC, only the correlation with TG was still present ($p = 0.024$) (Table 2). The values of 25-OHC were 84.2 (60.5, 96) ng/mL (Table 1) and did not differ according to sex (males 84.7 (72, 101) ng/mL; females 71.5 (54, 89) ng/mL) (Figure 2A), nor they correlated with age (Table 2). 25-OHC/TC levels (29 ng/mg (25, 38) (Table 1) were not different according to sex (males 32 (26, 41) ng/mg; females 27 (18, 32) ng/mg) (Figure 2B) and did not correlate with age (Table 2). Interestingly, both 25-OHC and 25-OHC/TC showed a significant positive correlation with TG ($p = 0.007$ and $p = 0.028$) (Table 2). Moreover, 24-OHC and 25-OHC levels were significantly correlated ($p = 0.0002$) (Table 2). The 27-OHC values (183.5 ng/mL (152, 211)), shown in Table 1, significantly ($p = 0.02$) diverged between males (209 (173, 230) ng/mL) and females (167.7 (126, 193) ng/mL) (Figure 2A) and did not correlate with age (Table 2). 27-OHC/TC values (72 ng/mg (51, 80)) were also different ($p = 0.008$) according to sex (males 77 (61, 90) ng/mg; females 56 (46, 72) ng/mg (Figure 2B) and did not correlate with age. In addition, 27-OHC showed a positive correlation trend with TG ($p = 0.056$) and non-HDL-C ($p = 0.065$), while it was significantly correlated with creatinine ($p = 0.017$). 27-OHC/TC was negatively correlated with HDL-C ($p = 0.006$) and apoAI ($p = 0.05$), whereas it was positively correlated with abdominal circumference ($p = 0.023$) and VFR ($p = 0.021$) (Table 2). Lp(a) was correlated with 24-OHC/TC ($p = 0.021$), 25-OHC ($p = 0.045$) and 25-OHC/TC ($p = 0.013$). Moreover, PCSK9 levels were negatively correlated with 27-OHC/TC ($p = 0.013$) (Table 2).

(A)

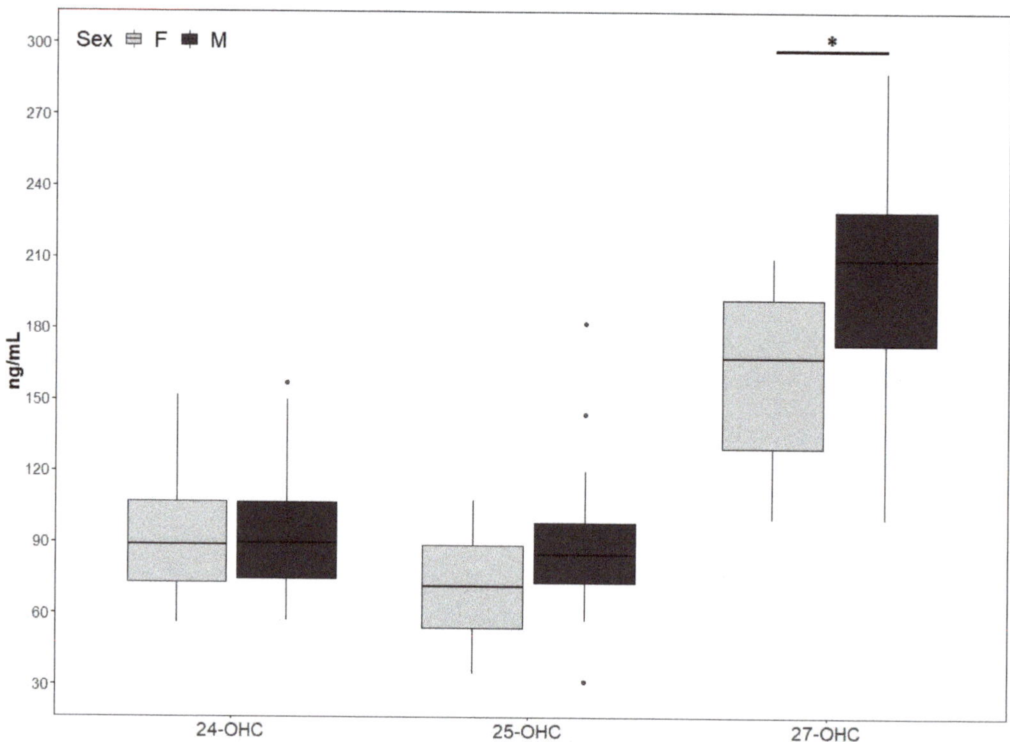

Figure 2. *Cont.*

(B)

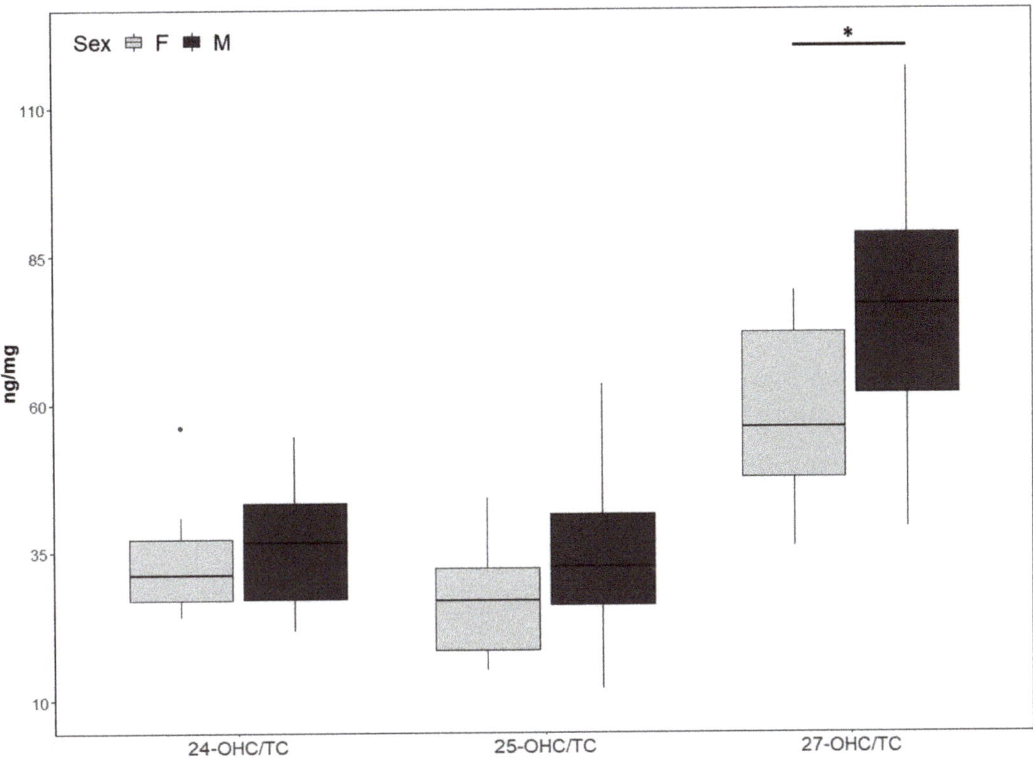

Figure 2. (**A**) 24-hydroxycholesterol (24-OHC), 25-hydroxycholesterol (25-OHC) and 27-hydroxycholesterol (27-OHC) plasma levels; (**B**) 24-OHC/total cholesterol (TC), 25-OHC/TC and 27-OHC/TC. TC values (mean $\pm$ SD) were 250.5 $\pm$ 75.6 mg/dL for males and 257.9 $\pm$ 80 mg/dL for females. M, males; F, females. * *p*-value < 0.01.

3.3. Effect of Nutraceutical Treatment on Oxysterols Plasma Levels

The participants of the cross-sectional study were also randomized to either nutraceutical combination or placebo (Figure 1). No differences between arms were found for the different variables, except age and apoB (Table 1). After nutraceutical intervention, compared to placebo, TC was significantly reduced (*p* < 0.0001; −16.7%), together with LDL-C (*p* < 0.0001; −25.7%), as previously reported [18] (please refer to this article for additional data on CVD biomarkers). After the normalization for TC (24-OHC/TC, 25-OHC/TC and 27-OHC/TC), oxysterol levels did not differ between the 2 groups. When considering the absolute value of circulating oxysterols, in the nutraceutical treatment arm 27-OHC concentrations were significantly (*p* = 0.008) decreased (−10.4%), whereas 24-OHC and 25-OHC levels did not change (Table 3).

Table 2. Correlation of circulating oxysterols levels and total cholesterol-normalized circulating oxysterols levels with clinical and biochemical characteristics of the study population.

	24-OHC	24-OHC/TC	25-OHC	25-OHC/TC	27-OHC	27-OHC/TC
Sex	0.68	0.42	0.14	0.07	0.02	0.008
Age	−0.269/0.151	−0.279/0.135	−0.29/0.121	−0.285/0.127	0.191/0.313	0.218/0.248
Weight	0.025/0.895	0.158/0.403	0.133/0.483	0.225/0.232	0.236/0.209	0.339/0.067
BMI	0.069/0.728	0.198/0.313	0.124/0.528	0.23/0.238	0.148/0.453	0.265/0.173
Abdominal Circumference	−0.034/0.859	0.118/0.536	0.06/0.754	0.159/0.401	0.272/0.146	0.414/0.023
Waist Circumference	0.037/0.847	0.109/0.567	−0.003/0.988	0.04/0.833	0.042/0.824	0.104/0.586
BIA	−0.054/0.776	0.06/0.753	−0.253/0.178	−0.171/0.367	0.026/0.89	0.136/0.475
VFR	0.021/0.912	0.168/0.376	0.138/0.468	0.253/0.177	0.292/0.118	0.42/0.021
SBP	−0.051/0.789	−0.094/0.62	0.172/0.365	0.182/0.336	−0.11/0.561	−0.135/0.477
DBP	−0.132/0.485	−0.194/0.304	−0.16/0.397	−0.175/0.354	−0.126/0.505	−0.18/0.34
HR	0.095/0.617	−0.012/0.95	0.128/0.499	0.073/0.7	−0.002/0.992	−0.105/0.569
TC	0.219/0.246	−0.193/0.307	0.23/0.221	−0.088/0.643	0.158/0.405	−0.295/0.115
LDL-C	0.022/0.909	−0.314/0.091	0.12/0.528	−0.151/0.426	0.23/0.221	−0.148/0.435
HDL-C	0.066/0.729	−0.153/0.419	−0.033/0.864	−0.188/0.319	−0.252/0.179	−0.489/0.006
non-HDL-C	0.215/0.253	−0.131/0.49	0.29/0.12	0.013/0.946	0.341/0.065	−0.041/0.828
TG	0.517/0.004	0.41/0.024	0.481/0.007	0.401/0.028	0.353/0.056	0.253/0.178
apoAI	0.104/0.586	−0.183/0.333	0.04/0.833	−0.164/0.387	−0.052/0.783	−0.361/0.05
apoB	0.455/0.012	0.306/0.1	0.155/0.415	0.025/0.897	0.201/0.287	0.032/0.866
oxLDL	0.202/0.285	0.155/0.414	−0.094/0.62	−0.136/0.472	−0.04/0.835	−0.045/0.815
24-OHC	-	0.912/<0.0001	0.626/0.00021	0.552/0.0015	−0.084/0.655	−0.177/0.347
24-OHC/TC	0.912/<0.0001	-	0.518/0.0033	0.580/0.00078	−0.145/0.443	−0.049/0.793
25-OHC	0.626/0.00021	0.518/0.0033	-	0.945/<0.0001	0.053/0.779	−0.060/0.749
25-OHC/TC	0.552/0.0015	0.580/0.00078	0.945/<0.0001	-	−0.0001/0.999	0.032/0.865
27-OHC	−0.084/0.655	−0.145/0.443	−0.053/0.779	−0.0001/0.999	-	0.892/<0.0001
27-OHC/TC	−0.117/0.347	−0.049/0.793	−0.06/0.749	0.032/0.865	0.892/<0.0001	-
Lp(a)	0.324/0.081	0.419/0.021	0.369/0.045	0.448/0.013	0.084/0.658	0.168/0.374
PCSK9	0.035/0.852	−0.188/0.321	0.067/0.724	−0.107/0.574	−0.223/0.236	−0.45/0.013
FPG	0.1/0.598	0.097/0.609	−0.015/0.938	−0.023/0.905	0.227/0.228	0.247/0.189
Insulin	0.019/0.92	0.92/0.419	0.073/0.7	0.176/0.352	0.14/0.46	0.272/0.146
HOMA-IR	0.029/0.879	0.153/0.42	0.085/0.654	0.181/0.339	0.166/0.38	0.292/0.117
FGF19	−0.099/0.637	−0.003/0.987	0.133/0.528	0.231/0.267	−0.225/0.279	−0.118/0.573
FGF21	−0.178/0.454	−0.208/0.379	0.218/0.357	0.197/0.405	0.013/0.956	0.043/0.857
Creatinine	−0.064/0.738	0.069/0.716	0.101/0.597	0.218/0.247	0.433/0.017	0.567/0.001

Pearson correlation coefficient and *P*-value are reported for each correlation, except for Sex (dichotomic variable). BMI: body mass index, BIA: bioelectrical impedance analysis/abdominal fat mass, VFR: visceral fat rating, SBP: systolic blood pressure, DBP: diastolic blood pressure, HR: heart rate, TC: total cholesterol, LDL-C: low-density lipoprotein. cholesterol, HDL-C: high-density lipoprotein cholesterol, TG: triglycerides, apoAI: apolipoprotein AI, apoB: apolipoprotein B, oxLDL: oxidized LDL, 24-OHC: 24-hydroxycholesterol, 25-OHC: 25-hydroxycholesterol, 27-OHC: 27-hydroxycholesterol, Lp(a): lipoprotein (a), PCSK9: proprotein convertase subtilisin/kexin type 9, FPG: fasting plasma glucose, HOMA-IR: Homeostatic Model Assessment of Insulin Resistance, FGF: fibroblast growth factor.

Table 3. Determination of serum levels of 24-hydroxycholesterol (24-OHC), 25-hydroxycholesterol (25-OHC) and 27-hydroxycholesterol (27-OHC) and their ratio to total cholesterol (TC).

	Placebo		Nutraceutical		Difference of Changes between Arms *p*-Value
	Baseline	**12 Weeks**	**Baseline**	**12 Weeks**	
24-OHC (ng/mL)	89 (73, 103)	97 (87, 118)	91 (72, 110)	94 (75, 139)	0.2
25-OHC (ng/mL)	86 (63, 106)	79 (52, 96)	81 (57, 96)	81 (62, 91)	0.91
27-OHC (ng/mL)	174 (115, 219)	179 (111, 232)	190 (166, 211)	170 (133, 187)	0.03
24-OHC/TC (ng/mg)	33 (27, 39)	34 (30, 41)	35 (26, 43)	40 (35, 51)	0.57
25-OHC/TC (ng/mg)	31 (25, 39)	25 (20, 34)	28 (22, 38)	35 (27, 41)	0.36
27-OHC/TC (ng/mg)	72 (47, 74)	67 (43, 82)	73 (51, 88)	74 (54, 94)	0.09

Data are shown as median (1st quartile, 3rd quartile), *p*-values are adjusted for age and apolipoprotein B.

4. Discussion

The present study was aimed at analyzing the relationship between circulating oxysterols, namely 24-, 25- and 27-OHC, with biomarkers related to atherosclerosis in subjects with moderate hypercholesterolemia. We also evaluated the effect of a nutraceutical combination containing *Bifidobacterium longum* BB536 and RYR and aimed to reduce hypercholesterolemia, on the circulating levels of these oxysterols. In our cohort of moderate hypercholesterolemic subjects, 24-OHC values were found to be within the range (33.2–227.0 ng/mL) previously reported for different populations [4]. On the contrary, the values of 25-OHC were almost 3 fold higher compared to those reported (range 2.0–31.0 ng/mL) previously [4]. Together with data indicating that hypercholesterolemic males have significantly higher 25-OHC levels compared to healthy males [10], our findings suggest that this elevation may be peculiar for this condition. 27-OHC values in our population were within the previously reported 27-OHC range (43.6–196.0 ng/mL) [4]. Different measurement techniques (high-performance liquid chromatography–mass spectrometry (HPLC-MS) without derivatization, charge-tagging with HPLC-MS analysis, dimethylglycine derivatization followed by HPLC-ESI-MS and GC–MShigh-performance liquid chromatography/electrospray ionization tandem mass spectrometry (HPLC-ESI-MS) and gas chromatography-mass spectrometry (GC–MS) analysis of oxys-terol trimethylsilyl derivatives), each having peculiar features might account for the variation range in oxysterols levels [27]. We found a significant correlation between the circulating levels of 24-OHC and 25-OHC. In this regard, CYP46, the enzyme mainly responsible for 24-OHC synthesis, was also shown to be capable of synthesizing 25-OHC (ratio: 4:1 (24-OHC:25-OHC)), in cell-based systems [28].

Interestingly, we observed that 27-OHC and 27-OHC/TC plasma levels were significantly higher in males than in females. These findings agree with previously reported data for 27-OHC [29], extending the observation to 27-OHC/TC. Circulating 27-OHC levels are lower in females than in males in both rodents and humans, and this is most likely related to the upregulation of CYP7B1 expression by estradiol and estrogen receptor activation [13,23]. Interestingly, 27-OHC itself has also been regarded has a peculiar selective estrogen receptor modulator (SERM), and has been shown to interfere with the atheroprotective activity of estrogens [30]. These findings may be relevant in the overall context of sex-related ASCVD risk and related response to drugs [31]. Moreover, experimental work indicates that 27-OHC also has an adverse impact on bone mineralization [32] and breast cancer proliferation [33].

In healthy subjects, 24-OHC and 27-OHC were found to correlate with TC, LDL-C and non-HDL-C [24]. In our cohort of moderate hypercholesterolemic patients, we observed a positive trend between plasma 27-OHC and non-HDL-C, but no correlations of oxysterols with TC and LDL-C, although 24-OHC levels positively correlated with apoB concentrations.

In our moderate hypercholesterolemic patients, a significant negative correlation was found between 27-OHC/TC and HDL-C and apoAI, its main lipoprotein component, circulating levels. This observation is in agreement with a previously reported inverse relationship between 27-OHC and HDL-C in normocholesterolemic subjects [34,35]. One possible explanation may be that 27-OHC, acting as liver X receptor (LXR) α ligand, upregulates the expression of cholesteryl ester transfer protein (CETP), which in turn transfers cholesteryl ester from HDL to other lipoproteins, leading to HDL-C reduction [34] and possibly to LDL-C increase [36]. It may be then speculated that the ineffective reverse cholesterol transport in individuals with very low levels of HDL-C may be compensated by this mechanism, which therefore could represent an alternative pathway in the context of the complex regulation of reverse cholesterol transport [37]. In our normo-triglyceridemic population, 24-, 25- and 27-OHC correlated with TG. While the 27-OHC and TG relationship has already been reported in healthy subjects [24], that between 24-, 25-OHC and TG is novel, to the best of our knowledge, and is still present upon normalization by TC

(24-, 25-OHC/TC and TG). This correlation could be the consequence of the known LXR modulation by these oxysterols [38] and the resulting stimulation of liver TG synthesis [39].

In the context of the studied population, featuring normal BMI and abdominal circumference, we observed that the 27-OHC/TC ratio positively correlated with abdominal circumference and VFR%. This finding is novel in the clinical setting, although the interrelationship between 27-OHC and adipose tissue has previously been addressed experimentally, leading to apparently controversial observations. In addition to circulate in serum, 27-OHC may also be locally produced by rodent and human adipocytes, where it may counteract adipogenesis [40]. 27-OHC content of white adipose tissue was negatively correlated with adipose mass in mice and exposure to 27-OHC suppressed intracellular TG accumulation by down-regulating lipogenic and adipogenic gene expression during adipocyte maturation of mouse 3T3-L1 cells [41]. However, in mice, 27-OHC administration has been shown to promote adipose tissue hyperplasia, independently from diet type, increasing visceral fat and local inflammation [42]. In light of the relevance of dysfunctional visceral and ectopic adipose for ASCVD [43,44], this intriguing connection requires further clarification.

To our knowledge, no data are available on the effects of hypocholesterolemic nutraceutical treatments on circulating oxysterol levels. Since, in this field, several nutraceutical combinations include RYR, whose main active component is monacolin K, notoriously structurally identical to the statin lovastatin, we may consider a comparison with the available data regarding the impact of statins treatment on 24-, 25- and 27-OHC in subjects with moderate hypercholesterolemia [10,45]. After nutraceutical intervention, 24-OHC level and 24-OHC/TC ratio were unchanged, differently with the reducing effect of simvastatin (80 mg/day) or atorvastatin (40 mg/day) [11,45]. The 25-OHC concentration and 25-OHC/TC ratio were also not affected in both placebo and active groups. Previously reported effects of statins showed decreased 25-OHC concentrations in hypercholesterolemic patients [10,46]. In agreement with the effect of simvastatin (80 mg/day) and atorvastatin (40 mg/day) [11], in our study, 27-OHC levels were significantly reduced in the nutraceutical group, whereas the 27-OHC/TC ratio was not different between arms. 27-OHC levels were also found to be decreased after treatment with atorvastatin or rosuvastatin in subjects with familial hypercholesterolemia or familial combined hyperlipidemia [47]. Due to the similarity of action of RYR extracts and statins, the observed 27-OHC reduction in our study may be the result of cholesterol synthesis inhibition, as also supported by the unchanged 27-OHC/TC ratio. Interestingly, such 27-OHC reduction may also contribute to the observed LDL-C decrease via downregulation of the CETP-pathway. As the nutraceutical combination used here, in addition to RYR, also contains the probiotic *Bifidobacterium longum* BB536, niacin, and coenzyme Q10, one should not exclude some contribution of these components to the effects on the observed reduction in 27-OHC level. One may hypothesize that *Bifidobacterium longum* BB536 may possibly contribute to this reduction by means of its biliary salt hydrolase activity, taking place in the ileum and consequently interfering with the enterohepatic circulation of cholesterol [18].

The evaluation of oxysterol levels after treatment with this nutraceutical combination provides further information on the safety of this product. Oxysterols seem to have a pathogenic role in hyperlipidemia and atherosclerosis both via modulating numerous systemic functions as well as with local actions, and the absolute reduction of 27-OHC concentration may result beneficial, in the context of this cohort of hypercholesterolemic subjects. Elevated circulating 27-OHC may have detrimental effects on the cardiovascular system through multiple mechanisms (SERM activity, LXRα and β ligand), promoting vascular inflammation, which is critically involved in atherogenesis [42,48–50]. The additional contribution of 27-OHC and CYP27A1, responsible for its synthesis, which are abundantly present in atherosclerotic plaques, needs further studies [29,51,52].

The findings of the present study may be relevant in terms of long-term safety in consideration of the potentially detrimental role of increased 24-, 25- and 27-OHC also in the context of neurodegenerative diseases such as mild cognitive impairment and dementia [53], as well as bone mineralization and breast cancer [54–56].

The main strengths of this study include (1). the extensive exploration of the relationships between 24-, 25- and 27-OHC with a relevant set of CVD risk biomarkers in subjects with moderate hypercholesterolemia, highlighting a series of novel correlations, and (2). the first evaluation of the impact of a nutraceutical combination, designed for the control of hypercholesterolemia, on the circulating levels of these oxysterols. The present study has some limitations. The lack of a control group, useful for comparison, is an intrinsic limitation of the cross-sectional study. Therefore, several observations cannot be extended to healthy subjects. A limitation of the interventional study is also that the dietary intake of the volunteers randomized to either placebo or nutraceutical intervention was not recorded.

Due to their relevant biological effects, the measurement of 24-, 25- and especially 27-OHC may be useful even when assessing the long-term safety of hypocholesterolemic treatments (statins, ezetimibe, bempedoic acid and PCSK9 inhibitors), indicating the need for further larger studies.

5. Conclusion

In conclusion, this study adds novel information on this hypocholesterolemic nutraceutical combination, regarding its efficacy and safety, according to oxysterol profile.

Author Contributions: Conceptualization, P.M. and S.C; methodology, M.D.P.; software, C.P.; validation, M.B. and G.M.; formal analysis, M.D.P.; investigation, M.B.; resources, M.B.; data curation, E.O.; writing—original draft preparation, P.M. and S.C.; writing, review and editing, A.L.C.; visualization, L.C.; supervision, A.L.C.; project administration, P.M.; funding acquisition, P.M. All authors have read and agreed to the published version of the manuscript.

Funding: The study was supported by unrestricted grants to Centro Dislipidemie (ASST Grande Ospedale Metropolitano Niguarda, Milan, Italy) and Università degli Studi di Milano (Milan, Italy) from Montefarmaco OTC S.p.A. (Bollate, MI, Italy). The founding sponsors had no role in the design of the study, in collection, analyses, or interpretation of data, in the manuscript writing, and in the decision to publish the results.

Institutional Review Board Statement: The study was conducted according to the guidelines of the Declaration of Helsinki, and approved by the Ethics Committee "Milano Area C" at ASST Grande Ospedale Metropolitano Niguarda (protocol code 598-112015 – 27 November 2015).

Informed Consent Statement: Informed consent was obtained from all subjects involved in the study.

Data Availability Statement: The data presented in this study are available on request from the corresponding author. The data are not publicly available due to privacy reasons.

Acknowledgments: The critical revision of the manuscript by G.D. Norata is gratefully acknowledged.

Conflicts of Interest: S.C., C.P., E.O., M.D.P., M.B., G.M. and L.C. have no conflict of interests. A.L.C. has received honoraria, lecture fees, or research grants from: Akcea, Amgen, Astrazeneca, Eli Lilly, Genzyme, Kowa, Mediolanum, Menarini, Merck, Pfizer, Recordati, Sanofi, Sigma Tau, Medco, Amryt. P.M.: has received honoraria, lecture fees, or research grants from: Montefarmaco OTC.

References

1. The Emerging Risk Factors Collaboration. Major Lipids, Apolipoproteins, and Risk of Vascular Disease. *JAMA* **2009**, *302*, 1993–2000. [CrossRef] [PubMed]
2. Gomaraschi, M.; Bonacina, F.; Norata, G.D. Lysosomal acid lipase: From cellular lipid handler to immunometabolic target. *Trends Pharmacol. Sci.* **2019**, *40*, 104–115. [CrossRef] [PubMed]
3. Zmysłowski, A.; Szterk, A. Oxysterols as a biomarker in diseases. *Clin. Chim. Acta* **2019**, *491*, 103–113. [CrossRef] [PubMed]
4. Mutemberezi, V.; Guillemot-Legris, O.; Muccioli, G.G. Oxysterols: From cholesterol metabolites to key mediators. *Prog. Lipid Res.* **2016**, *64*, 152–169. [CrossRef]
5. Yvan-Charvet, L.; Bonacina, F.; Guinamard, R.R.; Norata, G.D. Immunometabolic function of cholesterol in cardiovascular disease and beyond. *Cardiovasc. Res.* **2019**, *115*, 1393–1407. [CrossRef]

6. Kloudova, A.; Guengerich, F.P.; Soucek, P. The role of oxysterols in human cancer. *Trends Endocrinol. Metab.* **2017**, *28*, 485–496. [CrossRef]

7. Noguchi, N.; Saito, Y.; Urano, Y. Diverse functions of 24(S)-hydroxycholesterol in the brain. *Biochem. Biophys. Res. Commun.* **2014**, *446*, 692–696. [CrossRef]

8. De Backer, G.G. Prevention of cardiovascular disease: Much more is needed. *Eur. J. Prev. Cardiol.* **2018**, *25*, 1083–1086. [CrossRef]

9. Joseph, P.G.; Leong, D.; McKee, M.; Anand, S.S.; Schwalm, J.-D.; Teo, K.; Mente, A.; Yusuf, S. Reducing the global burden of cardiovascular disease, Part 1. *Circ. Res.* **2017**, *121*, 677–694. [CrossRef]

10. Dias, I.H.; Milic, I.; Lip, G.Y.; Devitt, A.; Polidori, M.C.; Griffiths, H.R. Simvastatin reduces circulating oxysterol levels in men with hypercholesterolaemia. *Redox Biol.* **2018**, *16*, 139–145. [CrossRef]

11. Thelen, K.M.; Laaksonen, R.; Päivä, H.; Lehtimäki, T.; Luetjohann, D. High-dose statin treatment does not alter plasma marker for brain cholesterol metabolism in patients with moderately elevated plasma cholesterol levels. *J. Clin. Pharmacol.* **2006**, *46*, 812–816. [CrossRef] [PubMed]

12. Ruscica, M.; Gomaraschi, M.; Mombelli, G.; Macchi, C.; Bosisio, R.; Pazzucconi, F.; Pavanello, C.; Calabresi, L.; Arnoldi, A.; Sirtori, C.R.; et al. Nutraceutical approach to moderate cardiometabolic risk: Results of a randomized, double-blind and crossover study with armolipid plus. *J. Clin. Lipidol.* **2014**, *8*, 61–68. [CrossRef] [PubMed]

13. Ruscica, M.; Pavanello, C.; Gandini, S.; Gomaraschi, M.; Vitali, C.; Macchi, C.; Morlotti, B.; Aiello, G.; Bosisio, R.; Calabresi, L.; et al. Effect of soy on metabolic syndrome and cardiovascular risk factors: A randomized controlled trial. *Eur. J. Nutr.* **2016**, *57*, 499–511. [CrossRef] [PubMed]

14. Pavanello, C.; Lammi, C.; Ruscica, M.; Bosisio, R.; Mombelli, G.; Zanoni, C.; Calabresi, L.; Sirtori, C.R.; Magni, P.; Arnoldi, A.; et al. Effects of a lupin protein concentrate on lipids, blood pressure and insulin resistance in moderately dyslipidaemic patients: A randomised controlled trial. *J. Funct. Foods* **2017**, *37*, 8–15. [CrossRef]

15. Ward, N.C.; Pang, J.; Ryan, J.D.; Watts, G.F. Nutraceuticals in the management of patients with statin-associated muscle symptoms, with a note on real-world experience. *Clin. Cardiol.* **2018**, *41*, 159–165. [CrossRef]

16. Magni, P.; Macchi, C.; Morlotti, B.; Sirtori, C.R.; Ruscica, M. Risk identification and possible countermeasures for muscle adverse effects during statin therapy. *Eur. J. Intern. Med.* **2015**, *26*, 82–88. [CrossRef]

17. Ruscica, M.; Macchi, C.; Morlotti, B.; Sirtori, C.R.; Magni, P. Statin therapy and related risk of new-onset type 2 diabetes mellitus. *Eur. J. Intern. Med.* **2014**, *25*, 401–406. [CrossRef]

18. Ruscica, M.; Pavanello, C.; Gandini, S.; Macchi, C.; Botta, M.; Dall'Orto, D.; Del Puppo, M.; Bertolotti, M.; Bosisio, R.; Mombelli, G.; et al. Nutraceutical approach for the management of cardiovascular risk-a combination containing the probiotic Bifidobacterium longum BB536 and red yeast rice extract: Results from a randomized, double-blind, placebo-controlled study. *Nutr. J.* **2019**, *18*, 13. [CrossRef]

19. HeartScore. Access HeartScore-Full Version. Available online: https://www.heartscore.org/en_GB/access (accessed on 10 September 2020).

20. Ruscica, M.; Ferri, N.; Macchi, C.; Meroni, M.; Lanti, C.; Ricci, C.; Maggioni, M.; Fracanzani, A.L.; Badiali, S.; Fargion, S.; et al. Liver fat accumulation is associated with circulating PCSK9. *Ann. Med.* **2016**, *48*, 384–391. [CrossRef]

21. Del Puppo, M.; Kienle, M.G.; Petroni, M.; Crosignani, A.; Podda, M. Serum 27-hydroxycholesterol in patients with primary biliary cirrhosis suggests alteration of cholesterol catabolism to bile acids via the acidic pathway. *J. Lipid Res.* **1998**, *39*, 2477–2482. [CrossRef]

22. Pandak, W.M.; Kakiyama, G. The acidic pathway of bile acid synthesis: Not just an alternative pathway. *Liver Res.* **2019**, *3*, 88–98. [CrossRef] [PubMed]

23. Fakheri, R.J.; Javitt, N.B. 27-Hydroxycholesterol, does it exist? On the nomenclature and stereochemistry of 26-hydroxylated sterols. *Steroids* **2012**, *77*, 575–577. [CrossRef] [PubMed]

24. Burkard, I.; Von Eckardstein, A.; Waeber, G.; Vollenweider, P.; Rentsch, K.M. Lipoprotein distribution and biological variation of 24S- and 27-hydroxycholesterol in healthy volunteers. *Atherosclerosis* **2007**, *194*, 71–78. [CrossRef] [PubMed]

25. Alberti, K.; Eckel, R.H.; Grundy, S.M.; Zimmet, P.Z.; Cleeman, J.I.; Donato, K.A.; Fruchart, J.-C.; James, W.P.T.; Loria, C.M.; Smith, S.C.; et al. Harmonizing the metabolic syndrome. *Circulation* **2009**, *120*, 1640–1645. [CrossRef]

26. Mach, F.; Baigent, C.; Catapano, A.L.; Koskinas, K.C.; Casula, M.; Badimon, L.; Chapman, M.J.; De Backer, G.G.; Delgado, V.; Ference, B.A.; et al. 2019 ESC/EAS Guidelines for the management of dyslipidaemia. *Eur. Heart J.* **2020**, *41*, 111–188. [CrossRef]

27. Dias, I.H.; Wilson, S.R.; Roberg-Larsen, H. Chromatography of oxysterols. *Biochimie* **2018**, *153*, 3–12. [CrossRef]

28. Lund, E.G.; Guileyardo, J.M.; Russell, D.W. cDNA cloning of cholesterol 24-hydroxylase, a mediator of cholesterol homeostasis in the brain. *Proc. Natl. Acad. Sci. USA* **1999**, *96*, 7238–7243. [CrossRef]

29. Brown, A.J.; Jessup, W. Oxysterols and atherosclerosis. *Atherosclerosis* **1999**, *142*, 1–28. [CrossRef]

30. Umetani, M.; Domoto, H.; Gormley, A.K.; Yuhanna, I.S.; Cummins, C.L.; Javitt, N.B.; Korach, K.S.; Shaul, P.W.; Mangelsdorf, D.J. 27-Hydroxycholesterol is an endogenous SERM that inhibits the cardiovascular effects of estrogen. *Nat. Med.* **2007**, *13*, 1185–1192. [CrossRef]

31. Mombelli, G.; Bosisio, R.; Calabresi, L.; Magni, P.; Pavanello, C.; Pazzucconi, F.; Sirtori, C.R. Gender-related lipid and/or lipoprotein responses to statins in subjects in primary and secondary prevention. *J. Clin. Lipidol.* **2015**, *9*, 226–233. [CrossRef]

32. Umetani, M.; Shaul, P.W. 27-Hydroxycholesterol: The first identified endogenous SERM. *Trends Endocrinol. Metab.* **2011**, *22*, 130–135. [CrossRef] [PubMed]

33. DuSell, C.D.; Umetani, M.; Shaul, P.W.; Mangelsdorf, D.J.; McDonnell, D.P. 27-Hydroxycholesterol is an endogenous selective estrogen receptor modulator. *Mol. Endocrinol.* **2008**, *22*, 65–77. [CrossRef] [PubMed]

34. Hirayama, T.; Mizokami, Y.; Honda, A.; Homma, Y.; Ikegami, T.; Saito, Y.; Miyazaki, T.; Matsuzaki, Y. Serum concentration of 27-hydroxycholesterol predicts the effects of high-cholesterol diet on plasma LDL cholesterol level. *Hepatol. Res.* **2009**, *39*, 149–156. [CrossRef] [PubMed]

35. Nunes, V.S.; Leança, C.C.; Panzoldo, N.B.; Parra, E.; Zago, V.S.; Cazita, P.M.; Nakandakare, E.R.; De Faria, E.C.; Quintão, E.C. Plasma 27-hydroxycholesterol/cholesterol ratio is increased in low high density lipoprotein-cholesterol healthy subjects. *Clin. Biochem.* **2013**, *46*, 1619–1621. [CrossRef]

36. Blauw, L.L.; Noordam, R.; Soidinsalo, S.; Blauw, C.A.; Li-Gao, R.; De Mutsert, R.; Berbée, J.F.P.; Wang, Y.; Van Heemst, D.; Rosendaal, F.R.; et al. Mendelian randomization reveals unexpected effects of CETP on the lipoprotein profile. *Eur. J. Hum. Genet.* **2019**, *27*, 422–431. [CrossRef]

37. Ferri, N.; Ruscica, M. Proprotein convertase subtilisin/kexin type 9 (PCSK9) and metabolic syndrome: Insights on insulin resistance, inflammation, and atherogenic dyslipidemia. *Endocrine* **2016**, *54*, 588–601. [CrossRef]

38. Bełtowski, J. Liver X receptors (LXR) as therapeutic targets in dyslipidemia. *Cardiovasc. Ther.* **2008**, *26*, 297–316. [CrossRef]

39. Schultz, J.R.; Tu, H.; Luk, A.; Repa, J.J.; Medina, J.C.; Li, L.; Schwendner, S.; Wang, S.; Thoolen, M.; Mangelsdorf, D.J.; et al. Role of LXRs in control of lipogenesis. *Genes Dev.* **2000**, *14*, 2831–2838. [CrossRef]

40. Li, J.; Daly, E.; Campioli, E.; Wabitsch, M.; Papadopoulos, V. De novosynthesis of steroids and oxysterols in adipocytes. *J. Biol. Chem.* **2014**, *289*, 747–764. [CrossRef]

41. Shirouchi, B.; Kashima, K.; Horiuchi, Y.; Nakamura, Y.; Fujimoto, Y.; Tong, L.-T.; Sato, M. 27-Hydroxycholesterol suppresses lipid accumulation by down-regulating lipogenic and adipogenic gene expression in 3T3-L1 cells. *Cytotechnology* **2016**, *69*, 485–492. [CrossRef]

42. Asghari, A.; Ishikawa, T.; Hiramitsu, S.; Lee, W.-R.; Umetani, J.; Bui, L.; Korach, K.S.; Umetani, M. 27-hydroxycholesterol promotes adiposity and mimics adipogenic diet-induced inflammatory signaling. *Endocrinology* **2019**, *160*, 2485–2494. [CrossRef] [PubMed]

43. Neeland, I.J.; Ross, R.; Després, J.-P.; Matsuzawa, Y.; Yamashita, S.; Shai, I.; Seidell, J.; Magni, P.; Santos, R.D.; Arsenault, B.; et al. Visceral and ectopic fat, atherosclerosis, and cardiometabolic disease: A position statement. *Lancet Diabetes Endocrinol.* **2019**, *7*, 715–725. [CrossRef]

44. Ross, R.; Neeland, I.J.; Yamashita, S.; Shai, I.; Seidell, J.; Magni, P.; Santos, R.D.; Arsenault, B.J.; Cuevas, A.; Hu, F.B.; et al. Waist circumference as a vital sign in clinical practice: A consensus statement from the IAS and ICCR working group on visceral obesity. *Nat. Rev. Endocrinol.* **2020**, *16*, 177–189. [CrossRef] [PubMed]

45. Locatelli, S.; Lütjohann, D.; Schmidt, H.H.-J.; Otto, C.; Beisiegel, U.; Von Bergmann, K. Reduction of plasma 24S-hydroxycholesterol (Cerebrosterol) levels using high-dosage simvastatin in patients with hypercholesterolemia. *Arch. Neurol.* **2002**, *59*, 213–216. [CrossRef] [PubMed]

46. Hukkanen, J.; Puurunen, J.; Hyötyläinen, T.; Savolainen, M.J.; Ruokonen, A.; Morin-Papunen, L.; Orešič, M.; Piltonen, T.T.; Tapanainen, J.S. The effect of atorvastatin treatment on serum oxysterol concentrations and cytochrome P450 3A4 activity. *Br. J. Clin. Pharmacol.* **2015**, *80*, 473–479. [CrossRef]

47. Bertolotti, M.; Del Puppo, M.; Corna, F.; Anzivino, C.; Gabbi, C.; Baldelli, E.; Carulli, L.; Loria, P.; Kienle, M.G.; Carulli, N.; et al. Increased appearance rate of 27-hydroxycholesterol in vivo in hypercholesterolemia: A possible compensatory mechanism. *Nutr. Metab. Cardiovasc. Dis.* **2012**, *22*, 823–830. [CrossRef]

48. Galkina, E.V.; Ley, K. Immune and inflammatory mechanisms of atherosclerosis. *Annu. Rev. Immunol.* **2009**, *27*, 165–197. [CrossRef]

49. Libby, P. Inflammation in atherosclerosis. *Arter. Thromb. Vasc. Biol.* **2012**, *32*, 2045–2051. [CrossRef]

50. Baragetti, A.; Catapano, A.L.; Magni, P. Multifactorial activation of NLRP3 inflammasome: Relevance for a precision approach to atherosclerotic cardiovascular risk and disease. *Int. J. Mol. Sci.* **2020**, *21*, 4459. [CrossRef]

51. Upston, J.M.; Niu, X.; Brown, A.J.; Mashima, R.; Wang, H.; Senthilmohan, R.; Kettle, A.J.; Dean, R.T.; Stocker, R. Disease stage-dependent accumulation of lipid and protein oxidation products in human atherosclerosis. *Am. J. Pathol.* **2002**, *160*, 701–710. [CrossRef]

52. Crisby, M.; Nilsson, J.; Kostulas, V.; Björkhem, I.; Diczfalusy, U. Localization of sterol 27-hydroxylase immuno-reactivity in human atherosclerotic plaques. *Biochim. Biophys. Acta Lipids Lipid Metab.* **1997**, *1344*, 278–285. [CrossRef]

53. Hughes, T.M.; Rosano, C.; Evans, R.W.; Kuller, L.H. Brain cholesterol metabolism, oxysterols, and dementia. *J. Alzheimer Dis.* **2013**, *33*, 891–911. [CrossRef] [PubMed]

54. Wu, Q.; Ishikawa, T.; Sirianni, R.; Tang, H.; McDonald, J.G.; Yuhanna, I.S.; Thompson, B.; Girard, L.; Mineo, C.; Brekken, R.A.; et al. 27-Hydroxycholesterol promotes cell-autonomous, ER-positive breast cancer growth. *Cell Rep.* **2013**, *5*, 637–645. [CrossRef]

55. Nelson, E.R.; Wardell, S.E.; Jasper, J.S.; Park, S.; Suchindran, S.; Howe, M.K.; Carver, N.J.; Pillai, R.V.; Sullivan, P.M.; Sondhi, V.; et al. 27-Hydroxycholesterol links hypercholesterolemia and breast cancer pathophysiology. *Science* **2013**, *342*, 1094–1098. [CrossRef] [PubMed]

56. DuSell, C.D.; Nelson, E.R.; Wang, X.; Abdo, J.; Mödder, U.I.; Umetani, M.; Gesty-Palmer, D.; Javitt, N.B.; Khosla, S.; McDonnell, D.P. The endogenous selective estrogen receptor modulator 27-hydroxycholesterol is a negative regulator of bone homeostasis. *Endocrinology* **2010**, *151*, 3675–3685. [CrossRef]

Review

Phytosterols, Cholesterol Control, and Cardiovascular Disease

Andrea Poli [1,*], Franca Marangoni [1], Alberto Corsini [2,3], Enzo Manzato [4], Walter Marrocco [5], Daniela Martini [6], Gerardo Medea [7] and Francesco Visioli [8,9]

1   Nutrition Foundation of Italy, 20124 Milan, Italy; marangoni@nutrition-foundation.it
2   Department of Pharmaceutical and Pharmacological Sciences, University of Milan, 20133 Milan, Italy; alberto.corsini@unimi.it
3   IRCCS MultiMedica, 20099 Sesto San Giovanni, Italy
4   Department of Medicine (DIMED), University of Padova, 35128 Padova, Italy; enzo.manzato@unipd.it
5   FIMMG—Italian Federation of General Medicine Doctors and SIMPeSV–Italian Society of Preventive and Lifestyle Medicine, 00144 Rome, Italy; wmarrocco54@gmail.com
6   Department of Food Environmental and Nutritional Sciences (DeFENS), University of Milan, 20133 Milan, Italy; daniela.martini@unimi.it
7   SIMG—Italian Society of General Medicine, 50142 Firenze, Italy; medea.gerardo@alice.it
8   Department of Molecular Medicine, University of Padova, 35121 Padova, Italy; francesco.visioli@unipd.it
9   IMDEA-Food, CEI UAM+CSIC, 28049 Madrid, Spain
*   Correspondence: poli@nutrition-foundation.it; Tel.: +39-02-7600-6271

Abstract: The use of phytosterols (or plant sterols) for the control of plasma cholesterol concentrations has recently gained traction because their efficacy is acknowledged by scientific authorities and leading guidelines. Phytosterols, marketed as supplements or functional foods, are formally classified as food in the European Union, are freely available for purchase, and are frequently used without any health professional advice; therefore, they are often self-prescribed, either inappropriately or in situations in which no significant advantage can be obtained. For this reason, a panel of experts with diverse medical and scientific backgrounds was convened by NFI—Nutrition Foundation of Italy—to critically evaluate and summarize the literature available on the topic, with the goal of providing medical doctors and all health professionals useful information to actively govern the use of phytosterols in the context of plasma cholesterol control. Some practical indications to help professionals identify subjects who will most likely benefit from the use of these products, optimizing the therapeutic outcomes, are also provided. The panel concluded that the use of phytosterols as supplements or functional foods to control Low Density Lipoprotein (LDL) cholesterol levels should be preceded by the assessment of some relevant individual characteristics: cardiovascular risk, lipid profile, correct understanding of how to use these products, and willingness to pay for the treatment.

Keywords: phytosterols; plant sterols; cholesterol; cardiovascular disease; supplements; functional foods

Citation: Poli, A.; Marangoni, F.; Corsini, A.; Manzato, E.; Marrocco, W.; Martini, D.; Medea, G.; Visioli, F. Phytosterols, Cholesterol Control, and Cardiovascular Disease. *Nutrients* **2021**, *13*, 2810. https://doi.org/10.3390/nu13082810

Academic Editor: Lindsay Brown

Received: 19 July 2021
Accepted: 13 August 2021
Published: 16 August 2021

Publisher's Note: MDPI stays neutral with regard to jurisdictional claims in published maps and institutional affiliations.

1. Introduction

The use of supplements or functional foods to keep plasma cholesterol concentrations under control is growing steadily in European countries [1–3] Among these products, phytosterols (or plant sterols) have recently gained traction because their cholesterol-lowering efficacy, within the frame of a healthy lifestyle, is acknowledged by authoritative guidelines [4] and, among others, by the European Food Safety Authority (EFSA) [5,6].

In the European Union, such products, formally classified as "food", can be freely purchased by the public under self-prescription; it is reasonable to believe that, if used after a professional prescription and under medical control, the appropriateness of their use and, consequently, their efficacy in improving plasma cholesterol concentrations and cardiovascular risk would significantly improve.

For this reason, NFI—Nutrition Foundation of Italy—has convened a group of experts with diverse medical and scientific backgrounds to critically evaluate and summarize the literature available on the topic, and to provide some practical indications to help health professionals identify persons who will most likely benefit from the use of phytosterols. The main goal of this effort is to entrust doctors who perform clinical activities and all health professionals with a proper use of these products, to improve cardiovascular prevention in the population.

2. Phytosterols' Chemistry

Phytosterols are fat-soluble compounds belonging to the triterpene's family, present in most plant cells where they contribute to membranes structure and stability. They are characterized by a tetracyclic structure, with a side chain in position 17 of the D ring [7]. Their structure is very similar to that of cholesterol, which is by far the most abundant sterol in animal cells, where it plays a similar structural role. Phytosterols differ from cholesterol in the side chain bound in their C-17 position; sitosterol, as an example, has an ethyl group linked in C-24 of the side chain, while campesterol has a methyl group in the same position, which is empty in cholesterol. Phytostanols are 5alpha-saturated derivatives of phytosterols [8]. Several hundred different phytosterol molecules have been identified in plant cells; the most common ones are beta-sitosterol, campesterol, stigmasterol, brassicasterol, and avenasterol [9,10].

The food content in phytosterols is highest in oily fruit, oil seeds, and in the oils obtained from them [11,12]. In particular, rapeseed oil, wheat germ oil, and corn oil are the oils richest in phytosterols, whereas among the various types of oily fruit the highest content is found in pistachios [13]. Phytosterols are also present in legumes and cereals, whilst fruit and vegetables contain much lower quantities. In general, the concentration of total phytosterols in vegetables varies from a few milligrams or tens of milligrams per 100 g of fruit and vegetables up to over 1000 mg per 100 g in some vegetable oils, with large differences among different foods [14].

In European countries, the overall dietary intake of phytosterols is around 250–400 mg/day, with a high variability [15]: a value quite similar to dietary cholesterol intake. The dietary intake may vary according to the prevalent dietary pattern; the highest content has been found in vegan diets (up to 500 mg/day). The most abundant dietary phytosterol is sitosterol (about 60–70% of total phytosterols in the diet), followed by campesterol (16%) and stigmasterol (10%) while sitostanol, campestanol, and Δ5-avenasterol collectively contribute <10% [16].

3. Human Metabolism and Metabolic Effects of Phytosterols

Due to their lipophilicity, phytosterols ingested with foods or supplements or enriched/functional foods, are absorbed by the human intestine after incorporation into the so-called "mixed micelles". These micelles derive from the emulsification of dietary fats by bile salts, and allow the entry of phytosterols into the enterocytes through a well characterized membrane transport protein called Niemann-Pick C1—Like 1 (NPC1L1). Most of absorbed phytosterols are immediately re-excreted in the intestinal lumen by efflux transporters of the ATP-binding cassettes (ABC) family, known as ABCG5 and ABCG8 [17]. These metabolic pathways are summarized in Figure 1 [18,19]. Limited amounts of phytosterols, instead esterified within enterocytes, incorporated in chylomicrons, and eventually captured by the liver, are to a large extent secreted into the bile through the ABCG5/G8 transporters present in the biliary pole of hepatocytes.

Plasma concentrations of phytosterols are, consequently, lower (usually by two orders of magnitude) than those of cholesterol, essentially due to the limited intestinal absorption (less than 5% of plant sterols and less than 0.5% of stanols are absorbed and enter the systemic circulation [11], versus about 50–60% of dietary cholesterol [20]) and to the rapid hepatic clearance through the bile.

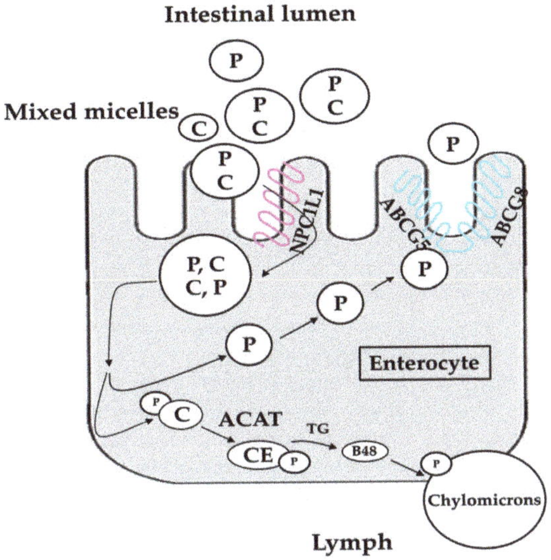

Figure 1. Main metabolic pathways of cholesterol and phytosterol in enterocytes. P, phytosterols.; C, cholesterol; CE, cholesteryl esters; ACAT, acylCoA cholesterol acyltransferase; NPC1L1, Niemann-Pick C1-Like 1; ABCG5, ATP-binding cassette G5; ABCG8, ATP-binding cassette G8; TG, triglycerides; B48, Apolipoprotein B-48. Modified from [18] and [19].

As cholesterol (deriving from ingested food or from the bile) is absorbed from the gut through the same pathway used by phytosterols, phytosterols compete with it for incorporation in the mixed-micelles and subsequent absorption in enterocytes, through the NPC1L1 transporter. Hence, cholesterol fractional absorption declines along with increasing amount of phytosterols present in the gut. The inhibition of cholesterol absorption by phytosterols ranges from about 5% for daily intakes of 300–400 mg (typical of most diets) up to 35–40% for intakes between 1500 and 2000 mg per day, which can only be achieved using enriched functional foods or specific supplements. In addition, phytosterols can also limit the absorption of cholesterol by directly co-crystallizing with cholesterol itself in the intestinal lumen and facilitating its elimination via the fecal route.

4. Effects of Phytosterols on Low Density Lipoprotein (LDL) Cholesterol: Characteristics and Clinical Relevance

The reduction in the amount of cholesterol absorbed from the intestine and reaching the liver through the chylomicron pathway triggers both a greater endogenous synthesis of cholesterol and a greater uptake of plasma LDL by hepatocytes, to maintain cholesterol homeostasis. The greater clearance of circulating LDL cholesterol yields the desired reduction in its plasma concentration. Such reduction is around 2–3% for the aforementioned dietary intakes of phytosterols (300–400 mg/day) [21] and reaches an average of 9% for supplementary dosages between 1500 and 2000 mg per day [22]. The effect can be as strong as a 12–12.5% reduction for dosages up to 3 g/day and tends to plateau for higher intakes [5].

Recent studies have confirmed such effects of phytosterols on plasma total and LDL cholesterol levels. Meta-analyses of published trials [23,24] indicate that the effect of the intake of phytosterols on plasma LDL cholesterol levels in humans falls within the range indicated by EFSA (Commission Regulation (EU) No 384/2010).

The efficacy of plant sterols on LDL cholesterol is independent of the initial LDL cholesterol concentrations; therefore, it can be useful in subjects with both low and high baseline LDL cholesterol levels [25]. Even in the presence of heterozygous familial hypercholesterolemia, the use of phytosterols can help reduce plasma LDL levels, as observed in, e.g., children [26].

High Density Lipoprotein (HDL) cholesterol is generally not significantly affected by phytosterols; triglycerides plasma levels are reduced to a minor extent, but the effect is larger when their levels exceed 150 mg/dL [27].

Because, as described above, the lipid-lowering effects of phytosterols are due to a competitive inhibition, such effects rapidly taper off upon discontinuation of intake and disappear after 7–10 days from the last dose of supplements/functional foods [6].

According to some authors [8], stanols are slightly more effective than the corresponding sterols, but a meta-analysis on the subject did not identify significant differences in the effect on LDL cholesterol levels between the two groups of molecules [23]. Furthermore, the effects do not appear to be influenced by the chemical form (free or esterified) in which sterols and stanols are ingested [28].

Experimental studies suggested that, perhaps due to the reduction of LDL cholesterol, phytosterols may perform a modest anti-inflammatory action. Nevertheless, according to a meta-analysis, regular intake of food enriched with phytosterols did not significantly impact levels of biomarkers of low-grade inflammation in obese subjects [29]. An anti-inflammatory effect, on the other hand, might also derive, at least in part, from an interaction between phytosterols and microbiota, improving the state of dysbiosis associated low-grade inflammation [30]. There is also some in vitro and in vivo experimental evidence suggestive of modulatory roles of phytosterols, namely through a reduction of selected bacterial species [10,31].

Endothelial function, evaluated as flow mediated dilation, would also improve after a treatment with phytosterols, but this is also controversial. Mechanistically, this effect could explain the mild reduction in blood pressure found in a recent meta-analysis [32].

Unfortunately, probably because of the large number of subjects needed and the challenge of controlling diets for a very long period, no data deriving from formal randomized clinical trials are available allowing to translate this well described effect of phytosterols on plasma LDL cholesterol levels into measurable direct clinical effects on cardiovascular morbidity and mortality [33].

Although such absence of clinical trials showing that phytosterols intake can reduce the incidence of clinical endpoints, such as myocardial infarction or coronary deaths, needs to be acknowledged, it is also necessary to remember that the accrued evidence clearly shows that lowering cholesterol concentrations by any means, e.g., via diet, ileal by-pass, or drugs with different mechanisms of action, is always accompanied by a proportional reduction in cardiovascular risk. Hence, both European Atherosclerosis Society (EAS) and EFSA [4,5] state that the plasma LDL cholesterol-reducing effects of phytosterols will proportionally reduce cardiovascular risk and related coronary events.

Interestingly, the selective effect of phytosterols on cholesterol absorption may have some positive preventive consequences.

It is well known that the balance between the rate of intestinal absorption or of hepatic synthesis of cholesterol in driving plasma LDL cholesterol levels may be different according to individual characteristics: in some individuals (often called "absorbers") a prevailing absorbing pattern from the gut can be observed, while in other subjects (often called "synthesizers"), hepatic synthesis is largely prevailing.

Subjects with genetic variants of NPC1L1 that limit cholesterol absorption have a much lower cardiovascular risk than their genetic counterparts with normal NPC1L1 activity, even if their cholesterol concentrations are only slightly lower [34]. Observational studies [35], and a meta-analysis [36], indeed, indicate that cardiovascular risk is higher in absorbers than in synthesizers, even if their plasma LDL cholesterol levels are comparable. Patients with chronic renal failure are usually absorbers and their higher cardiovascular risk and mortality seem to be related at least in part to their absorptive pattern [37,38].

This higher cardiovascular morbidity observed in absorbers as compared with synthesizers might be explained by the observation that the NPC1L1-mediated intestinal cholesterol absorption is poorly selective and takes up also molecules structurally similar to cholesterol, but potentially more atherogenic, such as oxysterols. Oxysterols are strongly atherogenic in experimental models and the blockage of NPC1L1 prevents the vascular damage exerted by these molecules [39].

These data, altogether, suggest that if an equal degree of LDL lowering is achieved, the effect obtained through inhibition of cholesterol absorption might be more advantageous than that obtained inhibiting cholesterol synthesis; the clinical benefits of phytosterols on the cardiovascular risk, consequently, might be larger than that solely expected by their impact on LDL cholesterol levels.

Preliminary data also suggest a possible preventive role of phytosterols in relation to the risk of some cancers and obesity, as well as a possible immunomodulatory role [4,40]. These associations, on the other hand, are more difficult to interpret from a mechanistic viewpoint, and could consequently be non-causal; they require further investigation.

5. Variables Affecting the Cholesterol-Lowering Effect of Phytosterols

A significant variability can be observed in the plasma cholesterol response of different individuals to phytosterols treatment.

The cholesterol-lowering effect of phytosterols supplementation in subjects with a typically synthetic pattern (for example, obese subjects, especially if insulin-resistant or frankly diabetic) will be smaller than that of persons with a profile more shifted towards the absorbing type (normal weight subjects with normal insulin sensitivity) [41].

The ApoE isoforms profile also appears to influence the efficacy of phytosterols, which may be higher in subjects with the E4 variant (at increased cardiovascular and cognitive decline risk) and lower in subjects with the more common E3/E3 isoform [42].

The possible effects of polymorphisms of other genes potentially influencing the plasma lipid response to the use of these products are also currently being evaluated.

On the other hand, age and sex do not appear to significantly affect the cholesterol-lowering response to phytosterols (which is perhaps slightly larger in males) [43].

A large number of studies has also considered the possible effect of variables that could affect the effectiveness of phytosterols supplementation on plasma LDL cholesterol concentrations [25,44,45]. In particular, the effect of the type of matrix (dairy products vs. other items, high fat vs. low fat foods, solid vs. liquid products), or of the type of administration (supplements vs. foods containing phytosterols), or of the method of supplementation (single dose vs. multiple doses), and the specific molecules used (sterols vs. stanols) have been considered [43].

In general, the data show greater efficacy when phytosterols are presented in solid rather than liquids foods [46]. However, this difference mainly develops at high dosages, whereas at the commonly used ones it appears to be negligible. A possible explanation of the lower efficacy of liquid foods can be linked to the faster gastric emptying, which results in a swifter transit time in the gastrointestinal tract where phytosterols play their cholesterol-lowering role [23].

In spite of the differences between solid and liquid matrices, the consumption of phytosterols incorporated in different foods does not appear to significantly influence their effects on plasma lipids. For example, the intake of phytosterols in milk-based or cereal-flour-based matrices has comparable effects on LDL cholesterol levels. Similarly, the cholesterol-lowering effects did not differ when comparing products rich in fat and non-fat foods [25,28].

Functional foods enriched with phytosterols and supplements based on phytosterols in capsules or tablets appear to have a similar effect on plasma LDL cholesterol concentrations [45]. Likewise, daily consumption in a single dose seems to be equally effective as the same quantity divided into three doses with meals. On the other hand, taking phytosterols at the end of one of the main meals, as compared with during fasting, amplifies the effect on plasma LDL cholesterol levels; consumption in a single dose at breakfast (especially after a small breakfast) is associated with a lower (less 30%) effect [47]. The explanation for this difference probably lies in the greater presence, after a meal, of cholesterol of food or biliary origin in the intestine, with which phytosterols can compete limiting its absorption [48]. Interestingly, since the cholesterol present in the intestine is largely, i.e., at least 75% of biliary and not food origin, supplements or foods enriched in phytosterols are also effective in vegetarians and vegans, who introduce low or negligible amounts of cholesterol with the diet.

Finally, no clear differences between supplementations with sterols or stanols, or comparing phytosterols in free or in esterified form have been described. According to a recent study, however, the extent of the achievable reduction depends on the specific mixture of sterols used and would increase (by a few percentage points) if at least 80% of the phytosterols used are composed of beta-sitosterol or the corresponding stanol [24].

6. Regulatory Framework

Some components of the regulatory framework of foods and food ingredients (as well as supplements and functional foods) aimed at controlling LDL cholesterol levels are relevant for their proper use and for a correct understanding of the interactions between manufacturers, medical doctors, and consumers.

In the European Union, phytosterols are classified among foods and are subject to the comprehensive food legislation. Communication of beneficial effects in relation to health and nutrition, on the product label or in the advertising of foods (as well as for food supplements), is defined by Regulation (EC) No 1924/2006 and is limited to "function health claims", pursuant to article 13(5) (for example: "product x contributes to the maintenance of normal blood cholesterol levels") and to "risk reduction claims" pursuant to art. 14(1)(a) of the same Regulation, on reducing a risk factor in the development of a disease (for example: "product y has been shown to lower/reduce blood cholesterol. High cholesterol is a risk factor in the development of coronary heart disease").

It is interesting to underline that only phytosterols (at a dose of 1.5–3 g/day) and beta-glucans (at a dose of 3 g/day) can benefit from a European Commission-authorized claim pursuant to art. 14(1)(a), which allows us, in the communication to the public, to refer to a cholesterol-lowering effect. Furthermore, for phytosterols, the precise magnitude of the effect to be expected has been defined (specifically from 7 to 10% if foods ensure a daily intake of 1.5–2.4 g/day and from 10 to 12.5% for 2.5–3 g/day), and the duration to obtain the effect, "in 2 to 3 weeks", must be specified.

This information can be used in promoting or advertising phytosterol enriched foods and supplements; notably, as already mentioned, it is reasonable to assume that a professional support, where possible medical, to the use of these formulations, formally classified as "food" could significantly improve the appropriateness of their use [49,50].

7. How to Identify Optimal Candidates for the Use of Phytosterols to Reduce LDL Cholesterol Levels

The recent Guidelines of the European Societies of Cardiology and Atherosclerosis (ESC/EAS) on treating cholesterol levels to lower cardiovascular risk are based on principles that can be considered the foundations of a correct approach to hypercholesterolemic patients [51]. These principles can be summarized as follows: (1) The correlation between increasing plasma LDL cholesterol and increasing cardiovascular risk is continuous; (2) there are no plasma levels below which a reduction in LDL cholesterol becomes ineffective, not being accompanied by a reduction in the risk of cardiovascular clinical events; (3) the decision to treat a patient, on the other hand, needs to be based on an estimate, as accurate and complete as possible, of his/her global cardiovascular risk, i.e., the probability to incur in a fatal or not fatal cardiovascular events over the following years; (4) in primary prevention, the future risk of cardiovascular events can be estimated using the SCORE algorithm [52]. This estimate can then be integrated by information regarding the specific characteristics of the patient, such as personal and family history, presence of other "classic" and "non-classic" risk indicators—socioeconomic status, level of individual stress, exposure to air pollution, sleep quality, etc.; and (5) as the estimated risk increases, therapeutic intervention on plasma cholesterol levels must progressively become more and more aggressive to reach lower LDL cholesterol values (target).

The aforementioned ESC/EAS Guidelines [51] recognize phytosterols with a significant and dose-dependent capacity to reduce LDL cholesterol (level of evidence A) without relevant effects on plasma HDL cholesterol and triglycerides levels. The Guidelines, also in light of the absence of significant side effects associated with their use [53], suggest considering phytosterols at doses up to 2 g/day after the main meal, leading to an average reduction of LDL cholesterol ranging from 7 to 10% in: (a) people with high cholesterol, low or intermediate overall cardiovascular risk, with no indication for drug treatment; or (b) patients at high or very high risk who do not reach their therapeutic goal in terms of LDL cholesterol despite treatment with statins (or who do not tolerate statins) to whom phytosterols can be administered in addition to drug therapy; or (c) adults and children (over 6 years of age) with familial hypercholesterolemia, within the Guidelines' framework.

As anticipated, although no data are available on the direct clinical effects of phytosterols on cardiovascular morbidity and mortality, clear evidence shows that lowering cholesterol concentrations by any mechanisms is always accompanied by a proportional reduction in cardiovascular risk [51].

Furthermore, as described in the EAS Guidelines [51], long-term monitoring studies indicate that phytosterols have a favorable safety profile which justifies their use as cholesterol-lowering agents both alone and in combination with drug therapy.

In this context, it is opinion of the expert group signing this document that the use of phytosterols as supplements or functional foods can be considered mainly in two different cases:

(1) People under the age of 40 years: in these subjects, estimating cardiovascular risk using the SCORE algorithm is formally not possible. Once patients with genetic hypercholesterolemia or with a previous cardiovascular event, whose plasma cholesterol levels must be treated according to the appropriate Guidelines indications, have been excluded, the cardiovascular risk of these people can be considered low by definition. However, on the basis of a thorough clinical evaluation including an accurate estimate of individual risk characteristics, a physician may decide to intervene on the cardiovascular risk of individual subjects by lowering their cholesterol; in these population groups, the use of a drug should be considered as off-label and, consequently, the use of supplements or functional foods is a valid alternative. Because, in these cases, the therapeutic goal for LDL cholesterol is set at 115 mg/dL, the optimal clinical target of phytosterols, as monotherapy, is represented by people who, following a correct diet, have a basal LDL cholesterol equal to or less than 130 mg/dL (Figure 2, flow chart A).

In individuals with higher plasma basal LDL cholesterol levels, the clinician may consider suggesting a combination of phytosterols with other food supplements indicated for cholesterol control.

(2) People over 40 years of age: in this age range, the prescription of the use of functional foods or supplements based on phytosterols should be considered for people with low or moderate risk, i.e., below 1% or in the 1–5% range at 10 years, respectively. For persons with a risk below 1% at 10 years, the target value for LDL cholesterol is set at 115 mg/dL. They can be also be treated as indicated in Figure 2, flow chart A.

For persons at moderate risk (1–5% at 10 years), the target value for LDL cholesterol is set at 100 mg/dL; phytosterols can be used in these persons as monotherapy when basal cholesterol levels are <110 mg/dL (Figure 3, flow chart B). Again, in individuals with higher plasma basal LDL cholesterol levels the clinician may consider to suggest a combination of phytosterols with other food supplements indicated for the cholesterol control.

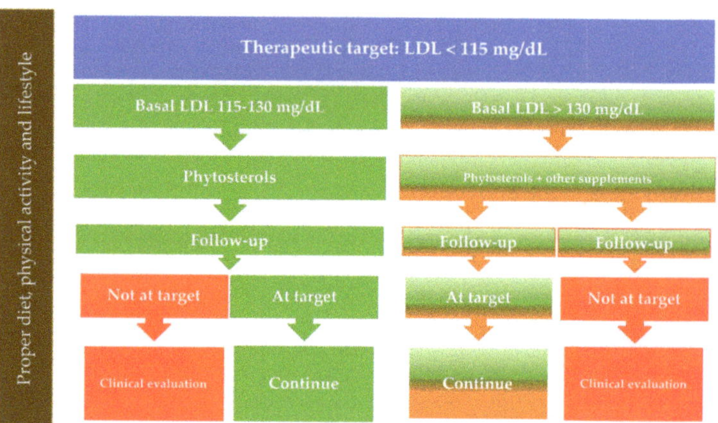

Figure 2. Flowchart A. Patient <40 years old or >40 years old, but with global cardiovascular risk <1%. This flow-chart is not appropriate for patients with genetic disorders of lipoprotein metabolism or with manifest cardiovascular disease, who need to be treated according to the Guidelines.

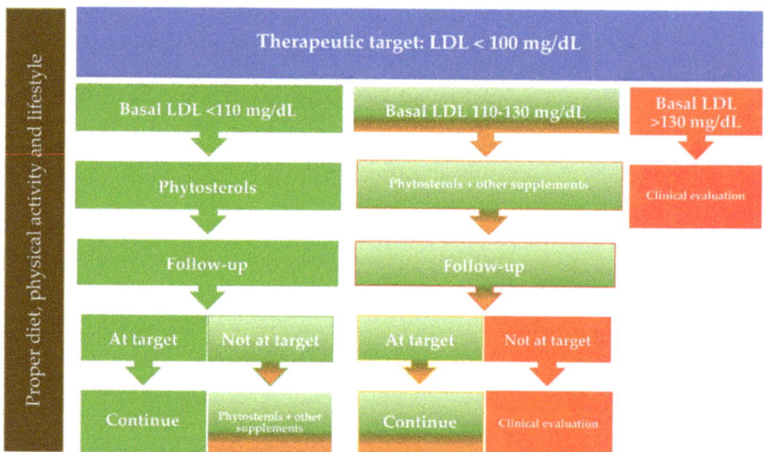

Figure 3. Flowchart B. Patient ≥40 years old with global cardiovascular risk 1–5%. This flow-chart is not appropriate for patients with genetic disorders of lipoprotein metabolism or with manifest cardiovascular disease, who need to be treated according to the Guidelines.

The expected LDL reductions discussed in previous paragraphs, on the other hand, reflect the average efficacy of phytosterols. Individual responses may also be significantly different and the reduction of LDL cholesterol concentrations that can be achieved can be larger, e.g., among absorbers.

The use of supplements or functional foods enriched in phytosterols in subjects with higher cardiovascular risk, in association with other drugs with a complementary mechanism of action, e.g., statins, as outlined by the ESC/EAS Guidelines mentioned above, can be considered after a careful personalized evaluation.

8. Side Effects of Phytosterols Use

The use of supplements or foods enriched in phytosterols, within the limit of 1.5–3.0 g per day, is not associated with relevant side effects [4,18].

Intestinal absorption of some carotenoids is moderately reduced by phytosterol intake, bringing their plasma levels to the low end of the oscillation range physiologically observed throughout the year (maximum in spring and summer and minimum in late winter). This reduction can be easily compensated by adopting a diet rich in these compounds, i.e., rich in colorful fruits and vegetables.

Some authors have proposed that an increase in plasma levels of phytosterols may represent a risk factor for cardiovascular events [54]; however, it is likely that, actually, their increase in circulating concentrations is rather an indicator of a high efficiency of the cholesterol absorption pathway, potentially atherogenic as previously discussed, and not a direct causal factor of atherosclerotic risk. In fact, no accumulation of phytosterols is observed in the tissues of subjects who take the recommended dosages of these compounds.

Post-marketing surveillance studies did not report any significant untoward effect of phytosterol use [55,56]. Of note, in patients with homozygous sitosterolaemia (in which the ABCG5 and/or ABCG8 transporters are not functional), the dietary intake of phytosterols greatly increases cardiovascular risk: the prevalence of this condition is, however, extremely low (about 1:10,000,000 of subjects) [57].

9. Use of Phytosterols in Addition to Other Supplements and Drugs

Other supplements and functional foods (or nutraceuticals), with different characteristics and mechanisms of action, are used worldwide for plasma LDL cholesterol control. Knowing the mechanisms underlying the effect of the aforementioned active ingredients on LDL cholesterol allows for their rational combinations, with the ultimate aim of optimizing their efficacy and safety.

For example, monacolin K, contained in fermented red rice, is chemically identical to lovastatin and inhibits the hepatic synthesis of cholesterol (the European Commission authorizes the claim "contributes to the maintenance of normal blood cholesterol levels" for products that provide at least 10 mg of monacolin K per day) [58]. The action of berberine is more articulated, and may include partial inhibition of PCSK9.

The mechanism of action of beta-glucans is quite similar to that of phytosterols. In this case, the European Commission has authorized the claim of reduction/maintenance of blood cholesterol levels for a daily intake of 3 g of beta glucans from oats, oat bran, barley, barley bran, or from mixtures of these sources [59].

A combination of phytosterols (inhibitors of cholesterol absorption) and statins (inhibitors of cholesterol synthesis) can be useful in subjects with more markedly altered lipid patterns (Figure 4) [60]. Phytosterols can, in fact, neutralize the compensatory increase in intestinal cholesterol absorption induced by statin. This combination therapy has been proposed based on the observation that the effect of phytosterols is additive to that of a diet low in saturated fat and statins [22,61]. Its efficacy is shown by a meta-analysis of 15 randomized clinical trials, including more than 500 subjects, which provides evidence that phytosterol enriched diets additionally lower total cholesterol and LDL cholesterol levels beyond that achieved by statins alone [60].

Theoretically, such effect might contribute to reduce the residual risk observed in patients treated with statins, which might be in part explained by the increased absorption pattern observed in these patients, but such interpretation must be considered merely speculative.

An additive effect on plasma LDL cholesterol levels can also by hypothesized for phytosterols and berberine [62]; the association of phytosterols with fiber and beta-glucans should be considered, on the opposite, less rationale [63]. Probably for the same reason, the evidence of add-on effects of phytosterols and ezetimibe is limited [64].

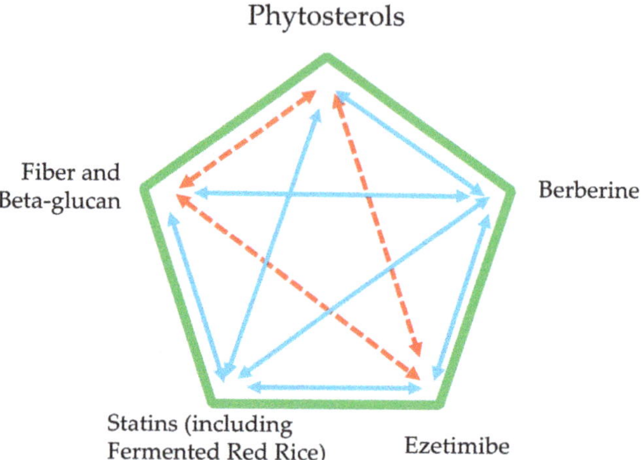

Figure 4. Combination of phytosterols with other food supplements or drugs active on plasma cholesterol levels. Blue (continuous) arrows: appropriate combination. Red (dotted) arrows: less appropriate combinations.

10. Conclusions and Practical Suggestions

Functional foods or supplements containing phytosterols are effective in controlling plasma LDL cholesterol levels if used appropriately.

These products must be taken on a daily basis; their cholesterol-lowering effect is rapid (noticeable after about three weeks); however, it is maintained only if the intake of phytosterols is routine. As mentioned above, the correct use of phytosterols induces an average reduction in plasma LDL cholesterol of 9–10%, which can reach up to 12.5% with higher dosages [65]; this reduction is added to that obtainable through appropriate diet and may be greater in some people, i.e., the absorbers.

Functional foods or supplements containing phytosterols should be taken in a single daily dose at the end of one of the main meals (lunch or dinner). Their administration on an empty stomach or after a small breakfast is not recommended [47].

The use of phytosterols is usually well tolerated, with no significant side effects. During the treatment, it is advisable to increase the intake of colored fruit and vegetables, especially the yellow, orange, and red ones.

The decision to propose the use of phytosterols as supplements or functional foods to control LDL cholesterol levels should be made by a physician or a qualified health professional after an assessment of individual cardiovascular risk, lipid profile, correct understanding of how taking these products, and willingness to pay for the treatment.

Author Contributions: Conceptualization, A.P. and F.M.; writing—original draft preparation, review and editing, A.P., F.M., A.C., W.M., D.M., E.M., G.M., F.V. All authors have read and agreed to the published version of the manuscript.

Funding: The preparation of this paper has been made possible by an unrestricted grant from Danone S.p.A. The sponsor had no role in the preparation and finalization of the manuscript, and in the decision to publish it.

Conflicts of Interest: A.P. and F.M. are the Chairman and Head of Research, respectively, of NFI—Nutrition Foundation of Italy, a non-profit organization partially supported by Italian and non-Italian Food Companies. All other authors declare no conflict of interest associated with this publication.

References

1. Baumgartner, S.; Bruckert, E.; Gallo, A.; Plat, J. The position of functional foods and supplements with a serum LDL-C lowering effect in the spectrum ranging from universal to care-related CVD risk management. *Atherosclerosis* **2020**, *311*, 116–123. [CrossRef]

2.   Poli, A.; Barbagallo, C.M.; Cicero, A.F.G.; Corsini, A.; Manzato, E.; Trimarco, B.; Bernini, F.; Visioli, F.; Bianchi, A.; Canzone, G.; et al. Nutraceuticals and functional foods for the control of plasma cholesterol levels. An intersociety position paper. *Pharm. Res.* **2018**, *134*, 51–60. [CrossRef] [PubMed]

3.   Cicero, A.F.G.; Fogacci, F.; Stoian, A.P.; Vrablik, M.; Al Rasadi, K.; Banach, M.; Toth, P.P.; Rizzo, M. Nutraceuticals in the Management of Dyslipidemia: Which, When, and for Whom? Could Nutraceuticals Help Low-Risk Individuals with Non-optimal Lipid Levels? *Curr. Atheroscler. Rep.* **2021**, *23*, 57. [CrossRef] [PubMed]

4.   Gylling, H.; Plat, J.; Turley, S.; Ginsberg, H.N.; Ellegard, L.; Jessup, W.; Jones, P.J.; Lutjohann, D.; Maerz, W.; Masana, L.; et al. Plant sterols and plant stanols in the management of dyslipidaemia and prevention of cardiovascular disease. *Atherosclerosis* **2014**, *232*, 346–360. [CrossRef] [PubMed]

5.   EFSA Panel on Dietetic Products Nutrition and Allergies. Scientific Opinion on the substantiation of a health claim related to 3 g/day plant sterols/stanols and lowering blood LDL-cholesterol and reduced risk of (coronary) heart disease pursuant to Article 19 of Regulation (EC) No 1924/2006. *EFSA J.* **2012**, *10*, 2693. [CrossRef]

6.   EFSA Panel on Dietetic Products Nutrition and Allergies. European Commission and a similar request from France in relation to the authorisation procedure for health claims on plant sterols/stanols and lowering/reducing blood LDL-cholesterol pursuant to Article 14 of Regulation (EC) No 1924/2006. *EFSA J.* **2009**, *1175*, 1–9.

7.   Moreau, R.A.; Nystrom, L.; Whitaker, B.D.; Winkler-Moser, J.K.; Baer, D.J.; Gebauer, S.K.; Hicks, K.B. Phytosterols and their derivatives: Structural diversity, distribution, metabolism, analysis, and health-promoting uses. *Prog. Lipid Res.* **2018**, *70*, 35–61. [CrossRef]

8.   Gylling, H.; Simonen, P. Phytosterols, Phytostanols, and Lipoprotein Metabolism. *Nutrients* **2015**, *7*, 7965–7977. [CrossRef]

9.   Feng, S.; Belwal, T.; Li, L.; Limwachiranon, J.; Liu, X.; Luo, Z. Phytosterols and their derivatives: Potential health-promoting uses against lipid metabolism and associated diseases, mechanism, and safety issues. *Compr. Rev. Food Sci. Food Saf.* **2020**, *19*, 1243–1267. [CrossRef] [PubMed]

10.   Vezza, T.; Canet, F.; de Maranon, A.M.; Banuls, C.; Rocha, M.; Victor, V.M. Phytosterols: Nutritional Health Players in the Management of Obesity and Its Related Disorders. *Antioxidants* **2020**, *9*, 1266. [CrossRef]

11.   Ostlund, R.E., Jr.; McGill, J.B.; Zeng, C.M.; Covey, D.F.; Stearns, J.; Stenson, W.F.; Spilburg, C.A. Gastrointestinal absorption and plasma kinetics of soy Delta(5)-phytosterols and phytostanols in humans. *Am. J. Physiol. Endocrinol. Metab.* **2002**, *282*, E911–E916. [CrossRef]

12.   Jimenez-Escrig, A.; Santos-Hidalgo, A.B.; Saura-Calixto, F. Common sources and estimated intake of plant sterols in the Spanish diet. *J. Agric. Food Chem.* **2006**, *54*, 3462–3471. [CrossRef]

13.   Esche, R.; Muller, L.; Engel, K.H. Online LC-GC-based analysis of minor lipids in various tree nuts and peanuts. *J. Agric. Food Chem.* **2013**, *61*, 11636–11644. [CrossRef] [PubMed]

14.   Wang, M.; Huang, W.; Hu, Y.; Zhang, L.; Shao, Y.; Wang, M.; Zhang, F.; Zhao, Z.; Mei, X.; Li, T.; et al. Phytosterol Profiles of Common Foods and Estimated Natural Intake of Different Structures and Forms in China. *J. Agric. Food Chem.* **2018**, *66*, 2669–2676. [CrossRef]

15.   Ras, R.T.; van der Schouw, Y.T.; Trautwein, E.A.; Sioen, I.; Dalmeijer, G.W.; Zock, P.L.; Beulens, J.W. Intake of phytosterols from natural sources and risk of cardiovascular disease in the European Prospective Investigation into Cancer and Nutrition-the Netherlands (EPIC-NL) population. *Eur. J. Prev. Cardiol.* **2015**, *22*, 1067–1075. [CrossRef] [PubMed]

16.   Racette, S.B.; Spearie, C.A.; Phillips, K.M.; Lin, X.; Ma, L.; Ostlund, R.E., Jr. Phytosterol-deficient and high-phytosterol diets developed for controlled feeding studies. *J. Am. Diet. Assoc.* **2009**, *109*, 2043–2051. [CrossRef] [PubMed]

17.   Berge, K.E.; Tian, H.; Graf, G.A.; Yu, L.; Grishin, N.V.; Schultz, J.; Kwiterovich, P.; Shan, B.; Barnes, R.; Hobbs, H.H. Accumulation of dietary cholesterol in sitosterolemia caused by mutations in adjacent ABC transporters. *Science* **2000**, *290*, 1771–1775. [CrossRef] [PubMed]

18.   Ostlund, R.E., Jr. Phytosterols and cholesterol metabolism. *Curr. Opin. Lipidol.* **2004**, *15*, 37–41. [CrossRef]

19.   Huff, M.W.; Pollex, R.L.; Hegele, R.A. NPC1L1: Evolution from pharmacological target to physiological sterol transporter. *Arter. Thromb. Vasc. Biol.* **2006**, *26*, 2433–2438. [CrossRef] [PubMed]

20.   Bosner, M.S.; Lange, L.G.; Stenson, W.F.; Ostlund, R.E., Jr. Percent cholesterol absorption in normal women and men quantified with dual stable isotopic tracers and negative ion mass spectrometry. *J. Lipid Res.* **1999**, *40*, 302–308. [CrossRef]

21.   Klingberg, S.; Ellegard, L.; Johansson, I.; Hallmans, G.; Weinehall, L.; Andersson, H.; Winkvist, A. Inverse relation between dietary intake of naturally occurring plant sterols and serum cholesterol in northern Sweden. *Am. J. Clin. Nutr.* **2008**, *87*, 993–1001. [CrossRef] [PubMed]

22.   Katan, M.B.; Grundy, S.M.; Jones, P.; Law, M.; Miettinen, T.; Paoletti, R.; Stresa Workshop, P. Efficacy and safety of plant stanols and sterols in the management of blood cholesterol levels. *Mayo Clin. Proc.* **2003**, *78*, 965–978. [CrossRef]

23.   Ras, R.T.; Geleijnse, J.M.; Trautwein, E.A. LDL-cholesterol-lowering effect of plant sterols and stanols across different dose ranges: A meta-analysis of randomised controlled studies. *Br. J. Nutr.* **2014**, *112*, 214–219. [CrossRef]

24.   Ying, J.; Zhang, Y.; Yu, K. Phytosterol compositions of enriched products influence their cholesterol-lowering efficacy: A meta-analysis of randomized controlled trials. *Eur. J. Clin. Nutr.* **2019**, *73*, 1579–1593. [CrossRef] [PubMed]

25.   Abumweis, S.S.; Barake, R.; Jones, P.J. Plant sterols/stanols as cholesterol lowering agents: A meta-analysis of randomized controlled trials. *Food Nutr. Res.* **2008**, *52*. [CrossRef]

26. Moruisi, K.G.; Oosthuizen, W.; Opperman, A.M. Phytosterols/stanols lower cholesterol concentrations in familial hypercholesterolemic subjects: A systematic review with meta-analysis. *J. Am. Coll. Nutr.* **2006**, *25*, 41–48. [CrossRef]

27. Demonty, I.; Ras, R.T.; van der Knaap, H.C.; Meijer, L.; Zock, P.L.; Geleijnse, J.M.; Trautwein, E.A. The effect of plant sterols on serum triglyceride concentrations is dependent on baseline concentrations: A pooled analysis of 12 randomised controlled trials. *Eur. J. Nutr.* **2013**, *52*, 153–160. [CrossRef]

28. Demonty, I.; Ras, R.T.; van der Knaap, H.C.; Duchateau, G.S.; Meijer, L.; Zock, P.L.; Geleijnse, J.M.; Trautwein, E.A. Continuous dose-response relationship of the LDL-cholesterol-lowering effect of phytosterol intake. *J. Nutr.* **2009**, *139*, 271–284. [CrossRef] [PubMed]

29. Rocha, V.Z.; Ras, R.T.; Gagliardi, A.C.; Mangili, L.C.; Trautwein, E.A.; Santos, R.D. Effects of phytosterols on markers of inflammation: A systematic review and meta-analysis. *Atherosclerosis* **2016**, *248*, 76–83. [CrossRef] [PubMed]

30. Scarmozzino, F.; Poli, A.; Visioli, F. Microbiota and cardiovascular disease risk: A scoping review. *Pharmacol. Res.* **2020**, *159*, 104952. [CrossRef]

31. Cuevas-Tena, M.; Bermudez, J.D.; Silvestre, R.L.A.; Alegria, A.; Lagarda, M.J. Impact of colonic fermentation on sterols after the intake of a plant sterol-enriched beverage: A randomized, double-blind crossover trial. *Clin. Nutr.* **2019**, *38*, 1549–1560. [CrossRef] [PubMed]

32. Gibney, E.R.; Milenkovic, D.; Combet, E.; Ruskovska, T.; Greyling, A.; Gonzalez-Sarrias, A.; de Roos, B.; Tomas-Barberan, F.; Morand, C.; Rodriguez-Mateos, A. Factors influencing the cardiometabolic response to (poly)phenols and phytosterols: A review of the COST Action POSITIVe activities. *Eur. J. Nutr.* **2019**, *58*, 37–47. [CrossRef]

33. Gylling, H.; Strandberg, T.E.; Kovanen, P.T.; Simonen, P. Lowering Low-Density Lipoprotein Cholesterol Concentration with Plant Stanol Esters to Reduce the Risk of Atherosclerotic Cardiovascular Disease Events at a Population Level: A Critical Discussion. *Nutrients* **2020**, *12*, 2346. [CrossRef]

34. Myocardial Infarction Genetics Consortium, I.; Stitziel, N.O.; Won, H.H.; Morrison, A.C.; Peloso, G.M.; Do, R.; Lange, L.A.; Fontanillas, P.; Gupta, N.; Duga, S.; et al. Inactivating mutations in NPC1L1 and protection from coronary heart disease. *N. Engl. J. Med.* **2014**, *371*, 2072–2082. [CrossRef] [PubMed]

35. Matthan, N.R.; Pencina, M.; LaRocque, J.M.; Jacques, P.F.; D'Agostino, R.B.; Schaefer, E.J.; Lichtenstein, A.H. Alterations in cholesterol absorption/synthesis markers characterize Framingham offspring study participants with CHD. *J. Lipid. Res.* **2009**, *50*, 1927–1935. [CrossRef]

36. Silbernagel, G.; Chapman, M.J.; Genser, B.; Kleber, M.E.; Fauler, G.; Scharnagl, H.; Grammer, T.B.; Boehm, B.O.; Makela, K.M.; Kahonen, M.; et al. High intestinal cholesterol absorption is associated with cardiovascular disease and risk alleles in ABCG8 and ABO: Evidence from the LURIC and YFS cohorts and from a meta-analysis. *J. Am. Coll. Cardiol.* **2013**, *62*, 291–299. [CrossRef]

37. Rogacev, K.S.; Pinsdorf, T.; Weingartner, O.; Gerhart, M.K.; Welzel, E.; van Bentum, K.; Popp, J.; Menzner, A.; Fliser, D.; Lutjohann, D.; et al. Cholesterol synthesis, cholesterol absorption, and mortality in hemodialysis patients. *Clin. J. Am. Soc. Nephrol.* **2012**, *7*, 943–948. [CrossRef]

38. Emrich, I.E.; Heine, G.H.; Schulze, P.C.; Rogacev, K.S.; Fliser, D.; Wagenpfeil, S.; Bohm, M.; Lutjohann, D.; Weingartner, O. Markers of cholesterol synthesis to cholesterol absorption across the spectrum of non-dialysis CKD: An observational study. *Pharm. Res. Perspect.* **2021**, *9*, e00801. [CrossRef]

39. Sato, K.; Nakano, K.; Katsuki, S.; Matoba, T.; Osada, K.; Sawamura, T.; Sunagawa, K.; Egashira, K. Dietary cholesterol oxidation products accelerate plaque destabilization and rupture associated with monocyte infiltration/activation via the MCP-1-CCR2 pathway in mouse brachiocephalic arteries: Therapeutic effects of ezetimibe. *J. Atheroscler. Thromb.* **2012**, *19*, 986–998. [CrossRef]

40. Huang, J.; Xu, M.; Fang, Y.J.; Lu, M.S.; Pan, Z.Z.; Huang, W.Q.; Chen, Y.M.; Zhang, C.X. Association between phytosterol intake and colorectal cancer risk: A case-control study. *Br. J. Nutr.* **2017**, *117*, 839–850. [CrossRef]

41. Rideout, T.C.; Harding, S.V.; Mackay, D.; Abumweis, S.S.; Jones, P.J. High basal fractional cholesterol synthesis is associated with nonresponse of plasma LDL cholesterol to plant sterol therapy. *Am. J. Clin. Nutr.* **2010**, *92*, 41–46. [CrossRef]

42. MacKay, D.S.; Eck, P.K.; Gebauer, S.K.; Baer, D.J.; Jones, P.J. CYP7A1-rs3808607 and APOE isoform associate with LDL cholesterol lowering after plant sterol consumption in a randomized clinical trial. *Am. J. Clin. Nutr.* **2015**, *102*, 951–957. [CrossRef]

43. Trautwein, E.A.; Vermeer, M.A.; Hiemstra, H.; Ras, R.T. LDL-Cholesterol Lowering of Plant Sterols and Stanols-Which Factors Influence Their Efficacy? *Nutrients* **2018**, *10*, 1262. [CrossRef]

44. Racette, S.B.; Lin, X.; Lefevre, M.; Spearie, C.A.; Most, M.M.; Ma, L.; Ostlund, R.E., Jr. Dose effects of dietary phytosterols on cholesterol metabolism: A controlled feeding study. *Am. J. Clin. Nutr.* **2010**, *91*, 32–38. [CrossRef]

45. Amir Shaghaghi, M.; Abumweis, S.S.; Jones, P.J. Cholesterol-lowering efficacy of plant sterols/stanols provided in capsule and tablet formats: Results of a systematic review and meta-analysis. *J. Acad. Nutr. Diet.* **2013**, *113*, 1494–1503. [CrossRef]

46. Cusack, L.K.; Fernandez, M.L.; Volek, J.S. The food matrix and sterol characteristics affect the plasma cholesterol lowering of phytosterol/phytostanol. *Adv. Nutr.* **2013**, *4*, 633–643. [CrossRef]

47. Plat, J.; van Onselen, E.N.; van Heugten, M.M.; Mensink, R.P. Effects on serum lipids, lipoproteins and fat soluble antioxidant concentrations of consumption frequency of margarines and shortenings enriched with plant stanol esters. *Eur. J. Clin. Nutr.* **2000**, *54*, 671–677. [CrossRef]

48. Doornbos, A.M.; Meynen, E.M.; Duchateau, G.S.; van der Knaap, H.C.; Trautwein, E.A. Intake occasion affects the serum cholesterol lowering of a plant sterol-enriched single-dose yoghurt drink in mildly hypercholesterolaemic subjects. *Eur. J. Clin. Nutr.* **2006**, *60*, 325–333. [CrossRef]

49. Lenssen, K.G.M.; Bast, A.; de Boer, A. Should botanical health claims be substantiated with evidence on traditional use? Reviewing the stakeholders' arguments. *PharmaNutrition* **2020**, *14*, 100232. [CrossRef]

50. Van der Waal, M.B.; Veldhuizen, C.K.; van der Waal, R.X.; Claassen, E.; van der Burgwal, L.H.M. A critical appreciation of intangible resources in PharmaNutrition. *PharmaNutrition* **2020**, *13*, 100208. [CrossRef]

51. Mach, F.; Baigent, C.; Catapano, A.L.; Koskinas, K.C.; Casula, M.; Badimon, L.; Chapman, M.J.; De Backer, G.G.; Delgado, V.; Ference, B.A.; et al. 2019 ESC/EAS Guidelines for the management of dyslipidaemias: Lipid modification to reduce cardiovascular risk. *Eur. Heart J.* **2020**, *41*, 111–188. [CrossRef]

52. Piepoli, M.F.; Hoes, A.W.; Agewall, S.; Albus, C.; Brotons, C.; Catapano, A.L.; Cooney, M.T.; Corra, U.; Cosyns, B.; Deaton, C.; et al. 2016 European Guidelines on cardiovascular disease prevention in clinical practice: The Sixth Joint Task Force of the European Society of Cardiology and Other Societies on Cardiovascular Disease Prevention in Clinical Practice (constituted by representatives of 10 societies and by invited experts) Developed with the special contribution of the European Association for Cardiovascular Prevention & Rehabilitation (EACPR). *Eur. Heart J.* **2016**, *37*, 2315–2381. [CrossRef] [PubMed]

53. Stock, J. Focus on lifestyle: EAS Consensus Panel Position Statement on Phytosterol-added Foods. *Atherosclerosis* **2014**, *234*, 142–145. [CrossRef]

54. Assmann, G.; Cullen, P.; Erbey, J.; Ramey, D.R.; Kannenberg, F.; Schulte, H. Plasma sitosterol elevations are associated with an increased incidence of coronary events in men: Results of a nested case-control analysis of the Prospective Cardiovascular Munster (PROCAM) study. *Nutr. Metab. Cardiovasc. Dis.* **2006**, *16*, 13–21. [CrossRef]

55. Willems, J.I.; Blommaert, M.A.; Trautwein, E.A. Results from a post-launch monitoring survey on consumer purchases of foods with added phytosterols in five European countries. *Food Chem. Toxicol.* **2013**, *62*, 48–53. [CrossRef]

56. Ras, R.T.; Trautwein, E.A. Consumer purchase behaviour of foods with added phytosterols in six European countries: Data from a post-launch monitoring survey. *Food Chem. Toxicol.* **2017**, *110*, 42–48. [CrossRef]

57. Bays, H.E.; Neff, D.; Tomassini, J.E.; Tershakovec, A.M. Ezetimibe: Cholesterol lowering and beyond. *Expert Rev. Cardiovasc. Ther.* **2008**, *6*, 447–470. [CrossRef]

58. EFSA Panel on Food Additives and Nutrient Sources added to Food (ANS); Younes, M.; Aggett, P.; Aguilar, F.; Crebelli, R.; Dusemund, B.; Filipic, M.; Frutos, M.J.; Galtier, P.; Gott, D.; et al. Scientific opinion on the safety of monacolins in red yeast rice. *EFSA J.* **2018**, *16*, e05368. [CrossRef]

59. Erguson, J.J.; Stojanovski, E.; MacDonald-Wicks, L.; Garg, M.L. High molecular weight oat beta-glucan enhances lipid-lowering effects of phytosterols. A randomised controlled trial. *Clin. Nutr.* **2020**, *39*, 80–89. [CrossRef]

60. Han, S.; Jiao, J.; Xu, J.; Zimmermann, D.; Actis-Goretta, L.; Guan, L.; Zhao, Y.; Qin, L. Effects of plant stanol or sterol-enriched diets on lipid profiles in patients treated with statins: Systematic review and meta-analysis. *Sci. Rep.* **2016**, *6*, 31337. [CrossRef] [PubMed]

61. Edwards, J.E.; Moore, R.A. Statins in hypercholesterolaemia: A dose-specific meta-analysis of lipid changes in randomised, double blind trials. *BMC Fam. Pract.* **2003**, *4*, 18. [CrossRef]

62. Wang, Y.; Jia, X.; Ghanam, K.; Beaurepaire, C.; Zidichouski, J.; Miller, L. Berberine and plant stanols synergistically inhibit cholesterol absorption in hamsters. *Atherosclerosis* **2010**, *209*, 111–117. [CrossRef]

63. Cicero, A.F.G.; Colletti, A.; Bajraktari, G.; Descamps, O.; Djuric, D.M.; Ezhov, M.; Fras, Z.; Katsiki, N.; Langlois, M.; Latkovskis, G.; et al. Lipid-lowering nutraceuticals in clinical practice: Position paper from an International Lipid Expert Panel. *Nutr. Rev.* **2017**, *75*, 731–767. [CrossRef] [PubMed]

64. Cofan, M.; Ros, E. Use of Plant Sterol and Stanol Fortified Foods in Clinical Practice. *Curr. Med. Chem.* **2019**, *26*, 6691–6703. [CrossRef] [PubMed]

65. Marangoni, F.; Poli, A. Phytosterols and cardiovascular health. *Pharm. Res.* **2010**, *61*, 193–199. [CrossRef] [PubMed]

Review

Nutraceuticals for Dyslipidaemia and Glucometabolic Diseases: What the Guidelines Tell Us (and Do Not Tell, Yet)

Manuela Casula [1,2], Alberico Luigi Catapano [1,2] and Paolo Magni [1,2,*]

[1] Dipartimento di Scienze Farmacologiche e Biomolecolari, Università degli Studi di Milano, 20133 Milan, Italy; manuela.casula@unimi.it (M.C.); alberico.catapano@unimi.it (A.L.C.)
[2] IRCCS MultiMedica, Sesto San Giovanni, 20099 Milan, Italy
* Correspondence: paolo.magni@unimi.it; Tel.: +39-02-50318229

Abstract: Background: The use of nutraceutical products and functional foods in the cardiovascular and metabolic field is rising in several countries. Preparation and implementation of guidelines are pivotal for translating research-derived knowledge and evidence-based medicine to the clinical practice. Based on these considerations, the aim of this paper is to explore if and how nutraceutical products are discussed by the most recent international guidelines related to cardio-metabolic diseases (dyslipidaemia, obesity, type 2 diabetes mellitus (T2DM) and cardiovascular disease (CVD) prevention). Some, but not all, guidelines for dyslipidaemia mention nutraceutical products as potential useful options for the treatment of mild dyslipidaemia, but also indicate the low level of evidence associated to their effects on hard endpoints (myocardial infarction, stroke, CVD-related death). In the most recent guidelines on obesity, it is mentioned that no safe and effective dietary supplement nor nutraceutical product is available for the management of weight loss in this condition, and more high-quality studies are necessary in this field. The examined guidelines for T2DM do not mention any specific nutraceutical approach to this disease, nor to milder forms, such as insulin resistance and pre-diabetes. Conclusions: The focus on nutraceutical products in the main international guidelines for cardio-metabolic disease management remains limited. Since robust scientific evidence is the background of useful and effective guidelines, the implementation of high-quality clinical research is strongly needed in the field of nutraceutical products for cardio-metabolic diseases.

Keywords: guidelines; nutraceutical product; cardiovascular disease; dyslipidaemia; metabolic disease; type 2 diabetes mellitus

Citation: Casula, M.; Catapano, A.L.; Magni, P. Nutraceuticals for Dyslipidaemia and Glucometabolic Diseases: What the Guidelines Tell Us (and Do Not Tell, Yet). *Nutrients* **2022**, *14*, 606. https://doi.org/10.3390/nu14030606

Academic Editors: Ina Bergheim and Stephen Anton

Received: 31 December 2021
Accepted: 28 January 2022
Published: 30 January 2022

Publisher's Note: MDPI stays neutral with regard to jurisdictional claims in published maps and institutional affiliations.

1. Introduction

Cardiovascular (CV) and metabolic diseases are the leading cause of morbidity, disability and mortality in the world [1] and their heterogenous causes are currently on the rise [2]. Their epidemiological and economic impact, thus, poses a significant challenge and requires a multifactorial approach, starting from the promotion of a healthy lifestyle and, when necessary, the use of preventive or treatment therapies, along with appropriate first-level and second-level diagnostic procedures. Indeed, over time, multiple diagnostic and treatment options have become available, including increasingly effective and safer medications, which also allow a more personalised approach. The current clinical practice in many fields and, in particular, in the CV disease (CVD) and metabolic disease area, is built on a body of knowledge and evidence, which is collected in guidelines focusing on specific medical issues [3,4]. These guidelines are constructed according to a specific development process [5,6], which takes into consideration several aspects, including a systematic review of the available evidence, the inclusion of a multidisciplinary panel of representatives and experts from key affected groups (such as patients), and the identification of potential conflicts of interest. Most important, they should periodically undergo reconsideration and revision, as appropriate according to any important novel evidence that arises. Among the treatment options, the use of nutraceutical products is currently growing in several biomedical areas, including the cardiometabolic one and, thus, requires a robust scientific validation. Nutraceuticals are nutritional compounds of natural origin, shown to be efficacious in preventive medicine or in the treatment or co-treatment of diseases. Specific to this review, several foods and dietary supplements have been shown to contribute to protection against the development of CVD [7]. However, within the landscape of intervention options, the role of nutraceutical products and functional foods, especially in the cardio-metabolic field, is still rather limited in the area of prescription medications, albeit variably depending on geographic context and disease type [8]. In any case, the overall nutraceutical market appears to have quite large proportions worldwide and the specific field of nutraceuticals for CVD and metabolic disease prevention may be expected to grow in the next years, as well. Thus, the discussion about the usefulness of nutraceutical options and the related potential critical issues in this area begin to be mentioned in the guidelines related to the metabolic field.

Based on these considerations, in this review, we aim at exploring if and how nutraceutical products are considered and discussed by the most recent guidelines devoted to the management of CV and metabolic diseases, with specific regards to dyslipidaemia, a major causative factor for atherosclerotic CVD (ASCVD), as well as obesity and type 2 diabetes mellitus (T2DM). This review is based upon the analysis of the international guidelines regarding CVDs, obesity, T2DM and metabolic syndrome, published in the last 20 years, in order to assess the relevance of this topic over time and in different regions of the world.

2. Nutraceutical Products and Guidelines for Dyslipidaemia

The use of nutraceutical products for the management of dyslipidaemia (and some features, such as insulin resistance, which are often associated to it) is a rather diffused option and the efficacy and safety of several such products, mostly combining two or more active phytoextracts or compounds, have been evaluated by good-quality RCTs [9–13].

Taking into consideration specific guidelines (Table 1), in the last version of the 2019 European Society of Cardiology (ESC)/European Society of Atherosclerosis (EAS) Guidelines for the management of dyslipidaemias [4], authors stressed the central role of nutrition for the prevention of ASCVD, but also the weakness of evidence and the lack of concordance among studies, suggesting caution in interpreting the results of randomised controlled trials (RCTs). These guidelines list a number of dietary supplements and functional foods to be considered for the treatment of dyslipidaemias, which are briefly discussed below. Plant sterols and stanols are phytosterols, and have been identified in several plant products, including various fruits and vegetables, cereals, nuts and seeds. Their biological activity is derived from their molecular structural similarity to cholesterol. According to the guidelines, in compliance with the LDL-C goal and without any safety contraindications, plant sterols/stanols "may be considered: (i) in individuals with high cholesterol levels at intermediate or low global CV risk who do not qualify for pharmacotherapy; (ii) as an adjunct to pharmacological therapy in high- and very-high-risk patients who fail to achieve LDL-C goals on statins or could not be treated with statins; and (iii) in adults and children (aged >6 years) with familial hypercholesterolemia (FH)" [4]. The role of phytosterols in controlling cholesterol levels according to the European Atherosclerosis Society is highlighted by the specific Consensus by Gylling et al. [14]. This critical appraisal of the evidence about the benefit-to-risk relationship of functional foods with added plant sterols and/or plant stanols underlined that "daily consumption of foods with added plant sterols and/or plant stanols in amounts of up to 2 g/day is equally effective in lowering plasma atherogenic LDL-C levels by up to 10%, and thus may be considered as an adjunct to lifestyle in subjects at all levels of CV risk" [14] and that "plant sterols and plant stanols can be efficaciously combined with statins" [14], while "very limited data suggest plant sterols/stanols may also lower LDL-C levels in combination with a fibrate or ezetimibe" [14].

Another product that these guidelines proposed for the management of hypercholesterolemia is red yeast rice (RYR) extract, thanks to its bioactive ingredient, monacolin k, that shows a statin-like mechanism. Different commercial preparations of RYR, with various concentrations of monacolins, are available on the market. However, safety issues, due to the possible presence of contaminants in some preparations, have been raised. The guidelines state that "Nutraceuticals containing purified RYR may be considered in people with elevated plasma cholesterol concentrations who do not qualify for treatment with statins in view of their global CV risk" [4], but also warrant for better regulation of this kind of supplements.

The role of dietary fibres in cholesterol lowering has been strengthened in the last version of these guidelines ("Foods enriched with these fibres or supplements are well tolerated, effective, and recommended for LDL-C lowering" [4]) after the publication of the Cochrane meta-analysis in 2016 [15]. In contrast, the role of soy, suggested in the 2011 version of the ESC/EAS Guidelines as "substitute for animal protein foods high in SFAs" [16], even if "expected LDL-C lowering may be modest (3–5%) and most likely in subjects with hypercholesterolaemia" [16], was then downscored: "LDL-C-lowering effect [. . .] was not confirmed when changes in other dietary components were taken into account" [4].

Finally, policosanol was described as showing "no significant effect on LDL-C, HDL-C, triglyceride (TG)" [4], and for berberine, despite a recent meta-analysis showing some lipid-lowering effect [17], "due to the lack of high-quality randomized clinical trials, its efficacy for treating dyslipidaemia needs to be further validated" [4].

Regarding the role of omega-3 polyunsaturated fatty acids (PUFA) in TG lowering, the recent guidelines reported that "Observational evidence indicates that consumption of fish (at least twice a week) and vegetable foods rich in omega-3 fatty acids [. . .] is associated with lower risk of CV death and stroke, but has no major effects on plasma lipoprotein metabolism" [4] and that "Pharmacological doses of long-chain omega-3 fatty acids (2–3 g/day) reduce TG levels by about 30%" [4]. Concerning hard CV endpoints, the 2019 guidelines had the opportunity to integrate results from the REDUCE-IT trial [18] (where a significantly lower risk of ischaemic events was reported in patients with elevated TG levels treated with 4 g of icosapent ethyl), but not from the most recent RCTs on omega-3 unsaturated fatty acids [19,20].

The US Guidelines approach is quite different. In the 2018 AHA/ACC/AACVPR/AAPA/ABC/ACPM/ADA/AGS/APhA/ASPC/NLA/PCNA Guideline on the Management of Blood Cholesterol [21], there is no mention of nutraceuticals (whether general or specific products) and only a generic indication to pay attention to the diet ("Lifestyle counseling: use principles of Mediterranean and DASH diets" [22]). In 2018, the International Lipid Expert Panel (ILEP) published a Position Paper [23] defining the use of nutraceuticals in the management of statin intolerance: "Nutraceuticals, such as red yeast rice, bergamot, berberine, artichoke, soluble fiber, and plant sterols and stanols alone or in combination with each other, as well as with ezetimibe, might be considered as an alternative or add-on therapy to statins" [23]. Among the listed products, berberine, omega-3 PUFA and RYR were deemed as class of recommendation I, and therefore, enough supported by evidence of benefit and safety to be recommended/indicated. Nevertheless, also in this case, authors recognise the limitation of non-statin studies. In the Japan Atherosclerosis Society (JAS) Guidelines for Prevention of Atherosclerotic Cardiovascular Diseases 2017 [24], the section dedicated to lifestyle modification clearly stated that "increasing the intake of omega-3 PUFA is effective in decreasing the TG level and may lead to suppression of coronary artery disease (CAD)" [24], and that "Consuming soy and soy products is recommended because they may decrease CAD and stroke risk [24]". Moreover, omega-3 PUFA are also indicated as treatment: "PUFAs are indicated for dyslipidaemia accompanied by an increased TG level, particularly for type IIb and type IV hyperlipidemia" [24].

The recently released 2021 ESC Guidelines on cardiovascular disease prevention in clinical practice [25] also comment on the potential usefulness of nutraceutical product to reduce lipid levels: "An evidence-based approach to the use of lipid-lowering nutraceuticals could improve the quality of the treatment, including therapy adherence, and achievement of the LDL-C goal in clinical practice" [25]. However, importantly, they warn about the lack of outcome studies proving that nutraceuticals can prevent CVD morbidity or mortality [26].

In conclusion, some, but not all, guidelines for dyslipidaemia mention nutraceutical products as potential useful options for the treatment of mild dyslipidaemia, but also indicate the low level of evidence associated to their effects on hard endpoints, such as prevention of myocardial infarction and stroke.

Table 1. Nutraceutical products and guidelines for dyslipidaemia.

Nutraceutical Products Mentioned: Plant Sterols and Stanols, Red Yeast Rice Extract, Dietary Fibres, Soy, Policosanol, Berberine, Omega-3 Polyunsaturated Fatty Acids, Bergamot, Artichoke		
Year	**Guideline (S) Name**	**Ref.**
2019	ESC/EAS Guidelines for the management of dyslipidaemias	[4]
2018	AHA/ACC/AACVPR/AAPA/ABC/ACPM/ADA/AGS/APhA/ASPC/	[21]
	NLA/PCNA Guideline on the Management of Blood Cholesterol	
2018	International Lipid Expert Panel Position Paper	[23]
2017	Japan Atherosclerosis Society (JAS) Guidelines for Prevention of Atherosclerotic Cardiovascular Diseases	[24]
2011	ESC/EAS Guidelines for the management of dyslipidaemias	[16]

3. Nutraceutical Products and Guidelines for Obesity

The increasing prevalence of obesity and the related cardio-metabolic complications is well established. However, the practical prevention and management of this pathological condition is complex and difficult, due to the interplay of multiple and individually specific causative factors. This situation led to the development of a large series of approaches, ranging from pharmaceutical products to approaches with specific dietary programs or using plant extracts and food supplements (Table 2). The effectiveness and safety of these approaches still appear to be determined in many instances. In 2018, the Endocrine Society (USA) published a scientific statement on obesity management [27]. In this document, a detailed table listed a series of dietary supplements proposed for obesity management, defined in the text as "Dietary supplements, over-the-counter products, and other treatments with unproven efficacy and unknown safety" [27], which thus, have not undergone any FDA evaluation. As a word of caution, the authors suggested "Evidence to support the effectiveness for weight loss or the safety of these preparations is usually nonexistent. Moreover, variability in the composition of these products adds an additional uncertainty to their use. We thus think that the public would be better served if the dietary supplements were held to a higher standard and were overseen by the FDA" [27]. These considerations then indicate that, at least in the USA, no specific validation of such products has been conducted. The American Association of Clinical Endocrinologists (AACE) guidelines for the practical management of obesity, published in 2016 [28], do not report any food supplement or nutraceutical approach to obesity, but recommend that "Combinations of FDA-approved weight loss medications should only be used in a manner approved by the FDA or when sufficient safety and efficacy data are available to assure informed judgment regarding a favorable benefit-to-risk ratio" [28], again suggesting caution in such weight loss strategies.

The 2015 European Guidelines for Obesity Management in Adults [29], by the European Association for the Study of Obesity (EASO), mentions, in the section "Alternative therapies" that "[…] unorthodox and unproven treatments flourish and are often offered. There is insufficient evidence to recommend in favour of herbal medicines, dietary supplements or homoeopathy for obesity management in the obese person. Physicians should advise patients to follow evidence-based treatments and recommend treatments only where evidence of safety and efficacy has been established" [29]. As an additional strategy, subjects with obesity may benefit of very-low-calorie diet (VLCD) approaches [30], which, however, may be detrimental to some patients and, thus, should be conducted under the supervision of a physician. An EASO Position Statement published in 2014 [31] mentioned the VLCD approach, which has been recently evaluated in depth, with the first definition of the European Guidelines on this topic [32].

In conclusion, according to the most relevant and recent guidelines on obesity, no safe and effective dietary supplement nor nutraceutical product is available for the management of weight loss in this condition, and more high-quality studies are necessary in this field.

Table 2. Nutraceutical products and guidelines for obesity.

Nutraceutical Products Mentioned: "Dietary Supplements, Over-the-Counter Products, and Other Treatments" Ref. [31], Herbal Medicines, Dietary Supplements, Very-Low-Calorie Diet		
Year	**Guideline (S) Name**	**Ref.**
2018	Endocrine Society (USA) scientific statement on obesity management	[27]
2016	American Association of Clinical Endocrinologists guidelines for the practical management of obesity	[28]
2015	EASO European Guidelines for Obesity Management in Adults	[29]
2014	EASO Position Statement	[31]

4. Nutraceutical Products and Guidelines for Type 2 Diabetes Mellitus

The pharmacological management of T2DM has recently been expanded to several different classes of drugs (dipeptidyl peptidase-4 (DPP4) inhibitors, sodium-glucose co-transporter 2 (SGLT2) inhibitors, glucagon-like peptide-1 receptor agonists (GLP1-RA), etc.) in addition to metformin, sulfonilureas and alpha-glucosidase inhibitors [33]. Interestingly, some (mainly SGLT2 inhibitors and GLP-1 RA) in addition to optimal control of glucose metabolism, also offer a relevant degree of cardiovascular and renovascular protection [34]. Within this context, especially for subjects with pre-diabetes, insulin resistance and moderate metabolic syndrome, or as an add-on to drug therapy, several nutraceutical products have been proposed for glycaemic control (Table 3). These products include, for example, berberine [35], *Morus alba* extract [36] and other herbal extracts [37].

In 2021 and in 2022, the American Diabetes Association (ADA) released the updated "Standards of Medical Care in Diabetes". The 2021 and 2022 sections devoted to pharmacologic approaches [33,38] report in great detail the pharmacological options for the management of T2DM, but do not mention the use of any supplement or nutraceutical product, either in a positive or negative way. In the section of these "Standards of Medical Care in Diabetes" devoted to cardiovascular disease and risk management [39,40], the authors include, in the 10.15 Recommendation, some dietary suggestions, including "[…] increase of dietary omega-3 fatty acids, viscous fiber, and plant stanols/sterols intake […]" [40], thus generically mentioning some nutraceuticals discussed in greater detail in the guidelines devoted to the management of dyslipidaemia (see above).

In the "2019 ESC Guidelines on diabetes, pre-diabetes, and cardiovascular diseases developed in collaboration with the European Association for the Study of Diabetes (EASD)" [34], a specific reference is present indicating that "Supplements with omega-3 fatty acids have not been shown to improve glycaemic control in individuals with DM and RCTs do not support recommending omega-3 supplements for the primary or secondary prevention of CVD" [34]. Moreover, a specific comment on the positive outcome of the REDUCE-IT trial [18], discussed above, on the primary endpoint of major adverse CV events (MACE) is also included. In 2018, a consensus report by the ADA and the EASD on "Management of hyperglycaemia in type 2 diabetes" was published [41]. This very detailed document does not mention, however, any food supplement or nutraceutical product proposed or even not recommended for T2DM management. Similarly, no reference to nutraceuticals or food supplements is present in the 2017 "Recommendations for Managing Type 2 Diabetes in Primary Care" by the International Diabetes Federation [42].

Obesity and T2DM may often be associated with non-alcoholic fatty liver disease (NAFLD). On this topic, in 2016, the European Association for the Study of the Liver (EASL), the EASD and the The European Association for the Study of Obesity (EASO) published the Clinical Practice Guidelines [43], which did not mention any nutraceutical product/food supplement.

In conclusion, guidelines for T2DM do not mention any specific nutraceutical approach to this disease, nor to milder forms such as insulin resistance and pre-diabetes, which may be observed in the early phases before proper T2DM development. In any case, if the validation of such products is considered important, robust clinical research will, thus, need to be implemented in this specific area.

Table 3. Nutraceutical products and guidelines for type 2 diabetes mellitus.

Nutraceutical Products Mentioned:
Berberine, Morus Alba Extract, other Herbal Extracts, Omega-3 Polyunsaturated Fatty Acids, Viscous Fibre, Plant Stanols/Sterols

Year	Guideline (S) Name	Ref.
2021–2022	American Diabetes Association "Standards of Medical Care in Diabetes"	
	Pharmacologic Approaches to Glycemic Treatment	[33,38]
	Cardiovascular Disease and Risk Management	[39,40]
2019	ESC Guidelines on diabetes, pre-diabetes and cardiovascular diseases developed in collaboration with the EASD	[34]
2018	Management of hyperglycaemia in type 2 diabetes	[41]
	A consensus report by ADA and EASD	
2017	IDF Recommendations for Managing Type 2 Diabetes in Primary Care	[42]
2016	EASL–EASD–EASO Clinical Practice Guidelines for the management of non-alcoholic fatty liver disease	[43]

5. Concluding Remarks

In recent years, the use of nutraceuticals has become increasingly substantial in several areas of medicine [44]. In parallel, the prescription of these products by physicians has also increased [45]. However, the focus on nutraceuticals and dietary supplements in the guidelines dedicated to disease management remains limited, both in general and, more specifically, regarding cardio-metabolic diseases and the related features, such as dyslipidaemia. The preparation and implementation of guidelines represent a pivotal aspect of the translation of research-derived knowledge and evidence-based medicine to the clinical practice. This process requires a set of subsequent steps that need to be accurately defined and implemented [46–48], starting from robust evidence that is the background of useful and effective guidelines. As indicated by the GRADE system, there are four levels of evidence. Evidence from RCTs is of the highest quality and, because of residual confounding, evidence that includes observational data is considered to be low quality. However, the literature has consistently claimed that there are few RCT data considering hard end-points to establish clinical benefits [49]. This may be partly due to the current methodological shortcomings of RCT study design and the complex characteristics of nutraceuticals. In contrast to pharmaceutical trials, clinical trials on nutraceuticals and on enriched dietary patterns show critical challenges in terms of study design and methodology, tiny effect sizes, high heterogeneity of the responses and limited translatability of the observed effect size. These observations prompt further studies, leading to an improvement of the methodology, aiming at building a stronger evidence-based data that supports the use of these products in cardiovascular and metabolic prevention and, possibly, management.

Author Contributions: Conceptualisation, M.C., P.M. and A.L.C.; methodology, M.C.; investigation, M.C. and P.M.; resources, P.M. and A.L.C.; data curation, P.M. and M.C.; writing—original draft preparation, M.C. and P.M.; writing, review and editing, P.M., M.C. and A.L.C.; project administration, P.M.; funding acquisition, P.M. and A.L.C. All authors have read and agreed to the published version of the manuscript.

Funding: The work of M.C. and P.M. is supported by the Ministry of Health-IRCCS MultiMedica GR-2016-02361198. The work of P.M. is also supported by the Università degli Studi di Milano (Transition grant PSR2015-1720PMAGN_01). The work of A.L.C. is supported by the Fondazione Cariplo 2015-0524 and 2015-0564; H2020 REPROGRAM PHC-03-2015/667837-2; ERANET ER-2017-2364981; PRIN 2017H5F943; Ministry of Health-IRCCS MultiMedica GR-2011-02346974; SISA Lombardia and Fondazione SISA. The authors are supported by the MIUR Progetto di Eccellenza and Università degli Studi di Milano through the APC initiative.

Institutional Review Board Statement: Not applicable.

Informed Consent Statement: Not applicable.

Data Availability Statement: Not applicable.

Conflicts of Interest: P.M. and M.C. have no conflict of interests. A.L.C. has received honoraria, lecture fees or research grants from: Akcea, Amgen, Astrazeneca, Eli Lilly, Genzyme, Kowa, Mediolanum, Menarini, Merck, Pfizer, Recordati, Sanofi, Sigma Tau, Amryt and Sandoz.

References

1. W.H.O. Available online: https://www.who.int/ (accessed on 30 December 2021).
2. Ralston, J.; Nugent, R. Toward a broader response to cardiometabolic disease. *Nat. Med.* **2019**, *25*, 1644–1646. [CrossRef] [PubMed]
3. Catapano, A.L.; Graham, I.; De Backer, G.; Wiklund, O.; Chapman, M.J.; Drexel, H.; Hoes, A.W.; Jennings, C.S.; Landmesser, U.; Pedersen, T.R.; et al. 2016 ESC/EAS Guidelines for the management of dyslipidaemias. *Eur. Heart J.* **2016**, *37*, 2999–3058. [CrossRef] [PubMed]
4. Mach, F.; Baigent, C.; Catapano, A.L.; Koskinas, K.C.; Casula, M.; Badimon, L.; Chapman, M.J.; De Backer, G.G.; Delgado, V.; Ference, B.A.; et al. 2019 ESC/EAS Guidelines for the management of dyslipidaemias: Lipid modification to reduce cardiovascular risk. *Eur. Heart J.* **2020**, *41*, 111–188. [CrossRef] [PubMed]
5. Schünemann, H.J.; Wiercioch, W.; Etxeandia, I.; Falavigna, M.; Santesso, N.; Mustafa, R.; Ventresca, M.; Brignardello-Petersen, R.; Laisaar, K.T.; Kowalski, S.; et al. Guidelines 2.0: Systematic development of a comprehensive checklist for a successful guideline enterprise. *CMAJ* **2014**, *186*, E123–E142. [CrossRef]
6. Magni, P.; Bier, D.M.; Pecorelli, S.; Agostoni, C.; Astrup, A.; Brighenti, F.; Cook, R.; Folco, E.; Fontana, L.; Gibson, R.A.; et al. Perspective: Improving nutritional guidelines for sustainable health policies: Current status and perspectives. *Adv. Nutr.* **2017**, *8*, 532–545. [CrossRef]
7. Alissa, E.M.; Ferns, G.A. Functional foods and nutraceuticals in the primary prevention of cardiovascular diseases. *J. Nutr. Metab.* **2012**, *2012*, 569486. [CrossRef]
8. Penson, P.E.; Banach, M. Nutraceuticals for the control of dyslipidaemias in clinical practice. *Nutrients* **2021**, *13*, 2957. [CrossRef]
9. Ruscica, M.; Gomaraschi, M.; Mombelli, G.; Macchi, C.; Bosisio, R.; Pazzucconi, F.; Pavanello, C.; Calabresi, L.; Arnoldi, A.; Sirtori, C.R.; et al. Nutraceutical approach to moderate cardiometabolic risk: Results of a randomized, double-blind and crossover study with armolipid plus. *J. Clin. Lipidol.* **2014**, *8*, 61–68. [CrossRef]
10. Pavanello, C.; Lammi, C.; Ruscica, M.; Bosisio, R.; Mombelli, G.; Zanoni, C.; Calabresi, L.; Sirtori, C.R.; Magni, P.; Arnoldi, A. Effects of a lupin protein concentrate on lipids, blood pressure and insulin resistance in moderately dyslipidaemic patients: A randomised controlled trial. *J. Funct. Foods* **2017**, *37*, 8–15. [CrossRef]
11. Ruscica, M.; Pavanello, C.; Gandini, S.; Gomaraschi, M.; Vitali, C.; Macchi, C.; Morlotti, B.; Aiello, G.; Bosisio, R.; Calabresi, L.; et al. Effect of soy on metabolic syndrome and cardiovascular risk factors: A randomized controlled trial. *Eur. J. Nutr.* **2018**, *57*, 499–511. [CrossRef]
12. Ruscica, M.; Pavanello, C.; Gandini, S.; Macchi, C.; Botta, M.; Dall'Orto, D.; Del Puppo, M.; Bertolotti, M.; Bosisio, R.; Mombelli, G.; et al. Nutraceutical approach for the management of cardiovascular risk—A combination containing the probiotic Bifidobacterium longum BB536 and red yeast rice extract: Results from a randomized, double-blind, placebo-controlled study. *Nutr. J.* **2019**, *18*, 13. [CrossRef]
13. Cicolari, S.; Pavanello, C.; Olmastroni, E.; Puppo, M.D.; Bertolotti, M.; Mombelli, G.; Catapano, A.L.; Calabresi, L.; Magni, P. Interactions of oxysterols with atherosclerosis biomarkers in subjects with moderate hypercholesterolemia and effects of a nutraceutical combination. *Nutrients* **2021**, *13*, 427. [CrossRef] [PubMed]

14. Gylling, H.; Plat, J.; Turley, S.; Ginsberg, H.N.; Ellegård, L.; Jessup, W.; Jones, P.J.; Lütjohann, D.; Maerz, W.; Masana, L.; et al. Plant sterols and plant stanols in the management of dyslipidaemia and prevention of cardiovascular disease. *Atherosclerosis* **2014**, *232*, 346–360. [CrossRef] [PubMed]

15. Hartley, L.; May, M.D.; Loveman, E.; Colquitt, J.L.; Rees, K. Dietary fibre for the primary prevention of cardiovascular disease. *Cochrane Database Syst. Rev.* **2016**, *1*, CD011472. [CrossRef] [PubMed]

16. Reiner, Z.; Catapano, A.; De Backer, G.; Graham, I.; Taskinen, M.; Wiklund, O.; Agewall, S.; Alegria, E.; Chapman, M.; Durrington, P.; et al. ESC/EAS guidelines for the management of dyslipidaemias: The task force for the management of dyslipidaemias of the European society of cardiology (ESC) and the European atherosclerosis society (EAS). *Eur. Heart J.* **2011**, *32*, 1769–1818. [CrossRef] [PubMed]

17. Ju, J.; Li, J.; Lin, Q.; Xu, H. Efficacy and safety of berberine for dyslipidaemias: A systematic review and meta-analysis of randomized clinical trials. *Phytomedicine* **2018**, *50*, 25–34. [CrossRef] [PubMed]

18. Bhatt, D.L.; Steg, P.G.; Miller, M.; Brinton, E.A.; Jacobson, T.A.; Ketchum, S.B.; Doyle, R.T.; Juliano, R.A.; Jiao, L.; Granowitz, C.; et al. Cardiovascular risk reduction with icosapent ethyl for hypertriglyceridemia. *N. Engl. J. Med.* **2019**, *380*, 11–22. [CrossRef] [PubMed]

19. Nicholls, S.J.; Lincoff, A.M.; Garcia, M.; Bash, D.; Ballantyne, C.M.; Barter, P.J.; Davidson, M.H.; Kastelein, J.J.P.; Koenig, W.; McGuire, D.K.; et al. Effect of high-dose omega-3 fatty acids vs corn oil on major adverse cardiovascular events in patients at high cardiovascular risk: The strength randomized clinical trial. *JAMA* **2020**, *324*, 2268–2280. [CrossRef]

20. Greenland, P.; Alpert, J.S.; Beller, G.A.; Benjamin, E.J.; Budoff, M.J.; Fayad, Z.A.; Foster, E.; Hlatky, M.A.; Hodgson, J.M.; Kushner, F.G.; et al. 2010 ACCF/AHA guideline for assessment of cardiovascular risk in asymptomatic adults: A report of the American college of cardiology foundation/american heart association task force on practice guidelines. *Circulation* **2010**, *122*, e584–e636. [CrossRef]

21. Grundy, S.M.; Stone, N.J.; Bailey, A.L.; Beam, C.; Birtcher, K.K.; Blumenthal, R.S.; Braun, L.T.; de Ferranti, S.; Faiella-Tommasino, J.; Forman, D.E.; et al. 2018 AHA/ACC/AACVPR/AAPA/ABC/ACPM/ADA/AGS/APhA/ASPC/NLA/PCNA guideline on the management of blood cholesterol: A report of the American college of cardiology/American heart association task force on clinical practice guidelines. *Circulation* **2019**, *139*, e1082–e1143. [CrossRef]

22. Grundy, S.; Stone, N.; Bailey, A.; Beam, C.; Birtcher, K.; Blumenthal, R.; Braun, L.; de Ferranti, S.; Faiella-Tommasino, J.; Forman, D.; et al. 2018 AHA/ACC/AACVPR/AAPA/ABC/ACPM/ADA/AGS/APhA/ASPC/NLA/PCNA guideline on the management of blood cholesterol: Executive summary: A report of the American college of cardiology/American heart association task force on clinical practice guidelines. *Circulation* **2019**, *139*, e1046–e1081. [CrossRef] [PubMed]

23. Banach, M.; Patti, A.M.; Giglio, R.V.; Cicero, A.F.G.; Atanasov, A.G.; Bajraktari, G.; Bruckert, E.; Descamps, O.; Djuric, D.M.; Ezhov, M.; et al. The role of nutraceuticals in statin intolerant patients. *J. Am. Coll. Cardiol.* **2018**, *72*, 96–118. [CrossRef] [PubMed]

24. Kinoshita, M.; Yokote, K.; Arai, H.; Iida, M.; Ishigaki, Y.; Ishibashi, S.; Umemoto, S.; Egusa, G.; Ohmura, H.; Okamura, T.; et al. Japan atherosclerosis society (JAS) guidelines for prevention of atherosclerotic cardiovascular diseases 2017. *J. Atheroscler. Thromb.* **2018**, *25*, 846–984. [CrossRef]

25. Visseren, F.L.J.; Mach, F.; Smulders, Y.M.; Carballo, D.; Koskinas, K.C.; Bäck, M.; Benetos, A.; Biffi, A.; Boavida, J.M.; Capodanno, D.; et al. 2021 ESC guidelines on cardiovascular disease prevention in clinical practice. *Eur. Heart J.* **2021**, *42*, 3227–3337. [CrossRef] [PubMed]

26. Cicero, A.F.G.; Colletti, A.; Bajraktari, G.; Descamps, O.; Djuric, D.M.; Ezhov, M.; Fras, Z.; Katsiki, N.; Langlois, M.; Latkovskis, G.; et al. Lipid-lowering nutraceuticals in clinical practice: Position paper from an international lipid expert panel. *Nutr. Rev.* **2017**, *75*, 731–767. [CrossRef]

27. Bray, G.A.; Heisel, W.E.; Afshin, A.; Jensen, M.D.; Dietz, W.H.; Long, M.; Kushner, R.F.; Daniels, S.R.; Wadden, T.A.; Tsai, A.G.; et al. The science of obesity management: An endocrine society scientific statement. *Endocr. Rev.* **2018**, *39*, 79–132. [CrossRef]

28. Garvey, W.T.; Mechanick, J.I.; Brett, E.M.; Garber, A.J.; Hurley, D.L.; Jastreboff, A.M.; Nadolsky, K.; Pessah-Pollack, R.; Plodkowski, R.; Reviewers of the AACE/ACE Obesity Clinical Practice Guidelines. American association of clinical endocrinologists and american college of endocrinology comprehensive clinical practice guidelines for medical care of patients with obesity. *Endocr. Pract.* **2016**, *22*, 842–884. Available online: https://www.aace.com/publications/guidelines (accessed on 10 December 2021). [CrossRef] [PubMed]

29. Yumuk, V.; Tsigos, C.; Fried, M.; Schindler, K.; Busetto, L.; Micic, D.; Toplak, H.; Obesity Management Task Force of the European Association for the Study of Obesity. European guidelines for obesity management in adults. *Obes. Facts* **2015**, *8*, 402–424. [CrossRef]

30. Tragni, E.; Vigna, L.; Ruscica, M.; Macchi, C.; Casula, M.; Santelia, A.; Catapano, A.L.; Magni, P. Reduction of cardio-metabolic risk and body weight through a multiphasic very-low calorie ketogenic diet program in women with overweight/obesity: A study in a real-world setting. *Nutrients* **2021**, *13*, 1804. [CrossRef]

31. Yumuk, V.; Frühbeck, G.; Oppert, J.M.; Woodward, E.; Toplak, H. An EASO position statement on multidisciplinary obesity management in adults. *Obes. Facts* **2014**, *7*, 96–101. [CrossRef]

32. Muscogiuri, G.; El Ghoch, M.; Colao, A.; Hassapidou, M.; Yumuk, V.; Busetto, L.; Obesity management task force (OMTF) of the European association for the study of obesity (EASO). European guidelines for obesity management in adults with a very low-calorie ketogenic diet: A systematic review and meta-analysis. *Obes. Facts* **2021**, *14*, 222–245. [CrossRef] [PubMed]

33. American Diabetes Association. 9. Pharmacologic approaches to glycemic treatment: Standards of medical care in diabetes—2021. *Diabetes Care* **2021**, *44*, S111–S124. [CrossRef] [PubMed]

34. Cosentino, F.; Grant, P.; Aboyans, V.; Bailey, C.; Ceriello, A.; Delgado, V.; Federici, M.; Filippatos, G.; Grobbee, D.; Hansen, T.; et al. 2019 ESC guidelines on diabetes, pre-diabetes, and cardiovascular diseases developed in collaboration with the EASD. *Eur. Heart J.* **2020**, *41*, 255–323. [CrossRef] [PubMed]

35. Guo, J.; Chen, H.; Zhang, X.; Lou, W.; Zhang, P.; Qiu, Y.; Zhang, C.; Wang, Y.; Liu, W.J. The effect of berberine on metabolic profiles in type 2 diabetic patients: A systematic review and meta-analysis of randomized controlled trials. *Oxid. Med. Cell. Longev.* **2021**, *2021*, 2074610. [CrossRef]

36. Morales Ramos, J.G.; Esteves Pairazamán, A.T.; Mocarro Willis, M.E.S.; Collantes Santisteban, S.; Caldas Herrera, E. Medicinal properties of *Morus alba* for the control of type 2 diabetes mellitus: A systematic review. *F1000Research* **2021**, *10*, 1022. [CrossRef]

37. Suksomboon, N.; Poolsup, N.; Boonkaew, S.; Suthisisang, C.C. Meta-analysis of the effect of herbal supplement on glycemic control in type 2 diabetes. *J. Ethnopharmacol.* **2011**, *137*, 1328–1333. [CrossRef]

38. Draznin, B.; Aroda, V.R.; Bakris, G.; Benson, G.; Brown, F.M.; Freeman, R.; Green, J.; Huang, E.; Isaacs, D.; Kahan, S.; et al. 9. Pharmacologic approaches to glycemic treatment: Standards of medical care in diabetes—2022. *Diabetes Care* **2022**, *45*, S125–S143. [CrossRef]

39. American Diabetes Association. 10. Cardiovascular disease and risk management: Standards of medical care in diabetes—2021. *Diabetes Care* **2021**, *44*, S125–S150. [CrossRef]

40. Draznin, B.; Aroda, V.R.; Bakris, G.; Benson, G.; Brown, F.M.; Freeman, R.; Green, J.; Huang, E.; Isaacs, D.; Kahan, S.; et al. 10. Cardiovascular disease and risk management: Standards of medical care in diabetes—2022. *Diabetes Care* **2022**, *45*, S144–S174. [CrossRef]

41. Davies, M.J.; D'Alessio, D.A.; Fradkin, J.; Kernan, W.N.; Mathieu, C.; Mingrone, G.; Rossing, P.; Tsapas, A.; Wexler, D.J.; Buse, J.B. Management of hyperglycemia in type 2 diabetes, 2018. A consensus report by the American diabetes association (ADA) and the European association for the study of diabetes (EASD). *Diabetes Care* **2018**, *41*, 2669–2701. [CrossRef]

42. Aschner, P. New IDF clinical practice recommendations for managing type 2 diabetes in primary care. *Diabetes Res. Clin. Pract.* **2017**, *132*, 169–170. [CrossRef] [PubMed]

43. European Association for the Study of the Liver (EASL); European Association for the Study of Diabetes (EASD); European Association for the Study of Obesity (EASO). EASL-EASD-EASO clinical practice guidelines for the management of non-alcoholic fatty liver disease. *Diabetologia* **2016**, *59*, 1121–1140. [CrossRef] [PubMed]

44. Williamson, E.M.; Liu, X.; Izzo, A.A. Trends in use, pharmacology, and clinical applications of emerging herbal nutraceuticals. *Br. J. Pharmacol.* **2020**, *177*, 1227–1240. [CrossRef]

45. Gosavi, S.; Subramanian, M.; Reddy, R.; Shet, B.L. A study of prescription pattern of neutraceuticals, knowledge of the patients and cost in a tertiary care hospital. *J. Clin. Diagn. Res.* **2016**, *10*, FC01–FC04. [CrossRef]

46. Guyatt, G.H.; Oxman, A.D.; Kunz, R.; Vist, G.E.; Falck-Ytter, Y.; Schünemann, H.J.; Group, G.W. What is "quality of evidence" and why is it important to clinicians? *BMJ* **2008**, *336*, 995–998. [CrossRef] [PubMed]

47. Guyatt, G.H.; Oxman, A.D.; Vist, G.E.; Kunz, R.; Falck-Ytter, Y.; Alonso-Coello, P.; Schünemann, H.J.; Group, G.W. GRADE: An emerging consensus on rating quality of evidence and strength of recommendations. *BMJ* **2008**, *336*, 924–926. [CrossRef] [PubMed]

48. Guyatt, G.; Oxman, A.D.; Akl, E.A.; Kunz, R.; Vist, G.; Brozek, J.; Norris, S.; Falck-Ytter, Y.; Glasziou, P.; DeBeer, H.; et al. GRADE guidelines: 1. Introduction-GRADE evidence profiles and summary of findings tables. *J. Clin. Epidemiol.* **2011**, *64*, 383–394. [CrossRef]

49. Mirmiran, P.; Bahadoran, Z.; Gaeini, Z. Common limitations and challenges of dietary clinical trials for translation into clinical practices. *Int. J. Endocrinol. Metab.* **2021**, *19*, e108170. [CrossRef]

MDPI
St. Alban-Anlage 66
4052 Basel
Switzerland
Tel. +41 61 683 77 34
Fax +41 61 302 89 18
www.mdpi.com

Nutrients Editorial Office
E-mail: nutrients@mdpi.com
www.mdpi.com/journal/nutrients

9 783036 571300